Global Economic and Environmental Aspects of
BIOFUELS

Advances in Agroecology

Series Editor: Clive A. Edwards

Global Economic and Environmental Aspects of

BIOFUELS

Edited by
David Pimentel

CRC Press
Taylor & Francis Group
Boca Raton   London   New York

CRC Press is an imprint of the
Taylor & Francis Group, an **informa** business

CRC Press
Taylor & Francis Group
6000 Broken Sound Parkway NW, Suite 300
Boca Raton, FL 33487-2742

First issued in paperback 2019

ISBN-13: 978-1-4398-3463-3 (hbk)
ISBN-13: 978-1-138-37436-2 (pbk)

Library of Congress Cataloging-in-Publication Data

Global economic and environmental aspects of biofuels / editor, David Pimentel.
    p. cm. -- (Advances in agroecology ; 17)
   Includes bibliographical references and index.
   ISBN 978-1-4398-3463-3
   1. Biomass energy--Economic aspects. 2. Biomass energy--Environmental aspects. 3. Agricultural ecology. I. Pimentel, David, 1925-

HD9502.5.B542G56 2012
333.95'39--dc23   2011035558

Visit the Taylor & Francis Web site at
http://www.taylorandfrancis.com

and the CRC Press Web site at
http://www.crcpress.com

Contents

v

Foreword

This volume is the latest in a series of books with the overall title *Advances in Agroecology*. The book focuses on an extremely important current environmental, ecological, and sociological issue: the use of biofuels produced from crops and organic waste as an alternative or supplement to oil. The U.S. Department of Energy and many national and international agencies predict that global supplies of petroleum will become increasingly expensive, and many experts predict that they may become exhausted by 2040. Discussions in the different chapters focus on key ecological and economic issues associated with the production of ethanol as a fuel from corn, sugarcane, crop residues, and other organic materials.

The book is edited by David Pimentel of Cornell University, who is an international authority on energy issues. He also contributes to several critical chapters in the book. The various chapter authors address the current and future impact of bioethanol technology on human food production, global malnutrition, human populations, water needs, and soil erosion and assess its global potential. Their overall conclusion is that the production of ethanol from grains, other crops, and crop residues has proven to be costly in both environmental and economic terms.

It has become clear that the potential energy production from bioethanol sources is small to negligible. For example, the 34 billion liters of ethanol being produced from corn provides only 1.3% of total oil consumption in the United States while using 33% of all the corn grown; if *all* U.S. corn were converted into alcohol, this would still provide the country with only 4% of its oil fuel needs. There are a few glimmers of hope for the technology reported in the book. First, the production of biodiesel and ethanol by saltwater algae holds possible future potential, but many technological problems with this process remain to be resolved. There is also a discussion of the economic potential of production of oil from tropical palms in Asia and elsewhere in the Far East, but again the technology needs to be developed.

The overall conclusion from most of the contributors is that green plants—corn, sugarcane, switchgrass, and other kinds of biomass—can convert only about 1% of solar energy into plant material annually. The authors unanimously conclude that the present and potential environmental and sociological impacts of ethanol production include global food shortages, increased carbon dioxide emissions, intensified soil erosion, and the consumption of enormous quantities of water.

This is a challenging book that brings together the opinions of a number of U.S. scientists with those of experts from Spain, Italy, the United Kingdom, and Brazil. There is remarkable agreement among the contributors that biofuels are not the answer to solving future oil shortages. The book will find a receptive audience among private readers, scientists, and government officials.

Prof. Clive A. Edwards
Editor-in-chief, Advances in Agroecology
The Ohio State University
Columbus, Ohio

Preface

People started using oil and coal about 150 years ago. Oil, natural gas, and coal have supported major advances in agriculture, industry, transport, and indeed all aspects of human life systems for more than a century. Fossil energy has allowed us to build up a human population of 7 billion and a massive world economy and agricultural system largely dependent on that energy. Now, world oil supplies have peaked and will slowly decline during the next 30 to 100 years. Some natural gas and coal will be used to replace oil, but this will be economically and environmentally costly. During the next 100 years, we will run out of *all* fossil energy. Then what?

People have been slow to recognize that we must reduce our population and change our economy as we move into a future with dwindling fossil fuel reserves. This book does not attempt to deal with the major changes that we must make, but focuses on the science and technology of biofuels. The focus is specifically on ethanol, biodiesel, and other biological materials that are possible substitutes for oil, natural gas, and coal resources.

Everyone recognizes that biofuels are made from green plants. However, depending on plants places a significant limitation on replacing oil, natural gas, and coal because green plants collect on average less than 0.1% of solar energy annually in the world. It took more than 700 million years for the green plants to collect all the energy stored in current oil, natural gas, and coal deposited in North America and elsewhere. As a replacement, a technology is needed that will collect solar energy at a rate about 200 times greater than what green plants currently do.

In this book, the authors examine various biofuel energy technologies and report on their potential to supply the United States and other nations with needed energy now and in the future. Some chapters examine several biofuel energy technologies and their potential to replace some fossil energy, while others focus on one or several technologies and their potential and limitations. The aim of the contributors to this volume is to share their analyses and for these analyses to be a basis for more research in biofuel energy technologies. Basic to all biofuel technologies is that they must attempt to minimize damage to the environment that supports all life.

Several of the chapters reflect the current lack of agreement in the field, as pressure mounts to explore and develop potential biofuel alternatives. The reader will notice considerable variability in the stated energy inputs and potential energy outputs in some of the studies. This is evidence of the complexity of assessing the large number of energy inputs that go into the production of a biofuel and the extraction of useful energy. As we collectively investigate the issues, we will discover if current analyses and biofuel outputs have adequately considered energy requirements, outputs, and environmental consequences. Hopefully, this research will help guide energy policy makers toward the most viable choices and away from energetically costly missteps.

I commend the authors of each of these chapters, who have done a superb job in presenting the most up-to-date perspectives of various biofuel energy technologies in a highly readable fashion.

David Pimentel

Acknowledgment

I wish to express my sincere gratitude to the Cornell Association of Professors Emeriti for the partial support of our research through the Albert Podell Grant Program. In addition, I wish to thank Mike Burgess for his valuable assistance in the preparation of our book.

About the Editor

David Pimentel is a professor of ecology and agricultural science at Cornell University. He received his bachelor of science in 1948 from the University of Massachusetts and his PhD from Cornell in 1951. From 1951 to 1954, he was chief of the Tropical Research Laboratory, U.S. Public Health Service, San Juan, Puerto Rico. From 1954 to 1955, he was a postdoctoral research fellow at the University of Chicago; in 1960–1961, an OEEC Fellow at Oxford University, and in 1961, an NSF scholar at the Massachusetts Institute of Technology. Dr. Pimentel was appointed assistant professor of insect ecology at Cornell University in 1955 and associate professor in 1961. In 1963, he was appointed professor and head of the Department of Entomology and Limnology. He served as department head until 1969, when he returned to full-time research and teaching as professor of ecology and agricultural sciences.

Nationally, Dr. Pimentel has served on the President's White House Science and Technology Program and the Presidential Commission on the Environment; numerous National Academy of Sciences committees and boards, including chairing the Board of Ecology; and committees in the Department of Health, Education, and Welfare, Department of Energy, Department of Agriculture, State Department, and the congressional Office of Technology Assessment. He also served as president of the Rachel Carson Council and as an elected member of the National Audubon Society and the American Institute of Biological Sciences.

His international honors and achievements have included his fellowship at Oxford University; an appointment as honorary professor of the Institute of Applied Ecology, China; receiving an honorary degree of science from the University of Massachusetts, receiving the Distinguished Service Award from the Rural Sociology Society; and serving on the board of directors of the International Institute of Ecological Economics, Royal Swedish Academy of Science.

Dr. Pimentel has authored nearly 700 scientific publications, written three books, and edited 31 others. He is now serving as editor-in-chief of the journal *Environment, Development and Sustainability*. His research spans the fields of energy, biological control, biotechnology, land and water conservation, and environmental policy.

Contributors

Claudinei Andreoli
Maize and Sorghum Research Unit
Brazilian Research Agricultural Corporation
Sete Lagoas, Brazil

Darcy Elizabeth Balcarce
College of Environmental Science and Forestry
State University of New York
Syracuse, New York

Stephen B. Balogh
College of Environmental Science and Forestry
State University of New York
Syracuse, New York

Sandra G. F. Bukkens
Institute of Environmental Science and Technology
Autonomous University of Barcelona
Barcelona, Spain

Sandra Fahd
Department of Sciences for the Environment
Parthenope University
Naples, Italy

Andrew R. B. Ferguson
Optimum Population Trust
Henley-on-Thames,
United Kingdom

Mario Giampietro
Institute of Environmental Science and Technology
Autonomous University of Barcelona
Bellaterra, Spain

Bari Greenfield
College of Agriculture and Life Sciences
Cornell University
Ithaca, New York

Aileen Maria Guzman
College of Environmental Science and Forestry
State University of New York
Syracuse, New York

Charles A. S. Hall
College of Environmental Science and Forestry
State University of New York
Syracuse, New York

Abbe Hamilton
College of Environmental Science
and Forestry
State University of New York
Syracuse, New York

Kate Hartman
College of Agriculture and Life
Sciences
Cornell University
Ithaca, New York

Edwin Kessler
Norman, Oklahoma

Danielle Kirshenblat
College of Agriculture and Life
Sciences
Cornell University
Ithaca, New York

Alan Kroeger
College of Agriculture and Life
Sciences
Cornell University
Ithaca, New York

R. Lal
Carbon Management and
Sequestration Center
The Ohio State University
Columbus, Ohio

Philip McMichael
Department of Development
Sociology
Cornell University
Ithaca, New York

Salvatore Mellino
Department of Sciences for the
Environment
Parthenope University
Naples, Italy

Emily Nash
College of Agriculture and Life
Sciences
Cornell University
Ithaca, New York

Sarah Palmer
College of Agriculture and Life
Sciences
Cornell University
Ithaca, New York

Tad W. Patzek
Department of Petroleum and
Geosystems Engineering
The University of Texas at
Austin
Austin, Texas

Simone Pereira de Souza
National Research Center of
Cassava and Tropical Fruits
Brazilian Research Agricultural
Corporation
Cruz das Almas, Brazil

David Pimentel
College of Agriculture and Life
Sciences
Cornell University
Ithaca, New York

Robert Rapier
Merica International
Kamuela, Hawaii

Jason Trager
Mechanical Engineering
Department
University of California, Berkeley
Berkeley, California

Sergio Ulgiati
Department of Sciences for the
Environment
Parthenope University
Naples, Italy

Jessica Zhang
College of Agriculture and Life
Sciences
Cornell University
Ithaca, New York

chapter one

Biofuels cause malnutrition in the world

David Pimentel

Contents

1.1 Introduction

Global shortages of fossil energy—especially of oil and natural gas—exist worldwide, and an emphasis on biofuels as a renewable energy source has developed globally (Pimentel and Pimentel, 2008), particularly those made from corn grain, soybeans, canola, rapeseed, palm oil, and sugarcane. In developing countries, about 2 kcal is required to cook 1 kcal of food, and most of this energy comes from wood or crop residues (Fujino et al., 1999). The use of crop residues is having a devastating impact on agriculture and food production (Lal, 2009). It is particularly serious because it increases soil erosion, water runoff, and loss of plant nutrients (Lal, 2002, 2009).

Conflicts have developed in the use of land, water, energy, and other environmental resources for food production versus biofuel production.

Both food and biofuels from corn, soybeans, canola, rapeseed, and sugarcane are dependent on the same resources for production. The objective of this chapter is to examine the interrelationship between malnutrition and the use of food crops for biofuel.

1.2   World malnutrition

Malnutrition—the inadequate intake of calories, protein, minerals, and vitamins—is a major disease related to the environment (Myers and Kent, 2001; Pimentel et al., 2007). The World Bank World Development Report estimated that deficiencies of vitamin A, iron, and iodine waste as much as 5% of global gross domestic product (GDP), while providing those needed nutrients would cost only 0.3% of the global GDP (World Bank, 1995; Maberly et al., 1998). Malnutrition exists in regions where the overall food supply is inadequate, where people lack adequate economic resources to purchase needed nutrients, and where political unrest and instability interrupt food supplies. The rapid growth in the use of food for biofuels in the United States, Europe, and other nations has increased world starvation and deaths (FAO, 2009; Pimentel, Marklein, et al., 2009). Food prices due to biofuels have also increased more than 140%, according to the World Bank (Mitchell, 2008).

In 1950, about 500 million people (20% of the world population) were malnourished (Grigg, 1993). Today, more than 4.8 billion people (70% of the world population) suffer from malnutrition (Pimentel and Satkiewicz, forthcoming). This is the largest number of malnourished people ever in history. Each year, an estimated 6 million children under the age of five die from malnutrition (FAO, 2002). Even in the United States, more than 14.5% of all households experienced food insecurity during 2010, and 5.4% of these households had at least one family member who went hungry (Coleman-Jensen et al., 2011). Malnutrition at an early age can lead to physical and mental underdevelopment as an adult, and this underdevelopment results in a poverty trap where people are stuck at a low level of productivity and is a major cost to society and the environment (WHO, 2000).

Vitamin A malnutrition diminishes and impairs the immune responses to infectious diseases in children. Vitamin A supplements have been shown to decrease mortality by 30% in vitamin A–deficient children between the ages of six months and five years (Stephensen, 2001). Each year, vitamin A deficiency causes 2.5 million deaths (Toenniessen, 2000). Vitamin A shortages can also cause mental disabilities in children (Uzendu, 2004; WHO, 2009). More than 13 million people suffer from night blindness or total blindness from a lack of vitamin A (WHO, 2009).

Similarly, iron intake per person has not appreciably improved during the past 10 years, especially in developing countries (WHO, 2000). Globally, from 4 billion to 5 billion people are iron deficient and 2 billion suffer from anemia (WHO, 2000). Iron deficiency decreases work capacity and decreases resistance to fatigue (WHO, 2000). In addition, about 1.6 billion people have iodine-deficient diets and may suffer from iodine deficiency disease (UN, 2004). Iron and iodine deficiencies cause mental disabilities in children as well (UN, 2004; UNICEF, 2008).

1.3   Energy resources

Approximately 50% of the solar energy captured worldwide by photosynthesis is used by humans for food, fiber, and forest products. Yet this amount of energy remains inadequate to meet all human food and other biomass needs (Pimentel et al., 2009). To make up for the shortfall, about 500 quads (1 quad = 1×10^{15} BTU) of energy is consumed, generated mostly from oil, natural gas, and coal (USDOE, 2006; IEA, 2008; Hoffman 2009). Of this total fossil energy, the United States, with only 4.5% of the world population, consumes 22% (USCB, 2008).

Each year, the U.S. population uses three times more fossil energy than the total solar energy captured by all the U.S. crops, forests, and pastures (Pimentel et al., 2009). Industry, transportation, home heating and cooling, and food production account for most of the fossil energy consumed in the United States (USCB, 2008). Per capita use of fossil energy in the United States amounts to 9,500 liters of oil equivalents (USCB, 2008).

Geologists estimate that worldwide oil supplies will be depleted by 2040 (IEA, 2008), natural gas within about 60 years (IEA, 2008), and coal hopefully in about 100 years (IEA, 2008). At present, the United States is importing about 70% of its oil, at an annual cost of about $700 billion (USCB, 2008).

1.4   Biomass resources

An estimated 26×10^{18} BTU of sunlight reaching the Earth (27.8×10^{15} BTU in the United States) is captured as biomass by green plants (crops, grasses, and forests) each year in the world, which suggests that green plants are collecting about 0.1% of the solar energy reaching the earth (Pimentel et al., 2009). The amount of biomass produced worldwide, $1,764 \times 10^9$ tons per year (see Table 1.1), is a thousand times greater than that in the United States (1.758×10^6 tons). Most of the biomass is crops and forest, totaling 1,000 billion tons (FAO, 2011). The best estimate for cooking and heating in developing countries is about 50×10^{15} kcal (Balat and Ayar, 2005).

The global forest area removed each year totals 15 million ha (Sundquist, 2007). Global forest biomass harvested each year is just over 1,430 billion kg, of which 60% is industrial roundwood and 40% fuelwood (FAO, 2011). About 90% of the fuelwood is utilized in developing countries (Parikka, 2004). Per capita consumption of woody biomass for heat in the United States amounts to 625 kg/yr, while the diverse biomass resources

Table 1.1 Total Amount of Biomass and Solar Energy Captured Each Year in the World

	Billion ha	Tons/ha	× 10⁹ tons	Total energy collected × 10¹⁸ BTU
Crops	1.5	5.0	750	3,000
Grasses	3.5	1.1	385	1,540
Forests	3.0	1.6	629	3,120
Total	*8.0*		*1,764*	*7,660*

(wood, crop residues, and dung) used in developing nations average about 630 kg/yr per capita (Kitani, 1999).

A significant portion (26%) of all forest wood—or about 65% of all fuelwood—is converted into charcoal (Williams, 1989). Production of charcoal causes between 30% and 50% of the wood energy to be lost (Demirbas, 2001) and produces enormous quantities of polluting smoke. On the other hand, charcoal is clean burning and produces little smoke when used for cooking (Williams, 1989). Charcoal is dirty to handle but is lightweight. In the production of charcoal, all the nutrients in the wood are lost; only the charcoal or biochar remains.

As mentioned, about 40% of the world forest wood is used for fuel. In developing countries, about 2 kcal of wood are utilized in cooking 1 kcal of food (Pimentel and Pimentel, 2008; Fujino et al., 1999). Thus, more biomass and more land and water are needed to produce the biofuel for cooking than are needed to produce the food (see Table 1.1).

Biomass can also be used to produce electricity. Assuming that an optimal yield of 3 dry tons per hectare per year can be harvested sustainably, as achieved in the northeastern United States (Ferguson, 2001), this would provide a gross energy yield of 13.5 million kcal/ha. Harvesting this wood biomass requires an energy expenditure of approximately 30,000 kcal/ha, plus the embodied energy for cutting and collecting wood for transport to an electric power plant (Pimentel et al., 1994). Thus, the energy input/output ratio for such a system is calculated to be 1:25. Woody biomass has the capacity to supply the United States with about 5 quads of its total gross energy supply by 2050, provided the amount of forest land remains constant (Pimentel, 2008). A city of 100,000 people using biomass from a sustainable forest producing 3 tonnes/ha/yr for electricity requires about 200,000 ha of sustainable forest area, based on an average electrical demand of slightly more than 1 billion kWh of electrical energy (Pimentel, 2008).

The air quality impact from burning biomass is less harmful than that associated with coal, but more harmful than that associated with natural gas (Pimentel, 2008). Biomass combustion releases more than 200 different chemicals, including 14 carcinogens, into the atmosphere (Burning Issues,

2006). As a result of cooking with biomass and the resultant continuous exposure to smoke, especially in developing countries, about 4 billion people suffer from various respiratory diseases (Smith, 2006) and 2.7 million to 3 million people worldwide die prematurely due to wood smoke (WHO, 1997). In the United States, wood smoke is reported to kill about 30,000 people each year (Marty, 2002; Burning Issues, 2003). Various pollution controls can be installed in wood-fired stoves. Of course, an owner of a small house could not afford to install such a unit.

1.5   Biofuels systems and net returns

An estimated 1.8 billion tons of biomass is produced per year on U.S. land area (see Table 1.2). This translates into nearly 7,040 quads of energy, which means that the solar energy captured by all the green plants in the United States each year equates to less than 30% of total fossil energy (Pimentel et al., 2009). Clearly, there is insufficient U.S. biomass to make the United States fossil-fuel independent.

1.5.1   Corn ethanol

Ethanol produced from corn grain is the largest source of biofuels in the world (USCB, 2008). However, even if we totally ignore corn ethanol's negative energy balance and high economic cost, we still find that it is absolutely impossible to use corn ethanol as a replacement for U.S. oil consumption (Pimentel and Patzek, 2008). If all 340 billion kilograms of corn produced in the United States each year (USDA, 2006) were converted into ethanol, only 130 billion liters of ethanol would be produced. This would provide only 4% of total oil consumption in the United States. Furthermore, in this situation, there would be no corn available for livestock production, food, or other human needs (Pimentel and Patzek, 2008). The World Bank's Food price index of corn increased in price 74% from the first quarter of 2010 to the first quarter of 2011, thus significantly affecting poor people (Chandrasekhar, 2011) plus sizable increases in United States soybeans and palm oil, both used to produce biodiesel.

Table 1.2 Total Amount of Biomass and Solar Energy Captured Each Year in the United States

	Million ha	Tons/ha	× 10⁶ tons	Total energy collected × 10¹⁵ BTU
Crops	160	5.5	901	3,600
Pasture	300	1.1	333	1,332
Forests	264	2.0	527	2,108
Total	724		*1,758*	*7,040*

1.5.2   Cellulosic ethanol

Tilman and colleagues (2006) assume that about 1,032 liters of ethanol can be produced through the conversion of 4 t/ha/yr of harvested grasses like switchgrass. However, Pimentel and Patzek (2008) report a negative 68% return in ethanol produced relative to the total fossil energy inputs. Furthermore, converting all 235 million hectares of U.S. grassland into ethanol, even at Tilman's optimistic rate, would provide the United States with only 12% of its annual consumption of oil (USCB, 2008).

Converting 235 million hectares of grassland into ethanol would mean that U.S. farmers would have to displace 97 million head of cattle, 6 million sheep, and 7.3 million horses (USDA, 2008; AVMA, 2011)—not to mention the millions of wild animals, including white-tailed deer, mule deer, elk, mice, rabbits, woodchucks, and numerous other native mammals, that depend on these grasses for food. Also, it has been reported that grasshoppers and various other insects that eat grasses consume as much or more grass than cattle (Moseley, 2009). For example, just six grasshoppers per square yard in a 10-acre hay field will eat as much as a cow in the same time period (Ratcliffe et al., 2004). In New York State pastures, it was reported that pasture insects consumed twice as much forage as cows per day (Wolcott, 1937). The cows ate about 180 kg during a study period, whereas the insects, for the same period, consumed 364 kg per hectare (USDA, 2006). Also, there is a serious overgrazing problem on U.S. grasslands (Brown, 2002; USDA, 2006). Thus, Tilman and colleagues are unduly optimistic.

1.5.3   Soybean biodiesel

The United States subsidizes biodiesel at about $11 billion per year for the production of about 900 million liters of biodiesel (Koplow and Steenblik, 2008), which is 75 times greater than the subsidies per liter of diesel fuel. The environmental impacts of producing soybean biodiesel fuel are second only to that corn ethanol production (Pimentel and Patzek, 2008). If all 71 billion tons of soybeans were converted into biodiesel, this would provide the United States with only 2.6% of total U.S. oil consumption (Pimentel and Patzek, 2008).

1.5.4   Rapeseed and canola

The European Biodiesel Board estimates total biodiesel production at nearly 9 million tons for 2009 (European Biodiesel Board, 2011). Well suited to the colder climates, rapeseed is the dominant crop in European biodiesel production. Often confused with canola, rapeseed is an inedible crop of the *Brassica* family yielding oil seeds high in erucic acid.

Rapeseed and canola require the application of fertilizers and pesticides in production. The energy required to produce these fertilizers and pesticides detracts from the overall net energy produced (Pimentel and Patzek, 2008). Although soybeans contain less oil than canola—about 18% soy oil versus 30% oil for canola seeds—soybeans can be grown with nearly zero nitrogen fertilizer inputs (Pimentel, Marklein, et al., 2009).

The biomass yield of rapeseed or canola per hectare is also lower than that of soybeans—about 1,600 kg/ha for rapeseed/canola compared with 2,900 kg/ha for soybeans. The production of 1,600 kg/ha of rapeseed requires the input of about 4.4 million kcal per hectare. About 3,333 kg of rapeseed/canola is required to produce 1,000 kg of biodiesel. The total energy input to produce the 1,000 liters of rapeseed/canola oil is 13 million kcal. This suggests a net loss of 58% of energy inputs. The cost per kilogram of biodiesel is also high: $1.52 or about $5.70/gallon. Rapeseed and canola are energy intensive and economically inefficient biodiesel crops (Pimentel, Marklein, et al., 2009).

1.5.5   Oil palm

A major effort worldwide exists to plant and harvest oil palms for biodiesel in several tropical developing countries, including Indonesia, Malaysia, Thailand, Colombia, and some West African countries (Thoenes, 2007). In the last 20 years, the production of vegetable oils has more than doubled. Palm oil makes up 33% of the biological oils produced worldwide (USDA Foreign Agricultural Service, 2011; USDA Foreign Agricultural Service, 2011a). In 2011–2012, more than 50.23 million tons of palm oil was produced worldwide, mostly in Indonesia and Malaysia (USDA Foreign Agricultural Service, 2011a).

The oil palm, once established (requiring four years for establishment), will produce about 4,000 kg of oil per hectare per year (Carter et al., 2007). Approximately 7.4 million kcal are required to produce 26,000 kg of oil palm bunches (Pimentel, Marklein, et al., 2009), which is sufficient to produce 4,000 kg palm oil. A total of 6.9 million kcal are required to process 6,500 kg of palm nuts to produce 1 ton of palm oil (Pimentel, Marklein, et al., 2009). This is clearly a better return than corn ethanol or soybean biodiesel. However, an estimated 200 ml (2,080 kcal) of methanol must be added to each 1,000 kg of palm oil for transesterfication, which results in an 8% negative net energy output for the palm oil (Pimentel et al., 2009).

There are several negative environmental and social issues associated with the oil palm plantations. First, the removal of tropical rain forests to plant the oil palm results in an increase in CO_2 (Thoenes, 2007). Second, this removal of tropical rain forests and planting of oil palms reduces biodiversity (Pimentel et al., 2006). Finally, using palm oil for biofuel reduces the availability of palm oil for human use and increases the price of the oil for everyone, including the poor (Thoenes, 2007). The World Bank's food price index of palm oil increased 55% in price from the first quarter of 2010 to the first quarter of 2011 (Chandrasekhar, 2011).

1.5.6 Algae for oil

Some cultures of microalgae may consist of 30% to 50% oil (Dimitrov, 2007). Thus, there is increasing interest in algae to increase U.S. biofuels. A study by Dimitrov (2007) reported that oil produced using algae costs an estimated $800 per barrel.

Another investigation calculated the cost per gallon of biodiesel using electricity in a bioreactor costing 7¢ per kWh to be $32.81 per gallon ($8.67 per liter) or about $1,378 per barrel (Kanellos, 2009). Yet another study reported that harvesting and dewatering costs for microalgae could be reduced by as much as 99% compared to centrifuges (Iyovo, 2009). These data suggest cost problems in producing algae for biodiesel.

However, Pimentel et al. (Forthcoming) report optimistically that it might be possible to produce 1.22 liters of algae oil for every 1 liter of oil equivalents invested in production. The cost of producing 1 liter of algae oil was $0.83 ($3.14 per gallon), which is encouraging.

1.6 Conclusion

The rapidly growing world population and rising consumption of fossil fuels is increasing the demand for both food and biofuels (Pimentel, Marklein, et al., 2009). This is exaggerating both food and fuel shortages. Producing biofuels today requires nearly 400 million tons of food per year.

Using food to produce ethanol and biodiesel raises major nutritional and ethical concerns. More than 70% of the world population is currently malnourished. Thus the need for grains and other basic food is critical. Producing crops for biofuels squanders land, water, and fossil energy resources vital for the production of food for people. Using food and feed crops has doubled or tripled the price of some foods, like bread. Overall, according to the U.S. Congressional Budget Office (2009), grocery bills have increased from $6 billion to $9 billion per year. Both the director general of the United Nations and the director general of the UN Food and Agricultural Organization report that using food crops to produce biofuel is increasing starvation worldwide (Diouf, 2008; Spillius, 2008). Note, the World Bank's commodity price index of corn and wheat increased 74% and 70%, respectively, from the first quarter of 2010 to the first quarter of 2011 (Chandrasekhar, 2011). Clearly, the use of food and feed crops is exacerbating the malnourishment problem worldwide by turning food into energy deficient biofuel.

Recent policy decisions have mandated increased production of biofuels in the United States and worldwide. For instance, the Energy

Independence and Security Act of 2007 set "a mandatory renewable fuel standard (RFS) requiring fuel producers to use at least 36 billion gallons of biofuel in 2022." Such a move would require harvesting of all biomass produced each year in the United States, including all crops. This would essentially be like taking a lawnmower to the total vegetation produced in the United States. Biodiversity and food supplies in the United States would be seriously decimated.

Most problems associated with biofuels have been ignored by many scientists and policy makers, both in the United States and abroad. The production of biofuels that are supposed to diminish dependence on imported fossil fuels is actually increasing U.S. dependence on fossil fuels. Most conversion of biomass into ethanol and biodiesel, except for algae oil, results in a negative energy return based on careful up-to-date analysis of all the fossil energy inputs. The negative energy returns in producing biofuels using food and feed crops range from 8% to 100%.

Clearly, the increased use of biofuels will further damage the global environment and especially the world food system. The serious world food malnutrition will further increase.

Acknowledgments

We wish to express our sincere gratitude to the Cornell Association of Professors Emeriti for the partial support of our research through the Albert Podell Grant Program.

References

AVMA. 2007. *U.S. Pet ownership and demographics sourcebook.* Schaumburg, IL: American Veterinary Medical Association, Member and Field Services.

Balat, M., and Ayar, G. 2005. Biomass energy in the world, use of biomass and potential trends. *Energy Sources,* 27:931–940.

Brown, L. R. 2002. Plan B updates: World's rangelands deteriorating under mounting pressure. Earth Policy Institute. http://www.earth-policy.org/plan_b_updates/2002/update6.

Burning Issues. 2003. References for wood smoke brochure. http://burningissues.org/car-www/science/wsbrochure-ref-3-03.htm.

Burning Issues. 2006. What are the medical effects of exposure to smoke particles? http://burningissues.org/health-effects.html.

Carter, C., Finley, W., Fry, J., Jackson, D., and Willis, L. 2007. Palm oil markets and future supply. *European Journal of Lipid Science and Technology,* 109:307–314.

Chandrasekhar, C. P. 2011. Finance, food and the poor. *Hindu,* May 1. http://www.thehindu.com/opinion/columns/Chandrasekhar/article1983450.ece.

Coleman-Jensen, A., Nord, M., Andrews, M. and Carlson, S. 2011. Household food security in the United States in 2010. U.S. Department of Agriculture, Economic Research Report Number 125. http://www.ers.usda.gov/Publications/ERR125/err125.pdf.

Demirbas, A. 2001. Biomass resource facilities and biomass conversion processing for fuels and chemicals. *Energy Conversion & Management*, 42(11): 1357–1378.

Dimitrov, K. 2007. Green fuel technologies: A case study for industrial photosynthetic energy capture. Brisbane, Australia. http://www.nanostring.net/Algae/CaseStudy.pdf.

Diouf, J. 2008. Statement by the director-general on the occasion of the launch of SOFA 2008. October 7. Food and Agriculture Organization of the United Nations, Rome. http://www.fao.org/newsroom/common/ecg/1000928/en/sofalaunch.pdf.

European Biodiesel Board. 2011. Statistics: The EU biofuel industry. European Biofuel Board. http://www.ebb-eu.org/stats.php.

FAO. 2002. Food insecurity in the world 2002. Food and Agriculture Organization of the United Nations, Rome. ftp://ftp.fao.org/docrep/fao/005/y7352e/y7352e00.pdf.

———. 2009. *The state of food insecurity in the world*. Rome: FAO. http://www.fao.org/docrep/012/i0876e/i0876e00.htm.

———. 2011. FAOSTAT data. [Online database accessed June 2, 2011.] http://faostat.fao.org/.

Ferguson, A. R. B. 2001. Biomass and energy. *Optimum Population Trust Journal*, 4(1): 14–18.

Fujino, J., Yamaji, K., and Yamamoto, H. 1999. Biomass-balance table for evaluating bioenergy resources. *Applied Energy*, 63(2): 75–89.

Grigg, D. B. 1993. *The world food problem*. Oxford, U.K.: Blackwell.

Hoffman, A. R. 2009. Energy poverty and security. *Journal of Energy Security* (April). http://www.ensec.org/index.php?view=article&catid=94%3A0409content&id=185%3Aenergy-poverty-and-security&tmpl=component&print=1&page=&option=com_content&Itemid=342.

IEA (International Energy Agency). 2008. *World energy outlook 2008*. Paris: OECD/IEA. http://www.iea.org/textbase/nppdf/free/2008/weo2008.pdf.

Iyovo, G. D. 2009. Top story: Algae biofuel again! *Sustainable Living with Sustainable Energy & Technology–Bioenergy* [blog], March 30. http://genedrekeke.blogspot.com/2009/03/top-storyalgae-biofuel-again.html.

Kanellos, M. 2009. Algae biodiesel: It's $33 a gallon. *Greentechmedia*. http://www.greentechmedia.com/articles/read/algae-biodiesel-its-33-a-gallon-5652/.

Kitani, O. 1999. Biomass resources. In *Energy and biomass engineering*, ed. Kitani, O., Jungbluth, T., Peart, R. M., and Ramdami, A., 6–11. St. Joseph, MI: American Society of Agricultural Engineers.

Koplow, D., and Steenblik, R. 2008. Subsidies to ethanol in the United States. In *Biofuels, solar and wind as renewable energy systems: Benefits and risks*, ed. Pimentel, D., 79–108. Dordrecht, The Netherlands: Springer.

Lal, R., ed. 2002. *Encyclopedia of soil science*. New York: Marcel Dekker.

Lal, R. 2009. Soil quality impacts of residue removal for bioethanol production. *Soil and Tillage Research*, 102(2): 233–241.

Maberly, G. F., Bagrinsky, J., and Parvanta, C. C. 1998. Forging partnerships among industry, government, and academic institutions for food fortification. *Food Nutrition Bulletin*, 19:122–130.

Marty, D. M. 2002. Smoke gets in your eyes (editorial). *E*, August 31. http://www.emagazine.com/archive/1107.

Mitchell, D. 2008. Note on rising food prices. World Bank Policy Research Working Paper: WPS no. 4682. http://www-wds.worldbank.org/external/default/

WDSContentServer/IW3P/IB/2008/07/28/000020439_20080728103002/
Rendered/PDF/WP4682.pdf.

Moseley, B. 2009. Profit from forage production strategies. *Eagle Land and Livestock Post.* http://www.landandlivestockpost.com/livestock/040109-cover.

Myers, N., and Kent, J. 2001. Food and hunger in the Sub-Saharan Africa. *Environmentalist,* 21(1): 41–69.

NAS (National Academy of Sciences). 2003. *Frontiers in agricultural research: Food, health, environment, and communities.* Washington, DC: National Research Council, National Academies Press.

Parikka, M. 2004. Global biomass fuel resources. *Biomass and Energy,* 27:613–620.

Pimentel, D. 2008. Renewable and solar energy technologies: Energy and environmental issues. In *Biofuels, solar and wind as renewable energy systems: Benefits and risks,* ed. Pimentel, D., 1–17. Dordrecht, The Netherlands: Springer.

Pimentel, D., Cooperstein, S., Randell, H., Filiberto, D., Sorrentino, S., Kaye, B., Yagi, C. J., et al. 2007. Ecology of increasing diseases: Population growth and environmental degradation. *Human Ecology,* 35(6): 653–668.

Pimentel, D., Marklein, A., Toth, M. A., Karpoff, M. N., Paul, G. S., McCormack, R., Kyriazis, J., and Krueger, T. 2009. Food versus biofuels: Environmental and economic costs. *Human Ecology,* 37(1): 1–12.

Pimentel, D., and Patzek, T. 2008. Ethanol production: Energy and economic issues related to U.S. and Brazilian sugarcane. In *Biofuels, solar and wind as renewable energy systems: Benefits and risks,* ed. Pimentel, D. 357–371. Dordrecht, The Netherlands: Springer.

Pimentel, D., Petrova, T., Riley, M., Jacquet, J., Ng, V., Honigman, J., and Valero, E. 2006. Conservation of biological diversity in agricultural, forestry, and marine systems. In *Focus on ecology research,* ed. Burk, A. R., 151–173. New York: Nova Science.

Pimentel, D., and Pimentel, M. 2008. *Food, energy, and society.* 3rd ed. Boca Raton, FL: CRC Press.

Pimentel, D., Rodrigues, G., Wang, T., Abrams, R., Goldberg, K., Staeckcr, H., Ma, E., et al. 1994. Renewable energy: Economic and environmental issues. *BioScience,* 44(8): 536–547.

Pimentel, D., and Satkiewicz, P. Forthcoming. World malnutrition. In *Encyclopedia of sustainability,* Vol. 4. Great Barrington, MA: Berkshire.

Pimentel, D., Trager, J., Palmer, S., Zhang, J., Greenfield, B., Nash, E., Hartman, K., Kirshenblatt, D., and Kroeger, A. Forthcoming. Energy production from corn, cellulosic, and algae biomass. In *Global economic and environmental aspects of biofuels,* ed. Pimentel, D. Boca Raton, FL: CRC Press.

Ratcliffe, S. T., Gray, M. E., and Steffey, K. L. 2004. Grasshoppers. *IPM: Integrated Pest Management.* University of Illinois Extension. http://ipm.illinois.edu/fieldcrops/insects/grasshoppers/index.html.

Smith, K. R. 2006. Health impacts of household fuelwood use in developing countries. *Unasylva* 224(57): 41–44. http://ehs.sph.berkeley.edu/krsmith/?p=113.

Spillius, A. 2008. IMF alert on starvation and civil unrest. *Brisbane Times,* April 15. http://www.brisbanetimes.com.au/news/world/imf-alert-on-starvation-and-civil-unrest/2008/04/14/1208025090977.html.

Stephensen, C. B. 2001. Vitamin A, infection, and immune function. *Annual Review of Nutrition,* 21:167–192.

Sundquist, B. 2007. Degradation data. In *Forest lands degradation: A global perspective.* http://home.windstream.net/bsundquist1/df4.html.

Tilman, D., Hill, J., and Lehman, C. 2006. Carbon negative biofuels from low-input, high-diversity grassland biomass. *Science*, 314:1598–1600.

Thoenes, P. 2007. Biofuels and commodity markets: Palm oil focus. Rome: FAO, Commodities and Trade Division. http://www.fao.org/es/ESC/common/ ecg/122/en/full_paper_English.pdf.

Toenniessen, G. H. 2000. Vitamin A deficiency and golden rice: The role of the Rockefeller Foundation. New York: Rockefeller Foundation. http:// www.rockefellerfoundation.org/uploads/files/4c1fc130-3db5-477d-a1b2-d2cfa60f5f65-111400ght.pdf.

UN. 2004. Independent expert on effects of structural adjustment. Special rapporteur on right to food present reports: Commission continues general debate on economic, social and cultural rights. Press Release HR/CN/1064. http://www.un.org/News/Press/docs/2004/hrcn1064.doc.htm.

UNICEF. 2008. UNICEF in action. Nutrition: Micronutrients – Iodine, iron and vitamin A. http://www.unicef.org/nutrition/index_iodine.html.

USCB (U.S. Census Bureau). 2008. *Statistical abstract of the United States, 2009*. Washington, DC: USCB.

U.S. Congressional Budget Office. 2009. The impact of ethanol use on food prices and greenhouse-gas emissions. http://www.cbo.gov/ftpdocs/100xx/ doc10057/04-08-Ethanol.pdf.

USDA (U.S. Department of Agriculture). 2006. *Agricultural statistics, 2006*. Washington, DC: USDA.

———. 2008. *Agricultural statistics, 2008*. Washington, DC: USDA.

USDA, Foreign Agricultural Service. 2011. Table 6. Major vegetable oils: World supply and distribution (Country view). United States Department of Agriculture, Foreign Agricultural Service, Circular Series, World Agricultural Production, WAP 09-11. http://www.fas.usda.gov/psdonline/psdreport .aspx?hidReportRetrievalName=BVS&hidReportRetrievalID=705&hidRepo rtRetrievalTemplateID=8.

———. 2011a. Table 16. Copra, palm kernel and palm oil production. United States Department of Agriculture, Foreign Agricultural Service, Circular Series, World Agricultural Production, WAP 09-11. http://www.fas.usda.gov/ psdonline/psdreport.aspx?hidReportRetrievalName=BVS&hidReportRet rievalID=454&hidReportRetrievalTemplateID=9.

USDOE (U.S. Department of Energy). 2006. *International energy annual, 2006*. U.S. Energy Information Agency. http://www.eia.gov/iea/.

Uzendu, M. 2004. Nigeria: 350,000 kids delivered with mental impairment yearly—Stella Obasnajo. *Daily Champion* (Lagos, Nigeria), October 29. http://allafrica.com/stories/200410290482.html.

Williams, M. 1989. Deforestation: Past and present. *Progress in Human Geography*, 13(2): 176–208.

Wolcott, G. N. 1937. An animal census of two pastures and a meadow in northern New York. *Ecological Monographs*, 7(1): 1–90.

World Bank. 1995. *Enriching lives: Overcoming vitamin and mineral malnutrition in developing countries*. Washington, DC: World Bank.

WHO (World Health Organization). 1997. *Health and environment in sustainable development: Five years after the Earth Summit*. Geneva: WHO.

———. 2000. *Nutrition for health and development: A global agenda for combating malnutrition*. Department of Nutrition for Health and Development. Geneva: WHO. http://whqlibdoc.who.int/hq/2000/WHO_NHD_00.6.pdf.

————. 2004. *Iodine status worldwide: WHO global database of iodine deficiency.* Department of Nutrition for Health and Development. Geneva: WHO. http://whqlibdoc.who.int/publications/2004/9241592001.pdf.

————. 2009. Global prevalence of vitamin A deficiency in populations at risk 1995–2005: WHO global database on vitamin A deficiency. World Health Organization of the United Nations, Geneva. http://whqlibdoc.who.int/publications/2009/9789241598019_eng.pdf.

chapter two

Biofuel and the world population problem

Mario Giampietro and Sandra G. F. Bukkens

Contents

2.1   Introduction

From the 1990s onward, the governments of the United States and the European Union have embraced the agro-biofuel concept as if it were the silver bullet for solving our problems of peak oil and greenhouse gas emissions, and ever since then they have focused on the implementation of agro-biofuel production by setting specific targets and policy strategies and providing massive subsidies (World Bank, 2008; Kutas et al., 2007). However, the consequent massive production of agro-biofuel in the late 1990s and early 21st century has only contributed to a world food crisis (BBC News, 2008; Cronin, 2008; Spiegel Online International, 2008; Chakrabortty, 2008) and did not generate any economic or reliable supply of liquid fuel capable of reducing either our dependence on fossil energy or the emission of CO_2 (Giampietro and Mayumi, 2009).

Food demand is expected to grow in the future because of both population growth and improvements in the standard of living. In fact, economic growth of less-developed countries is typically associated with changes in the traditional dietary pattern toward Western habits—changes that are likely to increase the consumption of animal products and hence augment the per capita grain consumption because of the double conversion involved in animal production (Pingali, 2007). To make matters worse, this additional increase in total food demand will take place in an already "full world" (Daly, 2005): the most fertile land is already in production or can no longer be used for crop production because of alternative uses (e.g., spread of urbanization). As a result, a larger food requirement will entail expanding food production on marginal land, a move that will intensify the use of technical inputs, increase the environmental impact per unit of cultivated area, cause additional destruction of natural habitat, and present an increased threat to biodiversity (MEA, 2005). Therefore, when looking at the future, the idea of a large-scale production of agro-biofuels implies establishing a direct competition for a limited supply of crops and hence for a limited supply of land, soil nutrients, and water, in relation to two distinct types of requirements: food for feeding people and liquid fuel for powering cars.

Also, in relation to fuels for transportation, we may expect that the world demand will continue to increase (EIA, 2009). The expected economic growth of low-income countries will be associated with structural changes in their economies and concomitant changes in lifestyle—both translating into a continuous increase in fuel consumption for commercial and private vehicles. Particularly important will be the increase in fuel consumption in Asia (led by China and India). The number of automobiles owned in Asia is expected to increase from 165 million in 2004 to 510 million in 2030 (IEEJ, 2006).

At present, the production of both food for people and fuel for transportation is completely dependent on the depletion of fossil energy stocks. Therefore, humankind has been able to satisfy these two demands thanks to the relentless plundering of fossil energy stocks.

In this chapter, we explain why large-scale production of liquid biofuel is an impractical idea in the face of a growing world population and a growing demand for liquid fuels. To make this point, we first explore the nature of the relation between food supply and fossil energy consumption, considering present trends in food consumption patterns. Then we shift the focus to liquid agro-biofuel production and debunk a systemic error in the assessment of the potential of agro-biofuel as a substitute for liquid fossil fuel. Indeed, in the literature, the systemic weakness of the performance of agro-biofuels as the primary energy source is often ignored and the potential supply of biofuels tends to be overestimated by several times. Our discussion on agro-biofuel production is necessarily of a technical nature, but let the reader not be discouraged. In order to see through the contrasting claims made by scientists and the different opinions expressed by politicians, corporations, and activists, it is important to adopt a more realistic method of assessment. At last, we illustrate with several numerical examples the impossibility of a large-scale implementation of liquid agro-biofuel production in the face of the competition between the requirement of agro-biomass for food and for fuels. In spite of the unavoidable simplifications used in the proposed scenarios, we strongly believe that these calculations show that the idea of substituting a significant fraction of fossil fuels with liquid agro-biofuels is totally impractical.

2.2   The dependence of our food supply on oil

> A great many people, including poorly informed political leaders, naively think that food production in Third World countries can be increased just by sending them seeds of high-yielding varieties of grains and a few "agricultural advisors." Cultivation of the "miracle" varieties requires expensive energy subsidies many underdeveloped countries cannot afford. (Odum, 1989, p. 83)

> The image is counterintuitive but true: survival of peasants in the rice fields of Hunan or Guadong—with their timeless clod-breaking hoes, docile buffaloes, and rice-cutting sickles—is now much more

> dependent on fossil fuels and modern chemical
> syntheses than the physical well-being of American
> city dwellers sustained by Iowa and Nebraska farm-
> ers cultivating sprawling grain fields with giant
> tractors. (Smil, 1991, p. 593)

Over the span of the 20th century, the world population has tripled, from about 2 billion in 1900 to more than 6 billion in 2000 (IAASTD, 2009). But the increase in the *pace* of population growth has been even more impressive. The world population increased from 3.5 billion at the begin- ning of the 1970s to 6.5 billion in 2005; that is to say that in a time span of only 35 years the population increase (of 3 billion people) exceeded that observed in the previous 35,000 years! In spite of this unprecedented increase in population growth and in the concomitant demand for food, humans were able to adjust food production: "Since 1960, population dou- bled while the economic activity is increased of 6 times, food production is increased 2½ times" (MEA, 2005).

The reason for this extraordinary success is well known: food secu- rity in modern times increasingly depends on heavy investments of fossil energy in agriculture and in the rest of the food system (Cottrell, 1955; Leach, 1976; Gever et al., 1991; Giampietro, 2003a; Hall et al., 1986; Odum, 1989; Odum, 1971; Pimentel and Pimentel, 1979, 1996; Smil, 1987, 1991; Steinhart and Steinhart, 1974; Stout, 1992). Indeed, an overall assessment of fossil energy used in both plant and animal production in developed and developing countries and at world level (see Table 2.1) shows that food security is fully dependent on a large consumption of fossil energy.

Three observations regarding these data are of particular relevance to our discussion. First, for the step of *agricultural production* alone, we use at world level an amount of fossil energy that is close to the food energy consumed in the diet. That is, the output/input ratio of food energy pro- duced to fossil energy consumed in agricultural production at world level is 1.2/1, meaning that 0.83 J of fossil energy are spent in world agriculture for each 1 J of food energy consumed at the household level. It should be noted that the direct consumption of the fossil energy input in agricul- tural production (the energy spent in agricultural production in Table 2.1) is just a fraction of the total amount of fossil energy spent in the entire food system. Especially in developed countries, the fossil energy con- sumed in postharvest activities is several times larger than the energy consumed in agricultural production (Pimentel and Pimentel, 1996; Heller and Keoleian, 2000).

The second observation concerns the marked differences in the ener- getics of food security between developed and developing countries, which make average world values meaningless. When considering the output/input ratio of food energy to fossil energy consumption, we see in

Table 2.1 Different Indicators of Fossil Energy Use in Agriculture

	Ratio output of food energy produced (MJ) to input of fossil energy consumed in agriculture (MJ)			Food energy consumption per capita (MJ/day)	
	Total food	Plant crops	Animal products	Total food	Animal food
World	1.2/1.0	1.8/1.0	1/2.5	11.6	1.8
Developed	1.0/2.0	1.0/1.2	1/4.3	13.6	3.6
Developing	2.2/1.0	2.6/1.0	1/1.0	11.1	1.3

	Total crop production	Fossil energy consumption in agriculture		Final food energy (in diet) production	
	per capita (MJ/day)	per hectare of arable land (GJ/ha)	per hour of farm labor (MJ/h)	per hectare of arable land (GJ/ha)	per hour of farm labor (MJ/h)
World	16.4	15.4	9	17.7	10
Developed	28.0	20.4	152	10.1	75
Developing	13.0	11.1	4	24.2	8

Source: Adapted from Giampietro, M., 2003a, Energy use in agriculture, in *Encyclopedia of life sciences* (ELS), http://onlinelibrary.wiley.com/doi/10.1038/npg.els.0003294/full.

Table 2.1 that developed countries consume 2 J of fossil energy in agriculture for each 1 J of food energy consumed in the diet, whereas developing countries get 2.2 J of food in the diet per each 1 J of fossil energy consumed in agriculture (0.45 J of fossil energy per 1 J of food). This reflects a different use of the fossil energy consumed in the production of crop plant per hectare: Developed countries consume 20.4 GJ/ha of fossil energy inputs to generate 10.1 GJ/ha of food energy, whereas developing countries consume 11.1 GJ/ha to generate 24.2 GJ/ha.

A more impressive difference is observed in the intensity of fossil energy use when considering the fossil energy consumed in crop production *per hour of labor*: Developed countries consume 152 MJ/h of fossil energy inputs to generate 75 MJ/h of food energy, whereas developing countries consume 4 MJ/h to generate 8 MJ/h of food energy. This is an important point, for economic development requires a dramatic increase in labor productivity, and this can only be obtained by heavy capitalization of the agricultural sector (more technical inputs per hour of labor). The difference in fossil energy consumed per hour of labor (used here as a proxy for the technical capital per worker) between developed and developing countries is of 38 times!

The third and last observation regards the share of animal products in the diet, which is an important determinant of the total consumption of fossil energy in food production (the overall output/input energy ratio). In fact, consumption of animal products in the diet implies a double conversion: first, crops are produced, which are then fed to animals. Table 2.1 shows that the overall intake of food energy in the diet (foods of plant and animal origin) is almost the same for developed and developing countries. In fact the difference (13.6 MJ/day per capita versus 11.1 MJ/day, respectively) may well be ascribed to differences in per capita *average* body mass (50 kg versus 30 kg, respectively), this in turn being related to different population structures (a larger share of children in developing countries). On the other hand, energy consumption from animal products is three times as high in developed countries compared to that in developing countries (3.6 versus 1.3 MJ/day per capita). This higher consumption of animal products translates into the need of producing more crops per capita (food and animal feed), depending on the types of animal products and technical coefficients of the production process. Note that, due to the low consumption of animal products in the diet, developing societies mainly use spare feeds (e.g. by-products and forages obtained on marginal land) rather than investing valuable crops in animal production. However, with this solution, the supply of animal products remains limited. If more animal products have to be produced, then dedicated crops have to be grown.

In the remainder of this section, we explore in detail the nature of the fossil energy dependence of the food supply in relation to the following three topics:

1. The two drivers requiring a continuous increase in the use of fossil energy inputs in agriculture: demographic and socioeconomic pressure
2. The relation between the growing phenomenon of urbanization and fossil energy consumption in the food system, especially with regard to the separation in space and time of the phase of food production and that of food consumption
3. The continuous increase in fossil energy consumption in the modern food system in relation to the widespread trend of an increasing share of animal products in the diet and in the level of convenience of food products

2.2.1   Demographic and socioeconomic pressure in relation to fossil energy consumption in agriculture

Technical progress in agriculture so far has resulted in a continuous increase in the use of external inputs (fertilizers, pesticides, irrigation, machinery, drying, and transport) and a corresponding increase in the consumption of fossil energy (Pimentel and Pimentel, 1979, 1996; Smil, 1987, 1991; Steinhart and Steinhart, 1974).

Technical "improvements" have been implemented to achieve two main objectives:

1. To use less agricultural land per capita while producing more food per capita to cope with demographic pressure
2. To use less hours of agricultural labor per capita while producing more food per capita to cope with socioeconomic pressure (i.e., a growing demand for labor hours in other sectors of the economy, especially the service sector) (Giampietro, 1997)

The relative presence of these two types of pressure entails a differential use of fossil-energy-based technical inputs in agricultural production: Irrigation and fertilizers are predominantly used to deal with demographic pressure, while machinery is employed to deal with socioeconomic pressure.

The implications of these two drivers on the consumption of fossil energy in agriculture have been investigated in detail using data referring to the 1990s (Giampietro, 1997; 2003a, b; Giampietro et al., 1999). A recent analysis (Arizpe et al., 2011) confirms the same findings using data from 2005. A sample of the results is given in Figure 2.1.

The upper two graphs in Figure 2.1 show that the higher the demographic pressure (i.e., the lower the arable land per capita) is, the greater is the use of irrigation and nitrogen fertilizer, technical inputs aimed at

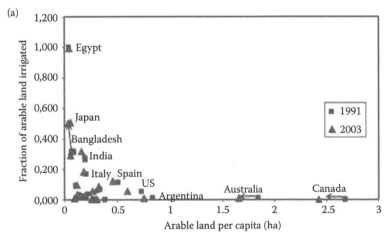

Figure 2.1 The use of fossil energy in world agriculture in relation to different inputs.

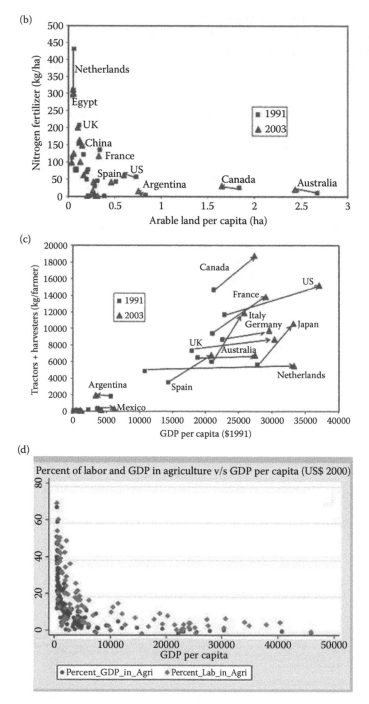

Figure 2.1 Continued

boosting land productivity. Indeed, these results confirm the findings that these two inputs are used more in densely populated countries, independently of the level of economic development (Giampietro, 1997; Conforti and Giampietro, 1997).

The two bottom graphs show that the higher the socioeconomic pressure (proxy: gross domestic product per capita) is, the lower is the percentage of the labor force allocated to agriculture (see Figure 2.1(d)) and the larger is the use of technical inputs like tractors and harvesters (see Figure 2.1(c)) aimed at boosting farm labor productivity. Indeed, the stunning performance of the industrial paradigm of agriculture is related not only to its ability to produce much more food on less land per capita but also to its ability to produce much more food using much less human labor. In fact, a prerequisite for economic development through industrialization and postindustrialization processes is the reduction of the percentage of the labor force engaged in agriculture so as to free labor for the industrial and service sectors. In the industrial era, countries with a gross domestic product higher than $10,000 per capita have less than 5% of the workforce allocated to agriculture. In the richest countries, the share of farmers in the workforce can fall as low as 2%. Just to give an idea, the entire amount of food consumed in one year by a U.S. citizen is produced using only 17 hours of work in the agricultural sector (Giampietro, 2003b).

2.2.2   Urbanization and the Copernican revolution in energy flows

According to the Population Division of the United Nations (*United Nations, Department of Economic and Social Affairs, Population Division*, 2010) more than 50% of the world population is now living in cities. This unprecedented phenomenon of urbanization at the global level indicates a clear discontinuity in human development: In the preindustrial era, the percentage of the population living in cities rarely exceeded 10% (Giampietro and Mayumi, 2009).

Massive urbanization entails important changes in the characteristics of the human food system, including a separation both in space and in time of the phase of food production (rural areas) and food consumption (urban areas) and an extreme specialization of the activities in the food system, from agricultural production to importation/exportation, industrial food processing, packaging, distribution, retailing, home storage, and eventually high-tech home preparation.

Urbanization is closely related to the Copernican revolution that took place in the energy flows in human society: without the massive switch to the use of fossil energy to carry out human affairs, massive urbanization simply would have been impossible. As illustrated in Figure 2.2a, in pre-industrial times, cities were the points where the diluted energy produced

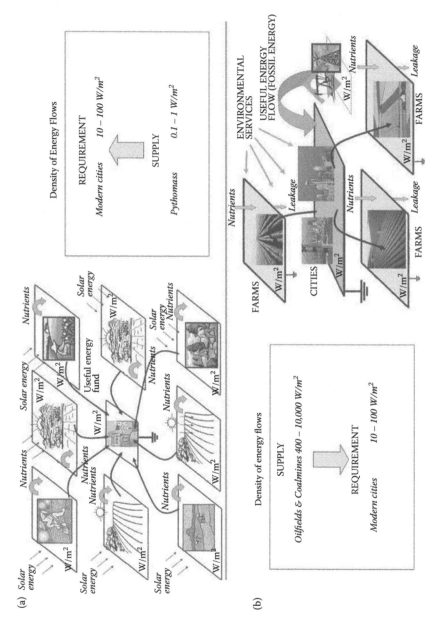

Figure 2.2 The different role of cities in the handling of energy flows.

by agriculture was concentrated in order to express a more diversified set of activities. To the contrary, modern cities are the points from which the concentrated flows of oil and other fossil energy fuels are distributed to the countryside to enable agricultural production (see Figure 2.2b). The driving force of this revolution is extremely clear if we look at the gradients in the density of energy flows in production and consumption (data on the density of energy flows in Figure 2.2 are from Smil, 2003, Figures 5.2 and 5.3).

A typical characteristic of the food system of modern urban society is that postharvest activities consume much more fossil energy than agricultural production itself. As illustrated in Figure 2.3, in 1995 in the United States, approximately 7.3 J of fossil energy was spent for each 1 J of food energy intake of U.S. residents. Only 20% of this fossil energy cost was accounted for by food production itself; the main share of the cost (80%) was incurred in postharvest activities.

As Figure 2.3 shows, the sources of fossil energy consumption in the food system are:

1. Energy required to move around the various food items imported and exported in world agriculture
2. Energy required by the food industry for processing and packaging of food products

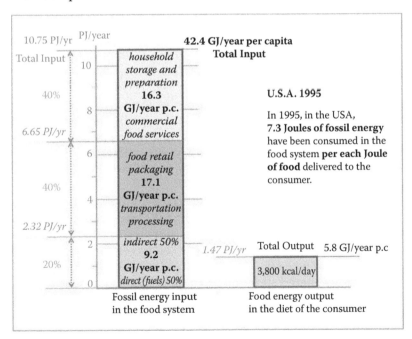

Figure 2.3 Breakdown of the total consumption of fossil energy in the food system.

3. Energy required for distribution, handling, and refrigeration of food items to and within shops
4. Energy required for conserving (refrigerator and freezer), preparing, cooking, and consuming food at the household level

The high energy cost in the postharvest sector is exacerbated by the significant food losses that necessarily occur during handling, processing, transport over long distances, and storage for long periods of time. Up to 50% of fresh fruits and vegetables and up to 20% of grains can be lost in the long journey from the field to the dinner table (Pimentel and Pimentel, 1996). This obviously augments the overall requirement of fossil energy per joule of food energy eventually consumed in the diet.

2.2.3   The effect of changes in the quality of the diet

As income increases, direct per capita food consumption of maize and coarse grains declines as consumers shift to wheat and rice. When incomes increase further and life style changes with urbanization, a secondary shift from rice to wheat takes place. In general, existing assessments project a continuation of these trends. The expected income growth in developing countries could become a strong driving force for increases in total meat consumption, in turn inducing strong growth in feed consumption of cereals. (IAASTD, 2009, p. 274)

This quote confirms the observation that economic growth is generally accompanied by marked changes in the quality of the diet. In this context, it is particularly useful to make a distinction between direct and indirect consumption of food grains. The breakdown of different end-uses of total grain consumption per capita in pre- and postindustrialized societies is illustrated in Figure 2.4.

As shown in Figure 2.4, in postindustrial countries, total grain consumption per capita is between 700 and 1,000 kg per year. But out of this amount only about 110–150 kg is consumed directly (in the form of bread, rice, pasta, breakfast cereals, etc.). The larger share goes into feed for animals (about 400–600 kg) and alcoholic beverages (depending on the country, this share can be more than 100 kg). Completely different is the situation in low-income countries, where grains generally represent the staple food (excluding some tropical countries where starchy roots play this role) and total consumption per capita amounts to about 300 kg per year. In these countries, direct grain consumption per capita is higher (up to 200 kg per year) than in wealthy countries, whereas indirect

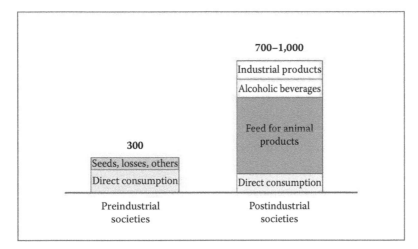

Figure 2.4 Direct and indirect consumption of grain for two different types of diet.

consumption, even when including losses and seed preservation, is much lower (around 100 kg per capita per year). Of course, the rough distinction between pre- and postindustrialized countries hides a much more articulated definition of consumption patterns.

2.2.4   Conclusions on the dependence of food security on fossil energy

At present, the food supply of humankind is fully dependent on fossil energy. This dependence is driven by a shortage of land and labor for use in food production. Demographic and economic pressure—the latter through changes in the quality of the diet—will further increase food requirements and hence the demand for fossil energy in food production. The continuing process of urbanization will inevitably augment the fossil energy costs in the postharvest system.

2.3   The energetics of agro-biofuel production

2.3.1   Debunking a systemic error in the appraisal of agro-biofuels

Nonhebel (2010) defines the conditions under which the use of biomass as an energy source could be considered a renewable resource with zero CO_2 emissions:

> One of the options to reduce emissions is the use of biomass as an energy source. The CO_2 emitted when using biomass as energy source equals the CO_2

> captured earlier that year in the photosynthetic pro-
> cess of the crops, on an annual basis no extra CO_2 is
> emitted to the atmosphere. (Nonhebel, 2010, p. 349)

Under these specific conditions, replacing the current use of oil with liquid agro-biofuel could represent a positive step toward sustainability, since it would mitigate both the depletion of oil stocks and the CO_2 accumulation in the atmosphere.

Unfortunately, in relation to the present production of liquid agro-biofuels in developed countries, the situation described by Nonhebel is mere wishful thinking. Indeed, this section shows that current liquid agro-biofuel production involves the consumption of massive quantities of fossil energy. Without fossil energy use (and without subsidies), there would be no liquid agro-biofuel production in developed countries.

We want to explain in this section why a fully renewable, zero-emission production of liquid biofuels, covering all of its own energy costs, is simply unfeasible. The widespread belief that the production of liquid agro-biofuels can solve sustainability problems is based on a sloppy analysis of the energetics of biofuel production (Giampietro and Mayumi, 2009). Since energy is a slippery concept, we point out below the major conceptual problems encountered in the assessment of biofuels in order to debunk the systemic error found in the vast majority of assessments of biofuel production.

2.3.2   Mandatory distinction between primary energy sources and energy carriers

Following the theoretical concepts developed within the field of energetics, it is essential to make a distinction between *primary energy sources* on the one hand and *energy carriers* on the other (Giampietro and Mayumi, 2009). Primary energy sources are physical gradients, such as solar energy, fossil energy stocks, or waterfalls, that enable the generation of a net supply of energy carriers. Primary energy sources measure the availability of energy in relation to external constraints—that is, the energy that society (seen as a black box) is taking/requiring from its environment. Energy carriers, in turn, are vectors, such as liquid fuels, electricity, or steam, that are *generated from a primary energy source by consuming (another) energy carrier.* Energy carriers represent the energy input required by the various compartments of the society to generate useful work; they refer to the functioning of the various parts of the society and the related internal constraints (inside the black box). Any energetic assessment requires both of these pieces of information.

The quantitative accounting of primary energy sources is done in terms of physical units (e.g., barrels of oil, tons of coal) rather than energy units (joules) to stress the function of this assessment. Indeed, the concept of primary energy sources is meant to assess the severity of *external constraints*, that is, the overall biophysical demand on the context implied by the energy consumption of a society. Those familiar with energy analysis have certainly come across the use of tons of oil equivalent (TOE) in national energy statistics (or tons of coal equivalent [TCE] in older statistics) as the reference primary energy source to express total energy consumption. These assessments of total energy consumption in TOE indicate the size of the required stocks or imports of primary energy sources associated with a given metabolic pattern (the total consumption of the black box in relation to its context).

On the contrary, assessments of energy consumption in terms of energy carriers refer to processes taking place inside the society (inside the black box), and are generally assessed in energy units (e.g., kWh for electricity or joules for liquid fuels). They indicate local requirements of a flow of specialized vectors of energy used to fulfill specific tasks. Examples of the latter are the consumption of liquid fuels for transportation, the electricity consumed in a house, and the steam used in an industrial plant.

2.3.3 The chicken-or-egg paradox in the exploitation of primary energy sources

The unrivaled success of reductionism in science has led to the blind adoption of linear analysis in the appraisal of virtually any process of production. Such linear analysis is typically represented by a default input/output system of accounting in which the efficiency of the process is captured by a simple number: the ratio between the output and the input flows. The assessment of the exploitation of primary energy sources has been no exception to this rule, and the linear approach of net energy analysis has invaded the scientific literature with the mantra "the higher the energy output per energy input, the higher the quality of the energy source."

This is very unfortunate, because a linear approach cannot be applied blindly to the assessment of the exploitation of primary energy sources. In fact, we must respect the laws of thermodynamics. The first law of thermodynamics rules out the possibility that the output is larger than the input in any energy transformation; the second law dictates that it is not even possible to break even. It is in this context in particular that the distinction between primary energy sources and energy carriers is of paramount importance, for it allows us to get out of the thermodynamic impasse.

Whenever we deal with an output/input energy ratio greater than one—an event rather commonly encountered in the literature on biofuels—we necessarily refer to the return (measured in joules) of energy carriers *per unit of energy carrier invested* (also measured in joules). That is, we necessarily deal *only with a part* of the whole set of energy conversions involved in the exploitation of primary energy sources (as in the simple linear view in Figure 2.5a). In fact, according to the laws of thermodynamics, the losses of conversions plus the net output of the process of exploitation are being paid for by a corresponding amount of primary energy sources lost in the process.

Thus, a *correct* application of net energy analysis to energy transformations requires a more sophisticated system of accounting (the "chicken or the egg" view in Figure 2.5b) than that commonly encountered in the literature. Indeed, in order to respect the rules of thermodynamics, we must make and use two distinctions:

- Between energy associated with primary energy sources (expressed in some physical unit other than joules) and energy associated with energy carriers (measured in joules)
- Between gross and net supply of energy carriers

Figure 2.5a shows the conventional linear representation of net energy analysis: the energy investment (input) goes into the process of exploitation of a primary energy source (PES) and gives us a return (output). As observed above, we are dealing here with an accounting done exclusively in terms of joules of energy carriers; the amount of primary energy sources consumed in the process is conveniently neglected. Hence, this method of accounting can result in an output/input ratio larger than one.

By using the distinction between primary energy sources and energy carriers, we can finally gather the two pieces of information relevant to the analysis of this conversion:

- *Internal constraint:* A quantitative assessment of the conversion of input of energy carrier into output of energy carrier mentioned

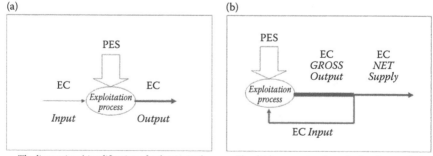

The linear view (simplification of reductionism).   The chicken-egg view (complexity of energetics).

Figure 2.5 Two systems of accounting for the output/input energy ratio of a primary energy source.

above; that is, how much input has to be invested by society in the exploitation?

- *External constraint:* The overall conversion of primary energy source into a net supply of energy carrier; in other words, how much primary energy source is needed to generate the required net supply of energy carrier?

If we want to generate useful information about this double task, there is a clear theoretical problem with the adoption of a linear representation. Indeed, in Figure 2.5a we must assume that the input of energy carriers invested in the process is already available to the exploiter before the exploitation takes place. Thus, when assuming a situation of steady state in a *self-sustaining* process, we have to adopt a circular view of the process, as illustrated in Figure 2.5b. In doing so, the direction of the arrow of the input of energy carrier represents a violation of the view used by reductionism. In this representation of energy flows, we have a chicken-or-egg dilemma: the output is needed by the system and entering into the process before the input!

2.3.4 *The output/input energy ratio and the definition of external and internal constraints*

According to net energy analysis, it is only the *net gain* of energy carriers delivered to society that matters in the exploitation of primary energy sources. Indeed, even if we have a huge primary energy source available—for example, the solar energy reaching our planet—its exploitation is not necessarily convenient. Like the pearls dispersed in the sea, if the cost of finding and gathering is higher than the net return, they do not represent an effective economic resource. Similarly, the assessment of the *gross output* of an energy carrier, as an independent piece of information, is irrelevant. We can generate a large supply of energy carriers from a given primary energy source, but then if 99.9% of this flow has to be reinvested in the production process, the resulting net supply for society can be negligible.

For all these reasons, a sound energetic analysis of the quality of energy sources should be able to assess the relationship between the primary energy source (associated with a definition of the external constraints) and the *net supply* of energy carrier (determining the conditions of feasibility in relation to internal constraints). Unfortunately, in the literature on biofuels, the only two quantitative assessments presented and debated are the output/input ratio of the process of exploitation and the gross output of the process. We claim that these two pieces of information do not provide any *direct* useful information by themselves. They are just two required ingredients to generate useful information.

We have to establish a relation between this quantitative information (the output/input ratio and the gross output) and the relevant information required to assess the viability and desirability of agro-biofuel. That is:

1. How much primary energy source is required to get the needed net supply?—to check external constraints.
2. What is the required investment of input to get the needed net supply?—to check the internal constraints (the availability in society of the required investment of capital, labor, and other internal resources).

When we try to frame the assessment of the exploitation of a primary energy source in terms of the chicken-or-egg process, as illustrated in Figure 2.5b, we soon realize that we have to acknowledge yet another complication: We have to study the implications of the distinction between the concepts of *gross* versus *net* supply of energy carriers. The gross supply of energy carriers is *the output* usually considered in the standard analysis. However, this output refers to the gross revenue of the process, which is different from the net supply, or the profit of the process after accounting for the expenses.

This conceptual distinction should always be addressed when studying the energetics of the exploitation of primary energy sources because, at low values of the output/input energy ratio, there is a *nonlinear* relation with the ratio of gross supply to net supply of energy carriers (joules). The mathematical relationship between the familiar output/input ratio used in net energy analysis and the concepts of gross and net supply of energy carriers is:

$$\text{Gross/Net} = (\text{Output/Input}) \times 1 / [(\text{Output/Input}) - 1]$$

The importance of making the distinction between gross and net supply of energy carrier using this mathematical relation is illustrated in Figure 2.6 (from Giampietro and Mayumi, 2009). As shown by the graph, when exploiting a primary energy source with a low output/input ratio of energy carriers (e.g. smaller than 2:1), the relation between the output/input ratio and the gross/net supply ratio falls within the range displaying nonlinearity.

This important point has been overlooked thus far in the literature on energy analysis, perhaps due to the high output/input ratio typical of fossil energy (larger than 12:1). At large values of the output/input ratio, there is a small difference between gross and net supply. Indeed, when dealing with a high-quality primary energy source, such as fossil energy, the internal loop of consumption of energy carriers in the

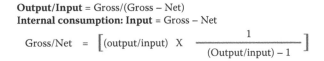

Output/Input = Gross/(Gross – Net)
Internal consumption: Input = Gross – Net

$$\text{Gross/Net} = \left[(\text{output/input}) \times \frac{1}{(\text{Output/input}) - 1} \right]$$

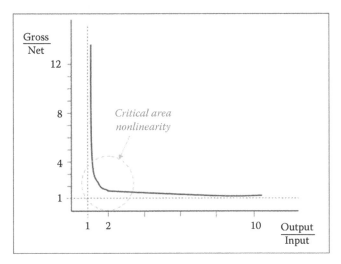

Figure 2.6 The nonlinear relation between output/input ratios and gross/net ratios.

exploitation process does not significantly affect the resulting net supply of energy carriers. Probably this explains why the conceptual distinctions (1) between primary energy source and gross energy carrier and (2) between gross and net supply of energy carrier were not recognized as being relevant in the recent past, not even by those working in the field of energetics.

In fact, in the 20th century, humans have been dealing almost exclusively with the exploitation of fossil fuel, an extremely high-quality primary energy source. When dealing with fossil energy, the quantitative estimate of the gross supply of the energy carrier is a fairly good estimate of the net supply of the energy carrier. A difference of about 15% in the value of latter flow may very well be within the error bar of the quantitative assessment itself. However, as soon as the output/input ratio of the energy carriers falls below the threshold of 2:1, which is the case with most liquid biofuels (when refraining from creative accounting), the internal losses of energy in the exploitation process cause a nonlinear decrease in the net supply of energy carriers.

Moreover, in the case of liquid agro-biofuels, the primary energy source can be considered to be the required amount of crops (feedstock of the process), which maps onto a set of requirements of land, water, soil, and nutrients (external constraints). As shown in Figure 2.6, when the output/input ratio falls below the threshold of 2:1, the internal consumption may become intolerably high, making the exploitation of the primary energy source very undesirable. This is why it is essential to consider these conceptual differences when dealing with the energetics of biofuels.

2.3.5 Relevance of these concepts to the assessment of biofuel production

Using the conceptual distinctions discussed so far, we can define a series of two energy conversions that are required when moving from a given amount of primary energy source (PES; e.g., corn or sugarcane) to the net supply (in joules) of biofuel energy carriers (EC; e.g., ethanol or biodiesel) to society. The relation between PES and net EC is determined by two conversions: PES to gross EC and gross EC to net EC (see Figure 2.5b). The integrated set of transformations involved in the production of ethanol from corn defined in this way is illustrated in Figure 2.7.

The data illustrated in Figure 2.7 show the poor quality of corn as a primary energy source when used to produce liquid energy carriers. In this example, taken from Giampietro et al. (2010), we use two different assessments of the process of ethanol production from corn: in the

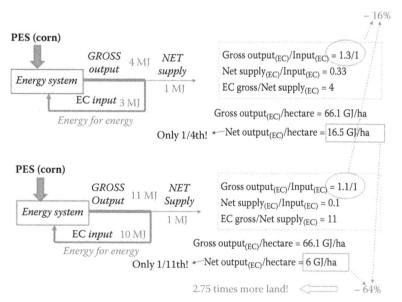

Figure 2.7 The nonlinear consequences of a change in output/input ratio in the production of a net supply of ethanol from corn.

upper graph, we adopt a generous assessment of the output/input ratio of 1.33:1 (calculated as MJ of EC ethanol output versus MJ of fossil EC going in), which includes energy credits for the possible use of the by-products of the process. In the lower graph, we adopt a more realistic assessment of the output/input ratio of 1.1:1. For a discussion on these values, see Giampietro and Mayumi (2009).

There are three basic points to be driven home from Figure 2.7. First, the assessments of biofuel published in the scientific and popular literature are based on *gross output* rather than *net supply* and therefore are misleading. It is simply untrue that by operating a biofuel system we can obtain 66 GJ/ha, the typical value found in the literature (Giampietro and Mayumi, 2009). If we are serious about considering this as a truly renewable source (catching the CO_2 of the energy consumption, including the internal consumption of the process), then the net supply of energy carriers that *can* be delivered by such a system to society is only about 6–12 GJ/ha.

Second, the performance of the system is subject to strong nonlinearity. In this example, a reduction of 16% in the output/input ratio (from 1.33 to 1.11) entails a much higher reduction (of 64%) of the density of the net supply. This nonlinearity implies huge increases in the requirement of primary energy resources per unit of net supply of energy delivered. When 11 liters of ethanol have to be produced to obtain a net supply of just one liter of ethanol for society, the corresponding land requirement (or other biophysical requirements or environmental costs such as water demand, soil erosion, pesticide pollution, nitrogen leakages, and so on) becomes intolerable. It simply translates into an unacceptable environmental/social impact per unit of net supply of energy delivered to society.

Third, the low value of the output/input ratio translates into the need of a large power level in the process of exploitation if we want to generate a large quantity of biofuel. That is, in order to keep the demand of labor in the energy sector low, the energy sector has to invest an incredibly high level of technical capital per hour of work (again when considering that the gross production of ethanol should be 11 times larger than the net supply to society).

The combination of these three points entails that large-scale production of corn ethanol in the United States is presently possible only because fossil energy is used as the input source in the process. In other words, it is *not* a self-sustained process. This being the case, one can only wonder why one should convert our scarce resources of oil, an extremely valuable energy carrier, into ethanol (a less valuable energy carrier, since it is less concentrated in terms of MJ per liter) if we gain hardly any joules in this process but only generate a drain on the economy and adverse environmental impacts (Giampietro and Mayumi, 2009).

2.4   Quantifying future competition for grains: Food for people versus fuel for cars

In this section, we establish a link between food and fuel. We use a quantitative assessment to illustrate the competition for grain crops if, in the future, produced crops will be used to supply both food for people and fuel for transportation. In particular, we run a scenario for the year 2030 assuming a population of 8 billion, changes in dietary habits toward increased meat consumption, and a level of production of agro-biofuel capable of replacing 10% of liquid fuels (referring to the targets implemented or discussed both in the European Union and United States). This analysis is inspired by a recent study carried out by Nonhebel (2010), even though several of our assumptions are different. In particular, we will not focus on bioenergy in general, as done by Nonhebel, but *only* on the performance of the production of liquid agro-biofuels. Furthermore, we use the analytical protocol explained in detail above, which is different from the protocol used by Nonhebel.

2.4.1   Assumptions about grain consumption for food in 2030

In 2000, world production of grain was about 2.3 billion tonnes (IAASTD, 2009, p. 277). Divided by the world population of 6 billion people, this translates into a world average consumption of about 380 kg per capita per year. This *world average* per capita consumption can roughly be explained using two rather distinct consumption levels (see Section 2.2):

1. An average consumption of 300 kg per capita per year for the 5 billion people living in developing countries (1.5 billion tonnes)
2. An average consumption of 800 kg per capita per year for the 1 billion people living in developed countries (0.8 billion tonnes)

The larger per capita consumption in developed countries reflects a diet with a large share of animal products.

For our scenario, in 2030 we assume a world population of 8 billion people (IAASTD, 2009, p. 262). Given the projection of economic development (IAASTD, 2009, p. 264), this total population can be divided into three income classes: high income (1 billion people), transition class (2 billion people), and low income (5 billion people). To these three groups, we assign three levels of per capita grain consumption: 800 kg/year for high income, 600 kg/year for transition, and 300 kg/year for low income. Indeed, according to Nonhebel (2010), the larger increase in consumption of animal products in the diet is associated with a change in per capita annual income from $2,000 to $10,000.

Table 2.2 Projections of Food Grain Consumption in 2030

| Income class | Population (billion) | Grain consumption per year | |
		kg per capita	total (million tonnes)
High income	1	800	800
Transition	2	600	1,200
Low income	5	300	1,500
World total	8	437.5	3,500

As shown in Table 2.2, use of these assumptions leads us to a total requirement of grain for food of:

(800 kg × 1 billion) + (600 kg × 2 billion) + (300 kg × 5 billion)
= 3.5 billion tonnes

Note that this amount implies a 50% increase in the level of grain production (of 2.3 billion tonnes) achieved in 2000.

2.4.2   *Assumptions about the consumption of liquid fuels to be replaced with biofuels*

Future consumption of liquid fuels is expected to steadily increase because of the effect of economic growth in developing countries.

> For both the non-OECD [Organization for Economic Cooperation and Development] and OECD economies, steadily increasing demand for personal travel is a primary factor underlying projected increases in energy demand for transportation. Increases in urbanization and in personal incomes have contributed to increases in air travel and motorization (more vehicles per capita) in the growing economies. Modal shifts in the transport of goods are expected to result from continued economic growth in both OECD and non-OECD economies. For freight transportation, trucking is expected to lead the growth in demand for transportation fuels. (EIA, 2009, p. 13)

Projections of the U.S. Energy Information Administration (EIA, 2009) indicate an increase in the world total consumption of liquid fuels from 170 EJ (1 exajoule [EJ] = 10^{18} joules) in 2005 to 217 EJ in 2030. Note that this projected increase is based *only* on an increase in population size. Changes in the per capita consumption are not taken into account in the

Table 2.3 Scenario 2035: Consumption of Only Gasoline and Diesel by
Class of Consumer

Consumer class	Population (billion)[a]	Liquid fuel consumption per capita (liters/year)[b]
Heavy consumers (USA, Canada, others)	0.5	2,000
High consumers (Europe, Japan, others)	0.5	1,000
Transition (China, Latin America, others)	3.0	300
Low consumers (Africa, rest of Asia)	4.0	150
World	*8.0*	*375*

Sources:

[a] International Assessment on Agriculture Science and Technology for Development, *Agriculture at a crossroads: Synthesis report*, Vol. 7 (Washington, DC: Island Press, 2009).
[b] U.S. Energy Information Administration, *International energy outlook, 2009* (Washington, DC: EIA, 2009).

EIA projection, and annual liquid fuel consumption per capita is sup-posed to remain about the same: 28 GJ in 2005 and 27 GJ in 2030.

Not all liquid fuel consumption can be readily replaced with agro-biofuel. An important share of liquid fuels (e.g., kerosene) is used for light and cooking in poor rural areas where the larger share of the world popu-lation is living. Other liquid fuels are used for air and water transporta-tion, power production, and heating in modern society. In our scenario of substitution, we will focus only on the consumption of liquid fuels such as gasoline and diesel. Therefore, our assumption of liquid fuel consumption for transportation for our heterogeneous world population of 8 billion in 2030 is based on consumption levels of gasoline and diesel reported by EIA (2009) and shown in Table 2.3.

Using the data in Table 2.3, we obtain a total demand for liquid fuels for transportation of 3.0 trillion liters. This demand, consisting of two-thirds gasoline (with an energy content of 32 MJ/liter) and one-third die-sel (with an energy content of 38 MJ/liter), is equivalent to a total demand in energy units of 100 EJ (EIA, 2009). Given that a liter of ethanol provides only 23.5 MJ of energy, this total demand of 100 EJ would correspond to 4.225 trillion liters of ethanol equivalent. Thus, when replacing fossil fuel with ethanol—with the consumption of both being expressed in units of volume (e.g., liters or barrels)—we have to multiply the demand by a factor of approximately 1.4 (4.225/3.0). In our calculations, we use data referring to ethanol because ethanol represents almost 90% of current biofuel pro-duction (Giampietro and Mayumi, 2009).

The implications of the target of satisfying 10% of the liquid fuel demand for transportation with biofuel are shown in Table 2.4. Calculations are based on the assumptions mentioned above. Thus, replacing 10% of

Table 2.4 Required Ethanol Production in 2030 to Meet the Target of 10% of Liquid Fuels

Consumer class	Population (billion)	10% of ethanol consumption (liters/year per capita)	Total ethanol demand (billion liters)
Heavy consumers	0.5	280	140
High consumers	0.5	140	70
Transition	3.0	42	126
Low consumers	4.0	21	84
World	*8.0*	*52*	*420*

the gasoline and diesel used for transportation with biofuel translates into the need of producing 420 billion liters of ethanol (or other types of biofuels to arrive at a total of 100 EJ) in 2030—keeping in mind that gasoline and diesel are only a part of the total liquid fuel demand.

2.4.3   Checking the external constraints for the required net supply

The relation between external constraints and the production of a flow of net supply of ethanol for society is illustrated in Figure 2.8. This is an application of the analytical framework presented earlier in Section 2.3

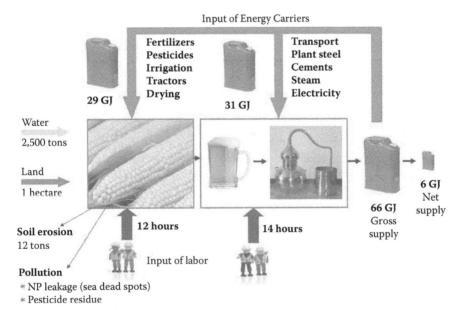

Figure 2.8 The link between external constraints, primary energy supply, and net supply of energy carrier for a corn-ethanol system.

(for more details, see Giampietro and Ulgiati, 2005; Giampietro and Mayumi, 2009). The quantitative assessments in Figure 2.8 refer to the flows associated with 1 hectare of corn production for ethanol in the United States. Hence, corn is considered the primary energy source, and its production process defines the external biophysical constraints—that is, the associated requirement of water and land and the resulting soil erosion and other stresses on the environment. The net supply of ethanol to society is considered here as the relevant output of the process.

As discussed earlier, the analysis of external constraints must be based on the requirement of primary energy source per unit of net supply of energy carrier. This relation is profoundly affected by the output/input ratio of energy carriers of the specific process of production. In this example (Figure 2.8), we do not include energy credit for the by-products of the process, which in principle can be used as animal feed, and simply adopt an output/input ratio of 1.1:1 based on the internal loop of energy carriers for energy carriers typical of corn-ethanol production in the United States. The reason for this choice has been explained in detail in Giampietro and Mayumi (2009): When enlarging the scale of production of ethanol from agro-biomass to cover a significant percentage of liquid fuels (e.g., 10%), the production of by-products would exceed the requirement of feed by orders of magnitude and, hence, rather than an energy credit, by-products should be considered an energy cost because of their high potential for pollution of water (biological oxygen demand).

Using the analytical approach explained in Section 2.3, we can say that the net supply of 255 liters of ethanol per hectare maps onto the production of 8 tonnes of corn, considered in this accounting as the primary energy source. That is, we can study the external constraints by using the ratio primary energy source (31 kg of grain) per unit of net supply (1 liter of ethanol energy carrier, equal to 23.5 MJ). Again, this is the only relevant information that ought to be considered when assessing the feasibility of a fully renewable biofuel production (covering all of its own energy costs and absorbing all the CO_2 emitted) in relation to external constraints.

This analysis of U.S. corn-ethanol production already gives us a first rough idea about the absurdity of the target of replacing 10% of liquid fuels with biofuels in postindustrial countries. When considering the replacement of 10% of the gasoline and diesel fuels consumed in the United States (which are only a part of the total liquid fuels consumed per capita in the United States) using a corn-ethanol system of production, the net supply of 280 liters of ethanol per capita per year (obtained by operating a fully renewable system, covering all its own energy consumptions) would require the production of 8,800 kg of grain/year per capita. That is, with this solution, U.S. citizens would require the production of 1 ton of grains for the diet (the current level of consumption) and almost 9 tons of grains for replacing 10% of their consumption of gasoline and diesel. In addition to the huge toll on the agricultural sector, people in the United States would still need a sufficient oil supply to provide the remaining 1,800 liters of fossil fuels consumed in the form of gasoline and diesel (which are only a small share of the total energy expenditure!). As a matter of fact, the 280 liters of ethanol per person per year (6.6 GJ) would represent about 2% of the *total* per capita fossil energy consumption of an average U.S. resident.

Obviously, this enormous demand for grain would translate in a tremendous impact on the environment. In fact, to replace this negligible fraction of fossil energy consumption, the chosen solution to produce ethanol from corn will also imply multiplying by nine times the soil erosion, the use of pesticides, the use of fertilizers, and the amount of land in corn production. This phenomenon will be amplified by the need to cultivate marginal land where the environmental impact (the biophysical costs indicated on the left of Figure 2.8) will be higher. Clearly, increasing the hectares in corn production nine times will have a tremendous impact on biodiversity and the preservation of traditional landscapes.

A second application of our system of accounting is illustrated in Figure 2.9 and refers to the production of ethanol from sugarcane in Brazil (data from Giampietro and Mayumi, 2009). The density of the net supply of ethanol per hectare, in this case, is much higher than that of the corn-ethanol system (but still an order of magnitude lower than the density achieved with fossil energy), and also the output/input ratio is much

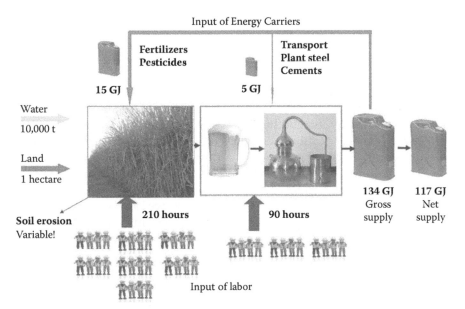

Figure 2.9 The link between external constraints, primary energy sources, and net supply of energy carrier for a sugarcane-ethanol system.

higher (7:1). However, as discussed below, this option remains problematic for its poor performance in relation to internal constraints (very low labor productivity) and for the dubious balance of CO_2 when the effects of land-use changes are considered.

2.4.4   Checking the internal constraints for the required net supply

The feasibility and desirability of a large-scale generation of agro-biofuels can only be assessed by comparing two sets of characteristics:

1. The characteristics of the energy system producing a net supply of energy carriers
2. The socioeconomic characteristics typical of the socioeconomic system using the energy carriers (Giampietro and Mayumi, 2009; Giampietro et al., 2010)

In fact, the metabolic pattern of modern societies imposes a series of internal constraints on the performance of specialized economic sectors. Economic development and a longer life span of the population determine important changes in the socioeconomic structure of a society:

- The dependency ratio increases because of the presence of more retired people and a higher level of education (longer education period).
- Not only do fewer people work in society but these workers also enjoy a lighter workload per year.
- A larger share of the paid work hours are allocated to the service and government sector (over 65% in the most developed countries).

All these changes imply that the primary sectors (agriculture, energy, and mining) must deliver the required huge flows of net supply to society, using only a tiny fraction of the available work hours and a small fraction of capital (see also section 2.2, where we discussed the implications of economic growth on the number of farmers in modern society).

This required high level of labor productivity of rich postindustrial countries can be used as benchmark to assess the quality of agro-biofuels as an alternative primary energy source. That is, we can compare the characteristics of a given *energy system*, which transforms a primary energy source into a net supply of energy carriers (the supply), with the characteristics of the *society* that will use the energy carriers and that, in return for this supply, has to provide the required investment of labor, capital, and energy carriers (the demand). Such a comparison is presented in Figure 2.10 using data of the two large-scale processes of agro-biofuel production available at the moment: ethanol from corn in the United States and ethanol from sugarcane in Brazil (from Giampietro and Mayumi, 2009).

The numerical values of Figure 2.10 clearly indicate a systemic lack of feasibility of these two options. The characteristics of the supply side—that is, the net supply of energy carriers from the energy sector to the society per hour of paid work—are simply not consistent with the characteristics of the demand side. Neither the U.S. corn-ethanol system nor the Brazilian sugarcane-ethanol system would be able to successfully operate in a postindustrial society because they cannot live up to the required net supply of energy carriers per hour of labor in the energy sector. In the case of Brazil, the output/input ratio of energy carriers is acceptable (7:1), but the extremely low power level (< 70 MJ/hr of energy invested per hour of work in the exploitation) makes this production system incompatible with the benchmarks associated with the metabolic pattern of modern society. The U.S. corn-ethanol system shows the reverse situation: the power level is low but acceptable (1,400 MJ/hr of energy invested per hour of work in the exploitation), but exactly because of this achievement, the output/input ratio of energy carrier (1.1:1) is way too low to make the system compatible with a modern society.

As explained before, this systemic problem can be associated with the low quality of biomass as primary energy source, since the values of the two variables—output/input ratio of energy carriers and power level on the input side—affect each other. If one wants to increase the power

---

Ethanol Production from Corn (USA)

Output/input = **1.1/1**	Power of worker (input) = **2,300 MJ/hour**
Net supply = **6 GJ/ha**	Net supply = **230 MJ/hour**
Land demand for energy	Work demand for energy

Ethanol Production from Sugarcane (Brazil)

Output/input = **7/1**	Power of worker = **67 MJ/hour**
Net supply = **117 GJ/ha**	Net supply = **390 MJ/hour**
Land demand for energy	Work demand for energy

Fossil Fuels Benchmarks

Output/input > **10/1**	Power of worker > **2,500 MJ/hour**
Net supply > **300,000 GJ/ha**	Net supply > **25,000 MJ/hour**
Land demand for energy	Work demand for energy

Figure 2.10 Analyzing the feasibility and desirability of agro-biofuels as an alternative primary energy source using benchmark values.

level in order to gather a dispersed flow of primary energy source using few labor hours, then one has to invest a large amount of energy carriers in the exploitation process (and hence the output/input ratio plummets). On the contrary, if one wants to keep the output/input ratio high by reducing the power level, then one has to dramatically increase the labor time invested in the energy sector, generating jobs with a miserable salary.

2.4.5   Assumptions used for illustrating the competition between food for people and fuel for cars

Even though Sections 2.4.3 and 2.4.4 have already illustrated the total lack of feasibility (let alone desirability or common sense) of the idea of replacing liquid fossil fuels with liquid agro-biofuels, we present in these final sections our quantitative assessments of the competition for agricultural production between food for people and fuel for cars. These assessments show that large-scale production of liquid agro-biofuels would represent an extravagant and completely unnecessary burden for the agricultural sector. As a matter of fact, humankind already struggles to augment the present level of food production to match the increasing demand due to population growth and changes in standard of living, while maintaining ecological compatibility (IAASTD, 2009; MEA, 2005).

We start with the requirement of liquid biofuel production reported in Table 2.4 (that is, 10% of only gasoline and diesel consumption), which translates into the need of producing 420 billion liters of ethanol. Just to provide a rough assessment of what this implies in terms of agricultural production, we assume that of this total, 320 billion liters will be generated using the corn-ethanol system in the United States, Europe, part of China, and other temperate countries, and the remainder 100 billion liters will be generated by the sugarcane-ethanol system in South America, Africa, and other tropical countries. Using the two assessments of the requirement of kilograms of grain (primary energy source) per net supply of ethanol energy carrier, referring to the corn-ethanol system discussed before, we can make two scenarios based on two coefficients:

- Realistic scenario: 31 kg of grain per liter of ethanol (output/input ratio 1.1:1)
- Optimistic scenario: 11 kg of grain per liter of ethanol (output/input ratio 1.3:1)

A net supply of 320 billion liters of ethanol would require the production of an additional 9.9 billion tons of grains (on top of that required for the food system) when considering the realistic coefficient, or 3.5 billion

Table 2.5 Overall Grain Requirement for the Two Scenarios in 2030
(in Metric Tons per Year)

Scenario	Grain requirement in 2030			Increase 2030/2000
	For fuel	For food	Total	
Realistic	9,900 million	3,500 million	13,400 million	5.8/1
Optimistic	3,500 million	3,500 million	7,000 million	3.0/1

tons when considering the optimistic coefficient (see Table 2.5). Even before considering the additional burden caused by the production of the remaining 100 billion liters of ethanol biofuels from tropical agriculture using sugarcane or oil palms as the primary energy source, we can safely state that the agro-biofuel feedstock required from temperate agriculture would represent an amount of grains exceeding that required for feeding all people.

We now also consider the implications of producing the additional 100 billion liters of ethanol equivalent from tropical agriculture. First of all, we should be aware that when converting tropical natural land covers into monocultures or plantations for biofuel production in Brazil, Africa, or Southeast Asia, one generates a *biofuel carbon debt* that can release from 17 to 240 times more CO_2 than the annual greenhouse gas (GHG) reduction claimed by biofuel proponents (Fargione et al., 2008). The same principle applies to the extension of land in production for corn-ethanol production within the United States. Searchinger and colleagues (2008) calculate that converting natural land covers into corn-ethanol hectares nearly doubles GHG emissions over 30 years and increases GHGs for 167 years.

The second issue to keep in mind is that the low productivity of labor of those working in sugarcane cultivations or palm plantations for biofuel production translates into the creation of poorly paid jobs.

> When dealing with the case of sugarcane-ethanol in Brazil we can better focus on the issue of "desirability." That is, after assuming that a fully renewable ethanol production from sugarcane is feasible in Brazil, how desirable is this choice for Brazilians? Is this providing an adequate a decent income to the workers making a living out of it? Is it delivering enough net supply of energy carrier (per hour of labor) to support the economic activity of other workers operating in different economic sectors? Would this choice increase the material standards of living of the Brazilian people, moving it closer to the level enjoyed by more developed countries?

> Or rather would it slow down the movement of the Brazilian economy over its trajectory of economic development? Countries such as China, India, Korea, competitors of Brazil in the international markets, are relying on oil as primary energy source and are at the moment using their work force, their capital, their finances and resources to boost other economic activities—to produce added value linked to the production of goods and services. On the contrary, Brazil, with the top priority given to agro-biofuel development is using an increasing fraction of its economic resources and work force just to produce an "oil equivalent" ethanol supply—an input for economic activities, but not an economic activity in itself, unless the energy carrier is exported to other countries. (Giampietro and Mayumi, 2009, pp. 189–190)

The economic development of modern societies can be easily explained by a dramatic increase in the use of oil to support labor and capital in the production of goods and services. The idea of a dramatic increase in the investments of existing labor, capital, goods, and services to generate oil simply seems to head in the wrong direction.

2.5 Conclusions

> It is said that we should use alcohol and vegetable oils after the petroleum energy has been exhausted. This reminds one of Marie Antoinette's advice to the Paris poor to eat cake when they had no bread. (Brody, 1945, p. 968)

This quote (from 1945!) about the folly of liquid agro-biofuels is highly relevant to our discussion, not only because it correctly exposes the idea of substituting oil with biofuels as wrong thinking but also for the implications of attributing this wrong thinking to Marie Antoinette. With regard to the thinking stigmatized by the quote, we can generalize that if you face a problem generated by scarcity (a situation in which the requirement of something is larger than the available supply), any proposed solution that will increase the original requirement will make the problem worse rather than solving it.

The incredible success of the industrial revolution and the progressive postindustrialization of the most developed countries is the direct consequence of an explosion in the use of fossil energy. For thousands of years, humankind relied on energy sources generated in a renewable way and did not lead to GHG accumulation in the atmosphere. However, exactly because of this, all preindustrial societies shared a common set of bio-physical constraints that limited their expansion and development, illustrated by the differences in the density of energy flows indicated in Figure 2.2. Powering a human society with a natural chain of energy conversions taking place within the system (the ultimate green solution) means facing the task of concentrating the energy surplus, produced at the typically low density associated with biomass production, in a limited number of privileged spots (Giampietro and Mayumi, 2009). In this context, the reference to the idea of replacing bread with cake is extremely apropos, since our civilization has been generated by the ability of using fossil energy stocks to save both land and labor that before had to be invested in producing energy and food.

Indeed, fossil energy enabled the emancipation from the constraints of land: Rather than working against the gradient to concentrate a diluted flow of primary energy source (biomass: 0.1–1.0 W/m^2) into high-density energy consumption within a limited number of primitive cities (10–100 W/m^2; see Figure 2.2a), fossil energy made it possible to create a favorable gradient, powering cities with a highly concentrated flow of primary energy source (400–10,000 W/m^2; see Figure 2.2b). When turning to the cultivation of grain to produce biofuel, the density of the energy supply becomes even lower than that of early-time biomass because of the internal loop of energy carrier for energy carrier (reducing the original density of biomass by an order of magnitude, 0.02–0.4 W/m^2). Thus, if we want to use biomass for energy, making beer with it does not seem a step in the right direction! As a matter of fact, under existing technologies, it is much better to use bioenergy directly—for example, fired for producing electricity—than for producing liquid fuels.

Fossil energy made it possible to eliminate the heavy social yoke of a large number of poor peasants. The differences in power level (MJ/hr) illustrated in the comparison over benchmark values in Figure 2.10 clearly show the key role of fossil energy in the development of our present modern human society. The idea of replacing a significant share of fossil energy with biofuels goes in the opposite direction of this process of evolution. As explained earlier, it is highly improbable that liquid biofuel production will ever generate a significant number of well-paid jobs.

Another limitation to be considered when dealing with any form of large-scale use of bioenergy is that biomass plays a key role in stabilizing ecological processes, including the food security of humankind.

Clearly, in its transition toward a low-carbon economy, humankind will have to find alternatives to fossil energy. The movement to alternative energy sources will have to start with a reduction of many of the excesses in extravagant lifestyles found in postindustrial societies. In this transition, biomass certainly will have an important role to play; however, this has nothing to do with producing dedicated crops within the model of high-external-input agriculture.

In light of the metaphor of the "Marie Antoinette reasoning," we can say that it indicates a situation of Ancien Régime in which decision makers are no longer in touch with reality. Such a situation, in general, should be expected after a long period of stability in which the power structure could afford to filter unpleasant information for too long. After a situation of continuous success, complex political institutions can arrive at a point where they are no longer able to understand the predicament they are facing. In the case of agro-biofuels, this awkward situation has been generated by an excess of funding allocated to those scientists telling to the power structure what the power structure wanted to hear. The fairy tale the power structure wanted to hear, which unfortunately was told by many scientists, is that biofuels were the solution capable of solving simultaneously many sustainability problems: peak-oil, climate change, the crisis of the agricultural sector, and the crisis of the car industry.

Unfortunately, the promise of a magic bullet solving all these problems is indeed no more than a fairy tale: A large-scale production of liquid agro-biofuels is not feasible in our modern society. It would entail severe competition between the production of crops for food and the production of fuel for cars. In this competition, it is sure that the poor will be the losers.

In a more detailed discussion, Giampietro and Mayumi (2009) also claim that a large-scale production of liquid agro-biofuels is not good for the environment and does not help rural development. In fact, it is marginalizing the rural poor and destroying traditional farming systems through the expansion of monocultures and plantations throughout Latin America, Africa, and Southeast Asia.

The policy of huge investments in the development of liquid agro-biofuels has been adopted simply for the following political reasons:

1. To give subsidies to an agricultural sector finding itself in great difficulty
2. To help out the companies producing inputs for high-tech agriculture, such as genetically modified organisms (GMO), pesticides, fertilizers, and tractors, who are facing a growing resistance to the adoption of GMO crops in many developed and developing countries
3. To help the car industry by reassuring possible buyers scared by peak-oil

We personally believe that given the serious problems that humankind is facing in the third millennium, this definition of priorities is highly questionable in relation to both practical and ethical considerations.

References

Arizpe, N., Giampietro, M., and Ramos-Martin, J. 2011. Food security and fossil energy dependence: An international comparison of the use of fossil energy in agriculture (1991–2003). In Sustainability of agriculture, ed. Paoletti, M., and Gomiero, T. Special issue. *Critical Review in Plant Sciences*, Vol. 30(1–2): 45–63.

BBC News. 2008. UN warns on biofuel crop reliance. July 18. http://news.bbc .o.uk/2/hi/7514677.stm.

Brody, S. 1945. *Bioenergetics and growth.* New York: Reinhold.

Chakrabortty, A. 2008. Secret report: Biofuel caused food crisis. *Guardian*, July 3. http://www.guardian.co.uk/environment/2008/jul/03/biofuels.renew ableenergy.

Conforti, P., and Giampietro, M. 1997. Fossil energy use in agriculture: An international comparison. *Agriculture, Ecosystems and Environment*, 65(3): 231–243.

Cottrell, W. F. 1955. *Energy and society: The relation between energy, social change, and economic development.* New York: McGraw-Hill.

Cronin, D. 2008. Europe: Warnings against biofuels get louder. IPS News, May 6. http://ipsnews.net/news.asp?idnews=42247.

Daly, H. E. 2005. Economics in a full world. *Scientific American*, 293(3): 78–85.

EIA (U.S. Energy Information Administration). 2009. *International energy outlook, 2009.* Washington, DC: EIA. http://www.eia.gov/forecasts/archive/ieo09/ index.html.

Fargione, J., Hill, J., Tilman, D., Polasky, S., and Hawthorne, P. 2008. Land clearing and the biofuel carbon debt. *Science*, 319:1235–1238.

Gever, J., Kaufmann, R., Skole, D., and Vörösmarty, C. 1991. *Beyond oil: The threat to food and fuel in the coming decades.* 3rd ed. Niwot: University Press of Colorado.

Giampietro, M. 1997. Socioeconomic pressure, demographic pressure, environmental loading and technological changes in agriculture. *Agriculture, Ecosystems and Environment*, 65(3): 201–229.

———. 2003a. Energy use in agriculture. Encyclopedia of Life Sciences (eLS). http://onlinelibrary.wiley.com/doi/10.1038/npg.els.0003294/full.

———. 2003b. *Multi-scale integrated analysis of ecosystems.* Boca Raton, FL: CRC Press.

Giampietro, M., Bukkens, S. G. F., and Pimentel, D. 1999. General trends of technological changes in agriculture. *Critical Reviews in Plant Sciences*, 18(3): 261–282.

Giampietro, M., and Mayumi, K. 2009. *The biofuel delusion: The fallacy of large-scale agro-biofuel production.* London: Earthscan.

Giampietro, M., Mayumi, K., and Sorman, H. A. 2010. Assessing the quality of alternative energy sources: Energy return on the investment (EROI), the metabolic pattern of societies and energy statistics. Report on Environmental Science no. 6. ICTA–Universitat Autònoma Barcelona.

Giampietro, M., and Ulgiati, S. 2005. An integrated assessment of large-scale biofuel production. *Critical Review in Plant Sciences*, 24:1–20.

Hall, C. A. S., Cleveland, C. J., and Kaufman, R. 1986. *Energy and resource quality.* New York: John Wiley & Sons.

Heller, M., and Keoleian, G. 2000. *Life-cycle based sustainability indicators for assessment of the U.S. food system.* Ann Arbor: Center for Sustainable Systems, University of Michigan. http://css.snre.umich.edu/css_doc/CSS00-04 .pdf.

IAASTD (International Assessment on Agriculture Science and Technology for Development). 2009. *Agriculture at a crossroads: Synthesis report.* Vol. 7. Washington, DC: Island Press.

IEEJ (Institute of Energy Economics of Japan). 2006. Asia/world energy outlook. Presented at the 395th Forum on Research Works, 21 September 2006, Tokyo, Japan, 36. http://eneken.ieej.or.jp/en/data/pdf/365.pdf.

Kutas, G., Lindberg, C., and Steenblik, R. 2007. *Biofuels—at what cost? Government support for ethanol and biodiesel in the European Union.* Geneva: Global Subsidies Initiative (GSI-IISD). http://www.globalsubsidies.org/files/assets/Subsidies_to_biofuels_in_the_EU_final.pdf.

Leach, G. 1976. *Energy and food production.* Guildford, England: IPC Science and Technology Press.

MEA (Millennium Ecosystem Assessment). 2005. *Ecosystems and human well-being: Synthesis.* Washington, DC: Island Press. http://www.maweb.org/documents/document.356.aspx.pdf.

Nonhebel, S. 2010. Biomass for energy and the impacts on food security. In *Energy option impact on regional security: Proceedings of the NATO Advanced Research Workshop on Energy Options Impact on Regional Security, Split, Croatia, 17–20 June 2009,* ed. Barbir, F., and Ulgiati, S., 341–361. Dordrecht, The Netherlands: Springer.

Odum, E. P. 1989. *Ecology and our endangered life-support systems.* Sunderland, MA: Sinauer Associates.

Odum, H. T. 1971. *Environment, power, and society.* New York: Wiley-Interscience.

Pimentel, D., and Pimentel, M. 1979. *Food, energy, and society.* London: Edward Arnold.

———. 1996. *Food, energy, and society.* Rev. ed. Niwot: University Press of Colorado.

Pingali, P. 2007. Westernization of Asian diets and the transformation of food systems: Implications for research and policy. *Food Policy,* 32(2): 281–298.

Searchinger, T., Heimlich, R., Houghton, R. A., Dong, F., Elobeid, A., Fabiosa, J., Tokgoz, S., Hayes, D., and Yu, T.-H. 2008. Use of U.S. croplands for biofuels increases greenhouse gases through emissions from land-use change. *Science,* 319:1238–1240.

Smil, V. 1987. *Energy, food, environment.* Oxford, England: Clarendon Press.

———. 1991. *General energetics: Energy in the biosphere and civilization.* New York: John Wiley & Sons.

———. 2003. *Energy at the crossroads: Global perspectives and uncertainties.* Cambridge, MA: MIT Press.

Spiegel Online International. 2008. World Bank leak: Biofuels may be even worse than first thought. July 4. http://www.spiegel.de/international/world/0,1518,563927,00.html.

Steinhart J. S., and Steinhart, C. E. 1974. Energy use in U.S. food system. *Science,* 184:307–316.

Stout, B. A., ed. 1992. *Energy in world agriculture.* 6 vols. Amsterdam: Elsevier.

United Nations. 2010. World urbanization prospects: The 2009 revision. UN Department of Economic and Social Affairs, Population Division. http://esa.un.org/unpd/wup/index.htm.

World Bank. 2008. *Biofuels: The promise and the risks.* World Development Report, 2008. Accessed July 6, 2011. http://econ.worldbank.org/WBSITE/EXTERNAL/EXTDEC/EXTRESEARCH/EXTWDRS/EXTWDR2008/0,,contentMDK:21501336~pagePK:64167689~piPK:64167673~theSitePK:2795143,00.html.

chapter three

Energy cropping in marginal land: Viable option or fairy tale?

Sandra Fahd, Salvatore Mellino, and Sergio Ulgiati

Contents

3.1   Introduction

The promotion of biofuel crops on marginal lands started a decade ago to avoid competition for arable land used for food production. Many specialty crops have already been investigated for their capacity to yield nonfood oils or high ligno-cellulosic biomass. The choice of specialty oil crops for further development depends on the geographical location and the potential use of products and by-products. Set-aside land or marginal land has become the focus for the production of nonfood crops. Idle, degraded, underutilized lands, wastelands, and abandoned croplands are all different terms to refer to marginal or set-aside lands.

To regulate land use for food production, the European Union Common Agricultural Policy reforms in 1992 introduced a rotational set-aside requirement for almost all arable land (CEEC, 1992). The proportion of set-aside varies considerably among the EU members and also at regional levels within each country. Some policy makers propose that agro-fuel crops be planted on land that is considered marginal in order not to compete with food production. Various policy measures have been set up to promote the use of biomass, including biofuels. The European Union has built up a large research program for the assessment of land suitability for biofuels. Therefore, we analyzed the extent of nonfood crop production within the general framework of set-aside lands in the European Union (Berger et al., 2006).

The Campania Region in southern Italy has a large amount of marginal land that is no longer cropped due to insufficient economic return or because the land was polluted by industrial and urban waste that was inadequately treated or illegally disposed of. As a consequence, such marginal land cannot be used for food production. Our selection of land use is based on the integration of GIS (Geographic Information Systems) data (spatially continuous) with data derived from the agricultural census (spatially discrete). The land resource inventory contains climate, topography, soil, and land use as the main thematic layers (Fischer et al., 2010). The energy value of the grain output was quantified for biodiesel use, and the agricultural residues and glycerin are considered as secondary by-products.

The aim of this study is to evaluate a scenario of energy production from dedicated crops given the characteristics of the local environment: geomorphology, climate, natural heritage, and current availability of marginal land use. The energy, material, and environmental input flows for each cropping system were calculated considering the operative times and specific fuel consumption for the different tractors used and the input of fertilizers and pesticides.

3.2   Materials and methods

3.2.1   The oilseed crops

The two main oilseed crops promoted for biodiesel in Europe are rapeseed (*Brassica napus oleifera*), providing 73% of the feedstock, and sunflower, with only 18% of the feedstock. The genus *Brassica* includes six cultivated crop species: *B. nigra (L)*, *B. rapa (L)*, *B. oleracea (L)*, *B. carinata*, and *B. juncea (L)*. A wide range of cultivars belonging to the various species were studied within a national research network. After years of experimentation in various locales, *B. carinata* yielded the best results in terms of seed production (Copani et al., 1999; De Mastro et al., 1999; Rosso et al., 1999).

For our case study, we selected *Brassica carinata* due to its agricultural characteristics and its potential use. *B. carinata* is a native plant of Ethiopia commonly known as Ethiopian mustard. According to Cardone et al. (2003), the agronomic performance and the energetic balance confirm that *B. carinata* adapts better and is more productive in adverse conditions than other Brassica species. *B. carinata* is considered a promising oil feedstock for cultivation in coastal areas of central and southern Italy, where it is more difficult to achieve the productivity potentials of *B. napus*. Some authors claim that it offers the possibility of exploiting the Mediterranean marginal areas for energetic purposes.

The adaptability of *B. carinata* to the various physical and chemical conditions of the upper layer of soil under the influence of climatic factors, as well as its low demands in terms of agricultural care and investment put it in a more favorable position compared with other biofuel crops. It could be grown on different types of soils (clay and sandy types) with lower nutritional levels, as well as in a semiarid temperate climate. Agricultural tests on *B. carinata* show a great capacity to adapt to different pedoclimatic conditions in Italy. It grows well during the autumn–spring cycle. The ideal climatic conditions for good productivity of *B. carinata* are characterized by mild winters and moderate humid springs. The plant responds poorly to waterlogging in winter and drought during summer. The prolonged absence of rainfall may be harmful for the crop during germination. *B. carinata* has several desirable agronomic characteristics, such as resistance to a wide range of diseases and pests. Moreover, its chemical requirements in terms of fertilizers and pesticides are very few.

To evaluate the agronomic and energy performances of *B. carinata* as an energy crop in southern Italy, we used data from experimental tests (De Mastro et al., 2009) in Puglia in 2007–2008 and compared these data to average data from the literature. The data from Puglia were used for estimates for Campania, which has a similar climate. *B. carinata* was cultivated in the experimental tests using different low- and high-cropping intensity systems (hereafter designated low input, LI, and high input, HI), that is, different expenditures of technical means such as soil tillage, fertilizers, and chemical treatments. In the case of LI production, the amount of fertilizer used was 54 kg/ha of nitrate and 20 kg/ha of phosphate. For HI production, the amount of fertilizers was doubled, and herbicides were also used to reduce weed competition. Under LI conditions, the total yield of seed in 2008 was 1.2 t/ha, while with the HI cropping system, the seed yield was 1.7 t/ha.

The energy, material, and environmental input flows for each cropping system were calculated considering the operative times and specific fuel consumption of the different types of machinery used and of the fertilizers and pesticides used. To evaluate the agricultural phase and its performances, the energy value of the seed output was quantified as a feedstock for biodiesel transformation and the agricultural residues as a secondary product to be used as a fuel for heating. The transport of seed to the biodiesel plant for processing was also included in the assessment.

The biodiesel is obtained as oil from the crushing of the seeds and then refined through a transesterification process to render it more compatible with combustion in a diesel engine. The whole process consists of three main phases—agricultural production, extraction of seed oil, and transesterification—to produce biodiesel, as shown in Figure 3.1.

3.2.2 The marginal land cropped

The selection of land use is based on the integration of GIS data (spatially continuous; EEA, 2000) with data derived from the agricultural census (spatially discrete). In particular, we overlaid the soil-use pattern from GIS with the areas declared polluted by the Agenzia Regionale Protezione Ambientale Campania (ARPAC, Agency for Environmental Protection in Campania Region; see Figure 3.2). From this inventory, we estimated a potential extended area of marginal land (within the boundary that denotes contaminated areas) suitable for *B. carinata* crops.

Figure 3.3 shows the total nonirrigated area where the oilseed crop could be planted within the contaminated land in Campania. Areas within the boundary are the contaminated ones (according to ARPAC), while darkened areas are nonirrigated. Our assumption was to crop for energy only the areas that are *both* contaminated and nonirrigated.

The study, first performed on a hectare basis, was then extended to the larger fraction of marginal land in the whole region, identified through GIS mapping.

3.3.3 The assessment method: Toward an integrated evaluation approach

Evaluation procedures proposed so far by many authors (energy analysis, ecological footprint, etc.) have been applied to different space-time perspectives and aimed at different investigations and policy goals. Many of these evaluation procedures offer valuable insights toward understanding and describing aspects of resource conversions and use. Bioenergy studies most often focus on the potential energy yield, with little or no attention to other cost such as water demand or topsoil erosion. However, without integration of different points of view, it is very unlikely that biophysical tools can be coupled to economic tools to support environmentally sound development policies.

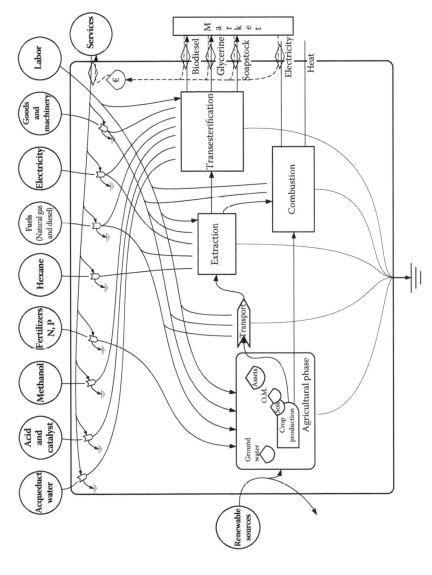

Figure 3.1 Diagram of biodiesel chain (systems symbols from Odum, 1996).

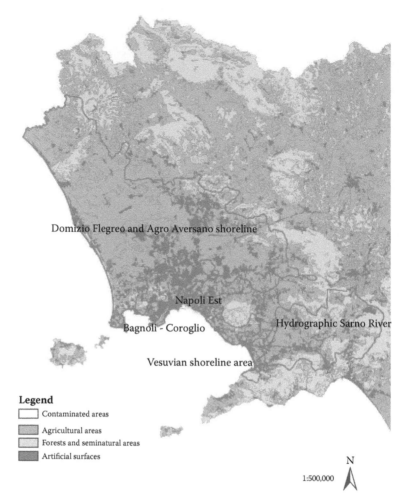

Figure 3.2 Land use data for the areas declared polluted in Campania.

We must be aware that the ecosystems and human economies are self-organizing systems, where processes are linked and affect each other at multiple scales. Investigating only the behavior of a single process and seeking maximization of one parameter (energy efficiency, production cost, or jobs, for instance) is unlikely to provide sufficient insight for sustainable policy making. Instead, if suitable approaches are selected—applicable at different scales and designed in such a way as to complement each other—integration would be feasible. Each of them may supply a piece of information about system performance at an appropriate scale, where the others do not fit, while their integration would supply an overall

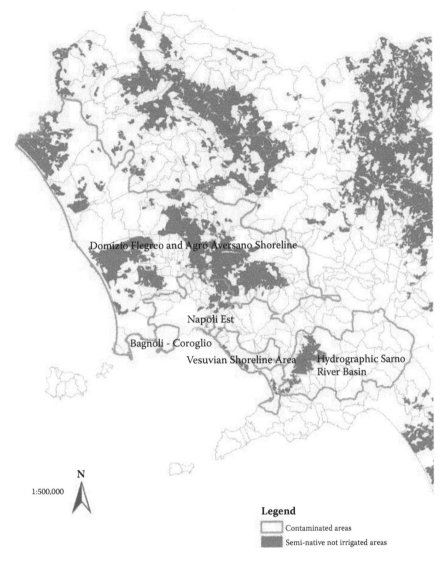

Figure 3.3 Semi-native areas (nonirrigated) within contaminated areas in Campania.

picture, characterized by an "added value" that could not be achieved by any approach individually.

To accomplish this goal, the multimethod, multiscale approach known as SUMMA (SUstainability Multiscale Multimethod Assessment; Ulgiati et al., 2006)—jointly based on assessing abiotic material, water, energy, emergy, airborne emissions, and economic flows—was applied to investigate the resource demand and use across space and time scales and to ensure a comprehensive understanding of the most important aspects of the agricultural system dynamics. In short, SUMMA is a conceptual framework for multicriteria decision making, consistent with life cycle assessment, in which the different perspectives complement each other but are not forced to combine. The approach is based on the parallel and joint application of a set of methods (material flow accounting, embodied energy analysis, emergy accounting, economic evaluation, and emission accounting) that provide complementary points of view on the complex issue of environmental impact and performance assessment. The methodology was used by Ulgiati and colleagues (2010) to assess different systems for cogeneration of heat and electricity, photovoltaic modules, alternative fuels and biofuels, transportation systems, waste management, and urban and agricultural systems.

3.2.4 *The analytical calculation procedure*

We first carried out a thorough inventory of all the input and output flows on the local scale of the farm, as well as of the other steps of the whole process (transport, industrial manufacture, etc.). It is important to underline that this inventory forms the common basis for all subsequent impact assessments, which are carried out in parallel, thus ensuring the maximum consistency of the input data and inherent assumptions.

The raw amounts of input and output flows from the inventory phase are multiplied by suitable conversion coefficients specific to each method applied. These express the "intensity" of the flow; that is, they quantify to what extent a material, energy, or environmental cost is directly or indirectly associated to the flow over its whole life cycle. Such coefficients are available in published life cycle assessment, energy, and environmental accounting literature. In so doing, the material, energy, and environmental "costs" associated with each flow are calculated for each step, according to the following generic equation:

$$C = \Sigma\, C_i = \Sigma\, f_i \times c_i, \text{ for } i = 1, \ldots, n,$$

where:

C = material, energy, or environmental cost associated with the investigated process

C_i = material, energy, or environmental cost associated with the i-th inflow or outflow of matter or energy

f_i = raw amount of the i-th flow of matter or energy

c_i = material, energy, or environmental unit cost coefficient of the i-th flow (from literature or calculated in this work).

The calculated material, energy, or environmental cost C is then divided by each step's product (seeds within farm, seeds at the plant gate, press oil, biodiesel) in order to yield the cost or impact indicator according to the method applied. When more than one product is generated, input and output flows are allocated to them according to traditional allocation procedures; in the emergy method only, the total emergy input is assigned to each coproduct according to the different "algebra" that characterizes the approach. Products from each intermediate step are fed to the next one, to which they contribute according to their material, energy, emergy, and emission intensities.

The process was divided into four steps: agricultural production, transport of seeds to the conversion plant, oil extraction, and finally biodiesel production (transmethylesterification). For each step, we performed an evaluation of the total abiotic material demand, total water demand, total embodied energy, total emergy, and finally main airborne emissions, over the whole life cycle of the product as well as of the matter and energy input flows. The tables presented in this section have been designed according to the equation above by multiplying each flow in the inventory by appropriate conversion coefficients in order to calculate the associated flow of embodied matter, water, energy, and emergy. Airborne emissions were calculated by multiplying the embodied energy flows by emission factors, thus generating the contribution of the energy used to the atmospheric impact.

3.3   Results

3.3.1   Material flows

The material flow accounting method (Schmidt-Bleek, 1993; Hinterberger and Stiller, 1998; Bargigli et al., 2004) evaluates the environmental disturbance associated with the withdrawal or diversion of material flows from their natural ecosystem pathways. The material inputs are divided into four different input categories: abiotic raw materials, water, air, and biotic raw materials (in this study, we account for only abiotic and water). Appropriate *material intensity* (MI) factors (g/unit) are multiplied by each input, respectively accounting for the total amount of matter that is directly or indirectly required to provide that very same input to the system. The resulting MIs of the individual inputs are then separately summed together for each environmental compartment and assigned to the system's output as a quantitative measure of its cumulative environmental burden from that compartment (often referred to as an "ecological rucksack").

All phases of a product have to be examined: production (including the extraction of raw materials, the manufacturing of preproducts, transport, and sales), use (including all consumption, transport, and repairs), recycling, and/or disposal. This extensive examination of the life cycle of a product is necessary, as it is not always apparent what environmental impact has occurred during manufacture or what impact is connected with the use of a product.

Table 3.1 converts the input flows to the agricultural and transport steps into embodied abiotic mass and embodied water, by means of abiotic and water intensity factors. The total amounts of abiotic matter and water are allocated to the seeds and straw produced according to their energy content. Then, the seeds enter the transport step and are delivered to the conversion plant. The total material and water flows for the seeds are added to the transportation inputs to yield cumulative flows of matter and water associated with the seeds in their final state at the gate of the plant. Table 3.1 clearly shows that the input flows to the transport step are negligible compared to the input flows to the agricultural step.

Brassica oil is extracted by pressing first and by solvent later, generating oil and a press cake as coproduct. The total abiotic material and water flows are allocated to these two products according to their energy content. The oil extracted is delivered to the transesterification phase, where it reacts with methanol and yields biodiesel and glycerin, plus other minor coproducts. Again, the input flows are allocated to the latter coproducts according to their energy content. Table 3.2 shows the input flows to each step and the results of the application of the equation in Section 3.2.4 to each flow.

As the result of all the calculations performed, Tables 3.1 and 3.2 provide the values of abiotic MIs (total demand of abiotic matter per unit of product) and water footprint (total demand of water per unit of product) in each intermediate and final step.

3.3.2   Embodied energy flows

Tables 3.3 and 3.4 show similar calculation procedures concerning the embodied energy, namely, the total energy investment calculated as oil equivalents and heat content in all the process steps and in all the processes that generated the input flows (Herendeen, 1998). More specifically, the embodied energy method focuses on fuels and electricity, fertilizers and other chemicals, machinery, and assets supplied to a process in terms of the oil-equivalent energy required to produce them, expressed in energy units per physical unit of good or service delivered (for instance, MJ per kg of steel). The method is concerned with the depletion of fossil energy, and therefore any process inputs of material and energy that do not require the use of fossil and fossil equivalent resources are not accounted for. Resources provided for free by the environment, such as topsoil and spring water, are likewise not accounted for by the embodied

Table 3.1 Mass Flow in the Agricultural and Transport Phases

Description of flow	Unit	Raw amount (ha⁻¹ yr⁻¹)	Abiotic intensity factor (g_{abiot} unit⁻¹)	Mass abiotic (unit ha⁻¹ yr⁻¹)	Water intensity factor (g unit⁻¹)	Mass water (unit ha⁻¹ yr⁻¹)	Ref
Agricultural phase							
Nonrenewable input (locally available)							
Top soil erosion	g	5.00E+05	0.76	3.80E+05	0.20	1.00E+05	[a]
Imported input							
Diesel and heavy fuel	g	7.74E+04	1.36	1.05E+05	9.70	7.50E+05	[b]
Fertilizers							
Nitrogen (N)	g	5.41E+04	24.98	1.35E+06	124.28	6.73E+06	[b]
Phosphate (PO₄)	g	2.00E+04	8.56	1.71E+05	59.52	1.19E+06	[b]
Agricultural machinery							
• steel and iron	g	3.15E+02	6.02	1.89E+03	11.41	3.59E+03	[c]
• aluminum	g	5.37E+01	10.27	5.51E+02	30.39	1.63E+03	[c]
• rubber and plastic material	g	3.84E+00	5.70	2.19E+01	146.00	5.60E+02	[d]
• copper	g	1.15E+01	179.07	2.06E+03	236.39	2.72E+03	[d]
Seeds	g	8.00E+03	0.73	5.81E+03	3.17	2.53E+04	[e]
Total material input				2.02E+06		8.80E+06	
Products and by-products (with allocation according to energy content)							
Seeds produced—dry matter	g	1.20E+06	0.73	.72E+05	3.17	3.80E+06	[e]
Straw (d.m., 30% left in soil)	g	2.38E+06	0.48	1.15E+06	2.10	5.00E+06	[e]

continued

Table 3.1 Mass Flow in the Agricultural and Transport Phases (Continued)

Description of flow	Unit	Raw amount (ha⁻¹ yr⁻¹)	Abiotic intensity factor (g_abiot unit⁻¹)	Mass abiotic (unit ha⁻¹ yr⁻¹)	Water intensity factor (g unit⁻¹)	Mass water (unit ha⁻¹ yr⁻¹)	Ref
Transportation phase							
Seeds produced—dry matter	g	1.20E+06	0.73	8.72E+05	3.17	3.80E+06	[e]
Steel for transport machinery	g	4.32E+02	5.44	2.35E+03	67.07	2.90E+04	[d]
Diesel for transport	g	9.00E+02	1.36	1.22E+03	9.70	8.73E+03	[b]
Products transported to the gate of the plant							
Seeds delivered	g	1.20E+06	0.73	8.76E+05	3.20	3.84E+06	[e]

Sources:

[a] Odum, H. T., *Environmental accounting: Emergy and environmental decision making* (New York: Wiley, 1996).
[b] Lettenmeier, M., Rohn, H., Liedtke, C., and Schmidt-Bleek, F., Resource productivity in 7 steps: How to develop eco-innovative products and services and improve their material footprint (Berlin: Wuppertal Institute for Climate, Environment and Energy, 2009).
[c] Bargigli, S., Analisi del ciclo di vita e valutazione di impatto ambientale della produzione ed uso di idrogeno combustibile, Ph.D. thesis, 2002.
[d] Average, Lettenmeier et al., 2009.
[e] Our calculation performed in this study.

Table 3.2 Mass Flow in the Industrial Phase

Description of flow	Unit	Raw amount (ha⁻¹ yr⁻¹)	Abiotic intensity factor (g_{abiot} unit⁻¹)	Mass abiotic (unit ha⁻¹ yr⁻¹)	Water intensity factor (g unit⁻¹)	Mass water (unit ha⁻¹ yr⁻¹)	Ref
Brassica oil extraction and refining phase							
Seeds delivered	g	1.20E+06	0.73	8.76E+05	3.20	3.84E+06	[a]
Electricity	J	1.04E+09	4.78E-04	4.99E+05	9.04E-03	9.44E+06	[b]
Natural gas	g	3.64E+04	1.22	4.44E+04	0.50	1.82E+04	[b]
Hexane	g	3.41E+03	1.36	4.63E+03	9.70	3.30E+04	[c]
Infrastructure							
• stainless steel	g	1.45E+02	16.18	2.35E-03	222.73	3.23E+04	[d]
• steel	g	2.77E+02	5.44	1.51E-03	67.07	1.86E+04	[d]
• cement	g	7.39E+02	2.75	2.03E-03	19.13	1.41E+04	[d]
Total material input				1.43E+06		1.34E+07	
Products and by-products (with allocation according to energy content)							
Brassica oil	g	3.96E+05	2.02	7.99E+05	18.91	7.49E+06	[a]
Brassica meal or pellets	g	6.96E+05	0.91	6.31E+05	8.49	5.91E+06	[a]
Transesterification biodiesel production							
Brassica oil	g	3.96E+05	2.02	7.99E+05	18.91	7.49E+06	[a]
Electricity	J	1.79E+08	4.78E-04	8.54E+04	9.04E-03	1.62E+06	[b]
Natural gas	g	1.36E+04	1.22	1.66E+04	0.50	6.79E+03	[b]
Methanol	g	5.66E+04	1.67	9.46E+04	4.46	2.53E+05	[b]
Acid (hydrochloric acid)	g	1.19E+04	3.03	3.61E+04	40.66	4.84E+05	[b]

continued

Table 3.2 Mass Flow in the Industrial Phase (Continued)

Description of flow	Unit	Raw amount $(ha^{-1} yr^{-1})$	Abiotic intensity factor $(g_{abiot}\ unit^{-1})$	Mass abiotic $(unit\ ha^{-1} yr^{-1})$	Water intensity factor $(g\ unit^{-1})$	Mass water $(unit\ ha^{-1} yr^{-1})$	Ref
Catalyst (sodium hydroxide NaOH)	g	8.73E+03	2.76	2.41E+04	90.31	7.88E+05	[b]
Infrastructure							
• iron	g	5.46E+00	16.69	9.12E+01	193.76	1.06E+03	[b]
• steel	g	9.11E+02	5.44	4.96E+03	67.07	6.11E+04	[d]
• cement	g	2.64E+02	2.75	7.25E+02	19.13	5.04E+03	[d]
Water	g	3.11E+05		0.00E+00	1.30	4.05E+05	[b]
Total material input				*1.06E+06*		*1.11E+07*	
Products and by-products (with allocation according to energy content)							
Biodiesel	g	3.88E+05	2.62	1.02E+06	27.41	1.06E+07	[a]
Glycerin	g	3.88E+04	1.16	4.50E+04	12.13	4.71E+05	[a]

Sources:

[a] Our calculation performed in this study.
[b] Lettenmeier, M., Rohn, H., Liedtke, C., and Schmidt-Bleek, F., Resource productivity in 7 steps: How to develop eco-innovative products and services and improve their material footprint (Berlin: Wuppertal Institute for Climate, Environment and Energy, 2009).
[c] Values taken as gasoline from Lettenmeier et al., 2009.
[d] Average, Lettenmeier et al., 2009.

Table 3.3 Energy Flows in the Agricultural and Transport Phases

Description of flow	Unit	Raw amount (ha^{-1} yr^{-1})	Energy intensity factor (g oil eq. $unit^{-1}$)	Ref	Oil equivalent demand (g oil eq. ha^{-1} yr^{-1})	Energy demand (J ha^{-1} yr^{-1})
Agricultural phase						
Imported input						
Diesel and heavy fuel	g	7.74E+04	1.23	[a]	8.48E+04	3.55E+09
Additional energy (heavy fuel oil) for diesel refining	g	1.61E+04	1.23		1.64E+04	6.88E+08
Fertilizers						
Nitrogen (N)	g	5.41E+04	1.75	[a]	9.48E+04	3.97E+09
Phosphate (PO_4)	g	2.00E+04	0.32	[a]	6.40E+03	2.68E+08
Agricultural machinery						
• steel and iron	g	3.15E+02	0.91	[b]	2.86E+02	1.20E+07
• aluminum	g	5.37E+01	5.36	[b]	2.88E+02	1.20E+07
• rubber and plastic material	g	3.84E+00	3.00	[a]	1.15E+01	4.82E+05
• fraction of copper	g	1.15E+01	2.21	[c]	2.54E+01	1.06E+06
Seeds	g	8.00E+03			5.86E+02	2.45E+07
Total energy demand					2.04E+05	8.52E+09

continued

Table 3.3 Energy Flows in the Agricultural and Transport Phases (Continued)

Description of flow	Unit	Raw amount (ha⁻¹ yr⁻¹)	Energy intensity factor (g oil eq. unit⁻¹)	Ref	Oil equivalent demand (g oil eq. ha⁻¹ yr⁻¹)	Energy demand (J ha⁻¹ yr⁻¹)
Products and by-products (with allocation according to energy content)						
Seeds produced—dry matter	g	1.20E+06	0.07	[d]		
Straw (d.m., 30% left in soil)	g	2.38E+06	0.05	[d]		
Transportation phase						
Seeds produced	g	1.20E+06	0.07	[d]	8.80E+04	3.68E+09
Steel for transport machinery	g	4.32E+02	1.91	[a]	8.25E+02	3.45E+07
Diesel for transport	g	9.00E+02	1.23	[a]	1.11E+03	4.64E+07
Total energy demand					8.99E+04	3.76E+09
Products transported to the gate of the plant						
Seeds delivered	g	1.20E+06	0.07			

Sources:

[a] Biondi, P., Panaro, V., and Pellizzi, G., *Le richieste di energia del sistema agricolo italiano*, Progetto Finalizzato Energetica, Sottoprogetto Biomasse ed Agricoltura, Report LB-20 (Rome: Consiglio Nazionale delle Ricerche, 1989).
[b] Bargigli, S., Analisi del ciclo di vita e valutazione di impatto ambientale della produzione ed uso di idrogeno combustibile, Ph.D. thesis, 2002.
[c] Our calculation from energy intensity data.
[d] Our calculation performed in this study.

Table 3.4 Energy Flows in the Industrial Phase

Description of flow	Unit	Raw amount (ha^{-1} yr^{-1})	Energy intensity factor (g oil eq. unit^{-1})	Ref	Oil equivalent demand (g oil eq. ha^{-1} yr^{-1})	Energy demand (J ha^{-1} yr^{-1})
Brassica oil extraction and refining phase						
Seeds delivered	g	1.20E+06	0.07	[a]	8.99E+04	3.76E+09
Electricity	J	1.04E+09	9.95E–05	[b]	6.24E+04	2.61E+09
Natural gas	g	3.64E+04	1.66	[c]	4.34E+04	1.82E+09
Additional energy for natural gas refining	g	5.20E+03	1.66	[c]	6.20E+03	2.59E+08
Texane	g	3.41E+03	1.32	[d]	4.50E+03	1.88E+08
Infrastructure						
• stainless steel	g	1.45E+02	1.91	[e]	2.77E+02	1.16E+07
• steel	g	2.77E+02	1.91	[e]	5.29E+02	2.22E+07
• cement	g	7.39E+02	0.11	[e]	8.13E+01	3.40E+06
Total energy demand					2.07E+05	8.68E+09
Products and by-products (with allocation according to energy content)						
Brassica oil	g	3.96E+05	0.29	[a]		
Brassica meal or pellets	g	6.96E+05	0.13	[a]		
Transesterification biodiesel production						
Brassica oil	g	3.96E+05	0.29	[a]	1.16E+05	4.85E+09
Electricity	J	1.79E+08	9.95E–05	[b]	1.07E+04	4.47E+08
Natural gas	g	1.36E+04	1.66	[c]	1.62E+04	6.77E+08
Additional energy for natural gas refining	g	1.94E+03	1.66	[c]	2.31E+03	9.67E+07

continued

Table 3.4 Energy Flows in the Industrial Phase (Continued)

Description of flow	Unit	Raw amount (ha⁻¹ yr⁻¹)	Energy intensity factor (g oil eq. unit⁻¹)	Ref	Oil equivalent demand (g oil eq. ha⁻¹ yr⁻¹)	Energy demand (J ha⁻¹ yr⁻¹)
Methanol	g	5.66E+04	1.03	[c]	5.82E+04	2.44E+09
Acid (hydrochloric acid)	g	1.19E+04	n.a.		0.00E+00	0.00E+00
Catalyst (sodium hydroxide NaOH)	g	8.73E+03	n.a.		0.00E+00	0.00E+00
Infrastructure						
• iron	g	5.46E+00	0.91	[f]	4.97E+00	2.08E+05
• steel	g	9.11E+02	1.91	[e]	1.74E+03	7.28E+07
• cement	g	2.64E+02	0.11	[e]	7.25E+02	3.04E+07
Water	g	3.11E+05	1.31E–04	[g]	4.09E+01	1.71E+06
Total energy demand					2.06E+05	8.61E+09
Products and by-products (with allocation according to energy content)						
Biodiesel	g	3.88E+05	0.51	[a]		
Glycerin	g	3.88E+04	0.22	[a]		

Sources:

[a] Our calculation performed in this study.
[b] Our calculation from Boustead, I., and Hancock, G. F., Handbook of industrial energy analysis (New York: John Wiley & Sons, 1979).
[c] Boustead and Hancock, 1979.
[d] Value equal to gasoline from Biondi, P., Panaro, V., and Pellizzi, G., Le richieste di energia del sistema agricolo italiano, Progetto Finalizzato Energetica, Sottoprogetto Biomasse ed Agricoltura, Report LB-20 (Rome: Consiglio Nazionale delle Ricerche, 1989).
[e] Biondi et al., 1989.
[f] Bargigli, S., Analisi del ciclo di vita e valutazione di impatto ambientale della produzione ed uso di idrogeno combustibile, Ph.D. thesis, 2002.
[g] Our calculation from energy costs of common materials.

energy method. Nor are human labor and economic services included in most evaluations.

Quantifying the total energy invested into a process allows an estimate of the total amount of primary energy invested and, as a consequence, the extent of the depletion of nonrenewable energy resources caused by the process. If the evaluation deals with an energy conversion process (e.g., conversion of oil or wind into electricity), the comparison of the energy output to the energy invested provides a measure of performance known as *energy return on investment* (EROI): The higher the EROI, the higher the energy benefit from the process.

Table 3.3 deals with the agricultural and transport steps, while Table 3.4 looks at the industrial phase. Each input flow is multiplied by a suitable oil equivalent intensity (amount of conventional commercial oil required to generate one unit of product) to yield the total oil-equivalent demand and its heat content. Oil equivalents are converted to heat units by multiplying the oil equivalents by $41{,}860$ J/g_{oil}. Table 3.4 yields the energy intensities (energy invested per unit of product) of intermediate and final products. It should be noted that Tables 3.3 and 3.4 are based on the same inventory as Tables 3.1 and 3.2, in order to provide consistent results.

3.3.3   Emergy flows

Tables 3.5 and 3.6 provide calculation and results of the emergy synthesis method applied to all process steps. The emergy accounting method, developed by Odum (1996), uses the thermodynamic basis of all forms of energy, materials, and human services provided to the process and converts them into equivalents of one form of energy: solar. This evaluation process has been termed *emergy synthesis*. Synthesis is the act of combining elements into coherent wholes. Rather than dissect and break apart systems and build understanding from the pieces upward, emergy synthesis strives for understanding by grasping the wholeness of systems. By evaluating complex systems using emergy methods, the major inputs from the human economy and those coming "free" from the environment can be integrated to analyze questions of public policy and environmental management holistically.

In Tables 3.5 and 3.6, input flows are multiplied by *unit emergy values* (UEVs, also known as *emergy intensities*) in order to convert them into the emergy flow associated with each item. The results of an emergy calculation are the UEVs of intermediate and final products as well as a set of performance indicators that shed light on the way the process uses the resources received by nature and by the economic system. The total emergy driving the process is assigned to each coproduct of each step, without allocation, following the emergy algebra (Brown and Herendeen, 1996).

Table 3.5 Emergy Flows in the Agricultural and Transport Phases

Description of flow	Unit	Raw amount (ha⁻¹ yr⁻¹)	Transformity (seJ/unit)	Ref	Emergy (seJ/yr)
Agricultural phase					
Renewable input (locally available)					
Sun	J/ha/yr	1.15E+13	1.00E+00	[a]	1.15E+13
Wind (kinetic energy of wind used at the surface)	J/ha/yr	1.73E+11	2.51E+03	[b]	4.34E+14
Rainfall (chemical potential)	J/ha/yr	1.59E+10	3.05E+04	[b]	4.84E+14
Deep heat (geothermal heat)	J/ha/yr	1.89E+10	1.20E+04	[b]	2.27E+14
Nonrenewable input (locally available)					
Top soil (erosion)	J/ha/yr	2.20E+08	1.24E+05	[b]	2.72E+13
Imported input					
Diesel and heavy fuel	J/ha/yr	3.55E+09	1.11E+05	[b]	3.93E+14
Fertilizers					
Nitrogen (N)	g/ha/yr	5.41E+04	6.37E+09	[b]	3.45E+14
Phosphate (PO_4)	g/ha/yr	2.00E+04	6.54E+09	[b]	1.31E+14
Agricultural Machinery					
• steel and iron	g/ha/yr	3.15E+02	5.31E+09	[f]	1.67E+12
• aluminum	g/ha/yr	5.37E+01	3.25E+10	[b]	1.75E+12
• rubber and plastic material	g/ha/yr	3.84E+00	3.69E+09	[g]	1.41E+10
• copper	g/ha/yr	1.15E+01	3.36E+09	[c]	3.87E+10
Seeds	g/ha/yr	8.00E+03	1.67E+09	[h]	1.33E+13
Human labor	€/ha/yr	4.76E+02	2.75E+12	[i]	1.31E+15
Indirect labor (services)	€/ha/yr	2.69E+02	2.75E+12	[i]	7.40E+14
Total emergy demand					3.44E+15

Products and by-products (without allocation)

Seeds produced—dry matter	g/ha/yr	1.20E+06	2.87E+09	[h]	3.44E+15
Straw (d.m., 30% left in soil)	g/ha/yr	3.40E+06	1.01E+09	[h]	3.44E+15
Transportation phase					
Seeds produced	g/ha/yr	1.20E+06	2.87E+09	[h]	3.44E+15
Steel for transport machinery	g/ha/yr	4.32E+02	5.31E+09	[f]	2.29E+12
Diesel for transport	g/ha/yr	9.00E+02	1.11E+05	[c]	9.96E+07
Labor	€/ha/yr	9.00E+00	2.75E+12	[i]	2.47E+13
Indirect labor (services)	€/ha/yr	1.47E+00	2.75E+12	[i]	4.03E+12
Total energy demand					3.48E+15
Products and by-products					
Seeds delivered	g/ha/yr	1.20E+06	2.90E+09	[h]	3.48E+15

Sources:

[a] By definition.
[b] Brown, M. T., and Ulgiati, S., 2004, Energy quality, emergy, and transformity: H. T. Odum's contributions to quantifying and understanding systems, *Ecological Modelling*, 178(1–2): 201–213.
[c] Odum, H. T., Brown, M. T., and Brandt-Williams, S., *Handbook of emergy evaluation: A compendium of data for emergy computation*, Folio #1, *Introduction and global budget* (Gainesville: Center for Environmental Policy, University of Florida, 2000).
[d] Buenfil, A., Sustainable use of potable water in Florida: An emergy analysis of water supply and treatment alternatives, in *Emergy synthesis: Theory and applications of emergy methodology*, Vol. 1, ed. Brown, M. T., Odum, H. T., Tilley, D., and Ulgiati, S., 107–116 (Gainesville: Center for Environmental Policy, University of Florida, 2000).
[e] Estimated from Biondi, P., Panaro, V., and Pellizzi, G., *Le richieste di energia del sistema agricolo italiano*, Progetto Finalizzato Energetica, Sottoprogetto Biomasse ed Agricoltura, Report LB-20 (Rome: Consiglio Nazionale delle Ricerche, 1989).
[f] Bargigli, S., and Ulgiati, S., Emergy and life-cycle assessment of steel production in Europe, in *Emergy synthesis: Theory and applications of emergy methodology*, Vol. 2, ed. Brown, M. T., Odum, H. T., Tilley, D., and Ulgiati, S. (Gainesville: Center for Environmental Policy, University of Florida, 2003).
[g] Our calculation after Odum, H. T., *Environmental accounting: Emergy and environmental decision making* (New York: Wiley, 1996).
[h] Our calculation performed in this work.
[i] Cialani, C., Russi, D., and Ulgiati, S., Investigating a 20-year national economic dynamics by means of emergy-based indicators, in *Emergy synthesis: Theory and applications of emergy methodology*, Vol. 3, ed. Brown, M. T., Odum, H. T., Tilley, D., and Ulgiati, S., 401–415 (Gainesville: Center for Environmental Policy, University of Florida, 2005).

Table 3.6 Emergy Flows in the Industrial Phase

Description of flow	Unit	Raw amount (ha⁻¹ yr⁻¹)	Transformity (seJ/unit)	Ref	Emergy (seJ/yr)
Brassica oil extraction and refining phase					
Seeds delivered	g/ha/yr	1.20E+06	2.90E+09	[a]	3.48E+15
Electricity	J/ha/yr	1.04E+09	2.81E+05	[b]	2.93E+14
Natural gas	g/ha/yr	3.64E+04	8.04E+04	[b]	2.93E+09
Hexane	g/ha/yr	3.41E+03	6.08E+09	[b]	2.07E+13
Infrastructure					
• stainless steel	g/ha/yr	1.45E+02	5.31E+09	[c]	7.71E+11
• steel	g/ha/yr	2.77E+02	5.31E+09	[c]	1.47E+12
• cement	g/ha/yr	7.39E+02	1.73E+09	[b]	1.28E+12
Labor	€/ha/yr	4.29E+00	2.75E+12	[d]	1.18E+13
Total services	€/ha/yr	5.92E+01	2.75E+12	[d]	1.63E+14
Total emergy demand					3.97E+15
Products and by-products (without allocation)					
Brassica oil	g/ha/yr	3.96E+05	1.00E+10	[a]	3.97E+15
Brassica meal or pellets	g/ha/yr	6.96E+05	5.70E+09	[a]	3.97E+15
Transesterification biodiesel production					
Brassica oil	g/ha/yr	3.96E+05	1.00E+10	[a]	3.97E+15
Electricity	J/ha/yr	1.79E+08	2.81E+05	[b]	5.02E+13
Natural gas	g/ha/yr	1.36E-04	8.04E+04	[b]	1.09E+09
Methanol	J/ha/yr	1.29E+09	1.89E+05	[e]	2.43E+14
Acid (hydrochloric acid)	g/ha/yr	1.19E+04	0.00E+00		0.00E+00
Catalyst (sodium hydroxide NaOH)	g/ha/yr	8.73E+03	0.00E+00		0.00E+00

Infrastructure				
• iron	g/ha/yr	5.46E+00	1.13E+10	6.18E+10	[f]
• steel	g/ha/yr	9.11E+02	5.31E+09	4.84E+12	[c]
• cement	g/ha/yr	2.64E+02	1.73E+09	4.56E+11	[b]
Water	g/ha/yr	3.11E+05	7.61E+05	2.37E+11	[g]
Human labor	€/ha/yr	6.43E+00	2.75E+12	1.77E+13	[d]
Annual services for industrial production	€/yr	3.84E+01	2.75E+12	1.05E+14	[d]
Total energy demand				4.39E+15	
Products and by-products (without allocation)					
Biodiesel	g/ha/yr	3.88E+05	1.13E+10	4.39E+15	[h]
Glycerin	g/ha/yr	3.88E+04	1.13E+11	4.39E+15	[h]

Sources:

[a] Our calculation performed in this work.
[b] Odum, H. T., Brown, M. T., and Brandt-Williams, S., *Handbook of emergy evaluation: A compendium of data for emergy computation*, Folio #1, *Introduction and global budget* (Gainesville: Center for Environmental Policy, University of Florida, 2000).
[c] Bargigli, S., and Ulgiati, S., Emergy and life-cycle assessment of steel production in Europe, in *Emergy synthesis: Theory and applications of emergy methodology*, Vol. 2, ed. Brown, M. T., Odum, H. T., Tilley, D., and Ulgiati, S. (Gainesville: Center for Environmental Policy, University of Florida, 2003).
[d] Cialani, C., Russi, D., and Ulgiati, S., Investigating a 20-year national economic dynamics by means of emergy-based indicators, in *Emergy synthesis: Theory and applications of emergy methodology*, Vol. 3, ed. Brown, M. T., Odum, H. T., Tilley, D., and Ulgiati, S., 401–416 (Gainesville: Center for Environmental Policy, University of Florida, 2005).
[e] Ulgiati, S., 2001, A comprehensive energy and economic assessment of biofuels: When "green" is not enough, *Critical Reviews in Plant Sciences*, 20(1): 71–106.
[f] Brown, M. T., and Ulgiati, S., 2004, Energy quality, emergy, and transformity: H. T. Odum's contributions to quantifying and understanding systems, *Ecological Modelling*, 178(1–2): 201–213.
[g] Buenfil, A., Sustainable use of potable water in Florida: An emergy analysis of water supply and treatment alternatives, in *Emergy synthesis: Theory and applications of emergy methodology*, Vol. 1, ed. Brown, M. T., Odum, H. T., Tilley, D., and Ulgiati, S., 107–116 (Gainesville: Center for Environmental Policy, University of Florida, 2000).
[h] Our calculation performed in this work.

3.3.4   Emissions

Finally, Tables 3.7 and 3.8 provide the global emissions released in each process step by all combustion processes involved. The embodied energy expenditures from Tables 3.3 and 3.4 are converted to average emissions according to emission factors from international literature (EPA, 2010; EEA, 2009). The flows listed in Tables 3.7 and 3.8 contribute to the global warming potential and acidification potential generated by the process. We also calculated the process contribution to other impact categories, including the eutrophication potential included in Tables 3.9 and 3.10.

3.4   Discussion

Values generated in Tables 3.1 through 3.8 for low-intensity production, as well as similar values for high-intensity production (not shown), are presented in Tables 3.9 and 3.10 in six columns, one column for each of the six intermediate and final products. Table 3.9 reports the intermediate and final product data for low-intensity *Brassica carinata* cropping in Puglia, while Table 3.10 shows the intermediate and final product data for high-intensity cropping, which uses more fertilizers, mechanization, and plant density (De Mastro et al., 2009).

The agricultural phase provides two coproducts: Brassica seeds and straw. Seeds are further processed into Brassica oil and cake meal. Oil is evaluated as such (for use as a chemical) and also further processed into biodiesel and glycerin. Of the Brassica straw, 30% is left on the fields, while 70% is assumed to be harvested and transported to a local boiler for combustion and heat generation. Cake meal in general practice would be used as a source of protein for livestock, but, due to its production in this case on polluted land, is delivered instead to the local boiler to be burned together with the straw. The assumption, described later in this study, is that the generated heat is converted into electricity, so that both electricity and residual cogenerated heat can be used by local firms.

Agricultural inputs are allocated to seeds and straw according to the percentage of their energy content, considering that the process is designed for the purpose of energy generation. (Other allocation procedures could have been adopted, such as an allocation based on the economic value of products, if the focus were on the market performance of the investigated process.) Then, the fraction of agricultural input allocated to seeds plus the resources for their transport and mechanical extraction are allocated to Brassica oil and cake meal, again according to their energy content. The fraction of inputs allocated to Brassica oil plus the inputs to chemical conversion are finally allocated to biodiesel and glycerin, still according to their energy content, although glycerin is not used as an energy source.

Table 3.7 Global Emissions in the Agricultural and Transport Phases

Description of flow	Global emissions (g CO_2)	Global emissions (g CO)	Global emissions (g NO_x)	Global emissions (g SO_2)
Agricultural phase				
Imported input				
Diesel and heavy fuel	2.44E+05	4.61E+02	1.05E+03	4.49E+01
Fertilizers				
• nitrogen (N)	2.91E+05	1.98E+01	8.53E+02	5.95E+02
• phosphate (PO_4)	1.96E+04	1.34E+00	5.76E+01	4.02E+01
Agricultural machinery				
• steel and iron	8.78E+02	5.99E-02	2.58E+00	1.80E+00
• aluminum	8.83E+02	6.02E-02	2.59E+00	1.81E+00
• rubber and plastic material	3.53E+01	2.41E-03	1.04E-01	7.23E-02
• copper	7.80E+01	5.32E-03	2.29E-01	1.60E-01
Seeds	1.80E+03	1.23E-01	5.28E+00	3.68E+00
Total global emission flow	5.59E+05	4.83E+02	1.97E+03	6.88E+02
Products and by-products (with allocation according to energy content)				
Emissions allocated to seeds produced	2.70E+05	2.10E+02	9.16E+02	3.42E+02
Emissions allocated to straw	3.55E+05	2.76E+02	1.20E+03	4.49E+02
Transportation phase				
Emissions allocated to seeds produced	2.70E+05	2.10E+02	9.16E+02	3.42E+02
Steel for transport machinery	2.53E+03	1.73E-01	7.43E+00	5.18E+00
Diesel for transport	3.40E+03	2.32E-01	9.97E+00	6.95E+00
Emissions from seeds delivering	2.76E+05	2.10E+02	9.33E+02	3.54E+02

Table 3.8 Global Emissions in the Industrial Phase

Description of flow	Global emissions ($g\ CO_2$)	Global emissions ($g\ CO$)	Global emissions ($g\ NO_x$)	Global emissions ($g\ SO_2$)
Brassica oil extraction and refining phase				
Emissions from seeds delivering	2.76E+05	2.10E+02	9.33E+02	3.54E+02
Electricity	1.61E+05	8.52E-01	3.99E+02	2.93E+02
Natural gas	1.02E+05	7.09E+01	1.62E+02	5.27E-01
Texane	1.38E+04	9.42E-01	4.05E+01	2.82E+01
Infrastructure				
• stainless steel	8.51E+02	5.80E-02	2.50E+00	1.74E+00
• steel	1.62E+03	1.11E-01	4.77E+00	3.32E+00
• cement	2.49E+02	1.70E-02	7.32E-01	5.11E-01
Total global emission flow	*5.56E+05*	*3.68E+02*	*1.54E+03*	*6.81E+02*
Products and by-products (with energy allocation)				
Brassica oil	5.19E+05	3.60E+02	1.57E+03	6.53E+02
Brassica meal or pellets	4.10E+05	2.85E+02	1.24E+03	5.16E+02
Transesterification biodiesel production				
Brassica oil	3.55E+05	2.42E+07	1.04E+09	7.27E+08
Electricity	2.76E+04	1.46E+01	6.82E+01	5.01E+01
Natural gas	3.80E+04	2.64E+01	6.03E+01	1.96E-01
Methanol	1.79E+05	1.22E+01	5.24E+02	3.65E+02
Acid (hydrochloric acid)	0.00E+00	0.00E+00	0.00E+00	0.00E+00

Catalyst (sodium hydroxide NaOH)	0.00E+00	0.00E+00	0.00E+00	0.00E+00
Infrastructure				
• iron	1.53E+07	1.04E+03	4.48E+04	3.12E+04
• steel	5.34E+09	3.64E+05	1.57E+07	1.09E+07
• cement	2.23E+09	1.52E+05	6.53E+06	4.55E+06
Water	1.25E+08	8.56E+03	3.68E+05	2.57E+05
Total global emission flow	7.70E+09	2.48E+07	1.07E+09	7.43E+08
Products and by-products (with energy allocation)				
Biodiesel	7.38E+09	2.37E+07	1.02E+09	7.12E+08
Glycerin	3.27E+08	1.05E+06	4.52E+07	3.15E+07

Table 3.9 Upstream and Downstream SUMMA Indicators for the Biodiesel Chain from Low-Input Farms for 2008

Indicator	Seeds	Straw	Brassica oil	Cake meal	Biodiesel	Glycerin
Material resource depletion						
MI_{abiot} (g/g)	0.73	0.48	2.02	0.91	2.62	1.16
MI_{abiot} (g/J)	2.63E–05	2.63E–05	5.39E–05	5.39E–05	6.99E–05	6.99E–05
MI_{abiot} (g/€)	2.08E+03	2.41E+04	4.59E+03	2.11E+03	3.35E+03	1.10E+04
MI_{water} (g/g) (water footprint)	3.17	2.10	18.91	8.49	27.41	12.13
MI_{water} (g/J) (water footprint)	1.15E–04	1.15E–04	5.06E–04	5.06E–04	7.31E–04	7.31E–04
MI_{water} (g/€) (water footprint)	9.05E+03	2.08E+02	4.30E+04	1.97E+04	3.50E+04	1.16E+05
Energy resource depletion						
GER per unit mass (J/g)	3.07E+03	2.03E+03	1.22E+04	5.50E+03	2.13E+04	9.41E+03
GER per energy (J/J)	1.11E–01	1.11E–01	3.27E–01	3.27E–01	5.67E–01	5.67E–01
GER per unit currency (J/€)	8.77E+06	1.02E+08	2.78E+07	1.28E+07	2.71E+07	8.96E+07
Oil_{eq} (g_{oil}/g)	0.07	0.05	0.29	0.13	0.51	0.22
Oil_{eq} (g_{oil}/J)	2.66E–06	2.66E–06	7.82E–06	7.82E–06	1.35E–05	1.35E–05
Oil_{eq} (g_{oil}/€)	2.09E+02	2.43E+03	6.65E+02	3.06E+02	6.48E+02	2.14E+03
EROI	9.00	9.00	3.05	3.05	1.76	1.76
Emergy (demand for environmental support)						
Specific emergy (seJ/g)	2.87E+09	1.01E+09	1.00E+10	5.70E+09	1.13E+10	1.13E+11
Transformity (seJ/J)	1.04E+05	5.54E+04	2.68E+05	3.39E+05	3.02E+05	6.81E+06
Emergy money ratio (seJ/€)	8.20E+12	7.24E+13	n.a	1.33E+13	1.44E+13	1.08E+15
EYR (Emergy Yield Ratio)	1.17	1.17	1.16	1.16	1.14	1.14

	6.12	6.12	7.13	7.13	8.01	8.01
ELR	6.12	6.12	7.13	7.13	8.01	8.01
ESI	0.19	0.19	0.16	0.16	0.14	0.14
% renewable	0.14	0.14	0.12	0.12	0.11	0.11
Climate change						
Global warming (g CO_2-equiv/g)	2.26E–01	1.50E–01	1.32E+00	5.92E–01	1.52E+00	6.73E–01
Global warming (g CO_2-equiv/J)	8.21E–06	8.21E–06	3.53E–05	3.53E–05	4.06E–05	4.06E–05
Global warming (g CO_2-equiv/€)	6.47E+02	7.46E+03	3.00E+03	1.38E+03	1.94E+03	6.41E+03
Acidification (g SO_2/g)	7.23E–04	4.80E–04	3.96E–03	1.78E–03	5.62E–03	2.49E–03
Acidification (g SO_2/J)	2.62E–08	2.62E–08	1.06E–07	1.06E–07	1.50E–07	1.50E–07
Acidification (g SO_2/€)	2.07E+00	2.40E+01	8.99E+00	4.13E+00	7.18E+00	2.37E+01
Eutrophication (g PO_4/g)	1.01E–04	6.67E–05	5.20E–04	2.34E–04	5.61E–04	2.48E–04
Eutrophication (g PO_4/J)	3.64E–09	3.64E–09	1.39E–08	1.39E–08	1.50E–08	1.50E–08
Eutrophication (g PO_4/€)	2.87E–01	3.33E+00	1.18E+00	5.43E–01	7.16E–01	2.36E+00

Table 3.10 Upstream and Downstream SUMMA Indicators for the Biodiesel Chain from High-Input Farms for 2008

Indicator	Seeds	Straw	Brassica oil	Cake meal	Biodiesel	Glycerin
Material resource depletion						
MI_{abiot} (g/g)	0.97	0.64	2.43	1.09	3.02	1.34
MI_{abiot} (g/J)	3.50E–05	3.50E–05	6.48E–05	6.48E–05	8.05E–05	8.05E–05
MI_{abiot} (g/€)	2.76E+03	3.21E+04	5.51E+03	2.53E+03	3.85E+03	1.27E+04
MI_{water} (g/g) (water footprint)	4.55	3.02	21.24	9.54	29.69	13.14
MI_{water} (g/J) (water footprint)	1.65E–04	1.65E–04	5.68E–04	5.68E–04	7.92E–04	7.92E–04
MI_{water} (g/€) (water footprint)	1.30E+04	2.83E+02	4.83E+04	2.22E+04	3.79E+04	1.25E+05
Energy resource depletion						
GER per unit mass (J/g)	3.58E+03	2.38E+03	1.31E+04	5.89E+03	2.21E+04	9.78E+03
GER per energy (J/J)	1.30E–01	1.30E–01	3.51E–01	3.51E–01	5.89E–01	5.89E–01
GER per unit currency (J/€)	1.02E+07	1.19E+08	2.98E+07	1.37E+07	2.82E+07	9.32E+07
Oil_{eq} (g_{oil}/g)	0.09	0.06	0.31	0.14	0.53	0.23
Oil_{eq} (g_{oil}/J)	3.10E–06	3.10E–06	8.38E–06	8.38E–06	1.41E–05	1.41E–05
Oil_{eq} (g_{oil}/€)	2.45E+02	2.84E+03	7.12E+02	3.27E+02	6.74E+02	2.23E+03
EROI	7.70	7.70	2.85	2.85	1.70	1.70
Emergy (demand for environmental support)						
Specific emergy (seJ/g)	2.60E+09	1.03E+09	9.18E+09	5.23E+09	1.05E+10	1.05E+11
Transformity (seJ/J)	9.41E+04	5.61E+04	2.46E+05	3.11E+05	2.79E+05	6.30E+06
Emergy money ratio (seJ/€)	7.42E+12	7.34E+13	n.a	1.22E+13	1.34E+13	9.96E+14

EYR	1.13	1.13	1.12	1.12	1.11	1.11
ELR	8.13	8.13	9.57	9.57	10.81	10.81
ESI	0.14	0.14	0.12	0.12	0.10	0.10
% renewable	0.11	0.11	0.09	0.09	0.08	0.08
Climate change						
Global warming (g CO_2-equiv/g)	2.64E–01	1.75E–01	1.40E+00	6.30E–01	1.58E+00	7.01E–01
Global warming (g CO_2-equiv/J)	9.57E–06	9.57E–06	3.75E–05	3.75E–05	4.22E–05	4.22E–05
Global warming (g CO_2-equiv/€)	7.54E+02	8.70E+03	3.19E+03	1.47E+03	2.02E+03	6.68E+03
Acidification (g SO_2/g)	9.05E–04	6.00E–04	4.45E–03	2.00E–03	5.86E–03	2.60E–03
Acidification (g SO_2/J)	3.28E–08	3.28E–08	1.19E–07	1.19E–07	1.56E–07	1.56E–07
Acidification (g SO_2/€)	2.58E+00	3.00E+01	1.01E+01	4.65E+00	7.49E+00	2.47E+01
Eutrophication (g PO_4/g)	1.12E–04	7.43E–05	5.39E–04	2.42E–04	5.85E–04	2.59E–04
Eutrophication (g PO_4/J)	4.06E–09	4.06E–09	1.44E–08	1.44E–08	1.56E–08	1.56E–08
Eutrophication (g PO_4/€)	3.20E–01	3.72E+00	1.22E+00	5.63E–01	7.47E–01	2.47E+00

A similar allocation procedure is applied to all kinds of impacts (material demand and emissions). However, the allocation procedure in the emergy evaluation follows a different algebra, specific to the emergy approach (Brown and Ulgiati, 2004), in which the total input emergy is allocated to each of the coproducts.

The indicators in Tables 3.9 and 3.10 are intensity indicators, that is, indicators per unit of product generated, where the product can be measured according to its energy content, mass, and economic value. The two tables are each divided into four main categories of indicators: *material resource depletion* (abiotic material depletion and water depletion, also indicated as water footprint); *energy resource depletion* (given as embodied energy use, MJ/unit of product, and oil equivalent, g_{oil}/unit of product); *emergy*, or *demand for environmental support* (measured as solar-equivalent joules per unit of product, seJ/unit), also called the *global environmental footprint*; and *climate change*, including the global warming potential (also called carbon footprint), acidification potential, and eutrophication potential per unit of product). Within the category of energy resource depletion, the EROI is provided. Similarly, within the category of emergy indicators, we provide the emergy yield ratio, the environmental loading ratio, the emergy sustainability index, and the percentage of emergy that is renewable.

Tables 3.9 and 3.10 show that cropping for energy is unlikely to be feasible and sustainable. Cropping only for biodiesel as final product has a high environmental impact in terms of local resources expenditure (top soil and land use). Even under the high-intensity cropping system, the indicators show relatively low yields and non-negligible impacts; these larger impacts are related to the agricultural phase.

Figure 3.4 provides a pictorial representation of the environmental impact for the three main products: seeds, oil, and biodiesel. The radar diagram was designed using the data in Table 3.9 (for low input). The values are normalized with reference to the total impact generated for biodiesel production. Values of indicators for seed and seed oil production are expressed as percentages of those calculated for biodiesel. In so doing, the relative impact of each process step is calculated as the difference between the percentages of the following and the preceeding steps.

Figure 3.4 shows increasing impacts as the agricultural product is processed. Results are more or less the same for high-intensity Brassica biodiesel, where the higher yield is offset by higher input of fertilizers and other agricultural resources. The low performance is made even worse in both systems because we have allocated input flows to all coproducts, thus indirectly considering a high-quality product like biodiesel comparable to low-quality energy products such as straw and cake meals.

Results in Tables 3.9 (low-intensity Brassica cropping) and 3.10 (high-intensity cropping) allow a very comprehensive evaluation of the

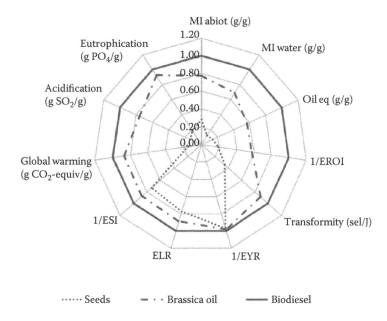

Figure 3.4 Environmental indicators for three main products (low-intensity cropping of *Brassica carinata* seeds and related oil and biodiesel). Values of indicators for seeds and seed oil are expressed as percentages of those calculated for biodiesel.

conversion chain—seeds to Brassica oil to biodiesel—with a focus on both the main products and coproducts. In short, the following aspects can be highlighted.

1. Impacts increase with manufacturing. All indicators increase with further processing steps, including those for coproducts, due to the allocated contribution from previous steps. The more a product is manufactured, the higher the upstream and downstream impacts, in terms of demand for resources and emissions. It must be noted that, since allocation of inputs was made according to the energy content in each coproduct, the performance indicators *per unit of energy delivered* are the same for the coproducts of each step, while the indicators *per unit of mass delivered* and *per unit economic value generated* are always different.

2. Figures 3.5 and 3.6 compare the contribution of each process step to the total material, water, and energy demand as well as to CO_2 emissions. Water use largely dominates all the steps (see Figure 3.5) and shows the highest demand in the industrial step of oil extraction. Considering that the crop is not irrigated and that the extraction is made via mechanical pressing, this is an indirect water demand,

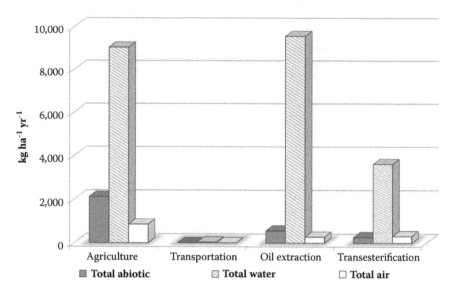

Figure 3.5 Total demand for abiotic material, water, and air for combustion, in each process step.

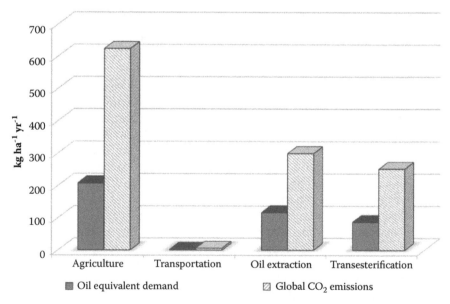

Figure 3.6 Embodied energy demand (as oil equivalent) and global-scale CO_2 emissions associated with each process step.

that is, water embodied in the processes that deliver fertilizers, electricity, fuels, machinery, and other production factors. The agricultural phase is responsible for the highest embodied energy demand (fertilizers and fuels) and, as a consequence, for the highest CO_2 emissions. Knowing where in the process the largest consumption of resources occurs helps to identify bottlenecks and requirements for technical improvement geared toward decreasing consumption and emissions.

3. *Water footprint:* One gram of seeds requires 3.17 g of water. When seeds are converted to oil, 1 g of oil requires 8.49 g of water. Finally, when oil is converted to biodiesel, 27.41 g of water are required for 1 g of biodiesel. Considering that 1 kg of biodiesel drives a car for about 10 km, this translates into about 3 kg of water per km run. If focus is placed on the added value of the process, we see that one euro of product in the form of seeds requires about 9 kg of water, in the form of Brassica oil requires 19 kg, and in the form of biodiesel requires 35 kg.

4. *Energy demand:* It takes only 0.07 g oil equivalent to make 1 g of seeds. The requirement is much higher for Brassica oil (0.29 g oil equiv) and for biodiesel (0.51 g). The final result is that half a gram of oil is invested to generate a gram of oil-equivalent energy. This means that the net energy delivered to society is only 50% of the total yield and that two hectares are needed in order to make one "net hectare." In terms of EROI, it goes from a high 9:1 for seeds to a low of 1.76:1 for biodiesel. Although the energy delivered seems higher than the energy invested (also due to the allocation procedure that splits the input and assigns a fraction to the coproducts), such a result is largely inadequate to support an energy-intensive production process or society, compared to the EROI of fossil fuels (in the range 10:1 or 20:1). Moreover, according to Ulgiati (2001), we identify here an amplifying loop due to the low energy return. Even if energy inputs are allocated to the main product as well as to the residues, the 50% net energy delivered means that for the process to become independent of fossil fuels, one hectare must be cropped to replace the fossil energy needed for production. This means that all upstream and downstream impacts (water demand, abiotic material demand, emergy) and related impacts (soil erosion, carbon footprint, acidification, etc.) are doubled, thus increasing by 100% the overall pressure on the environment.

5. *Carbon footprint:* Emissions of CO_2 and other greenhouse gases are relatively low in the first step, seed production (0.26 g of CO_2 per gram of dry seed). In the following steps, oil extraction and biodiesel production, emissions grow to 1.32 and 1.52 g of CO_2 per gram of oil and biodiesel, respectively, as a consequence of the high direct and

indirect demand for energy in the production process. These emissions do not include the emissions released by the use of the fuel, because that is assumed to be recycled by the new biomass growing season. Instead, emissions are linked to the 0.51 g of oil equivalent used per gram of biodiesel production. Considering that the allocation procedure also assigns part of the emissions to the straw in the agricultural phase and to the press cake and glycerin in the industrial phase, the total amount of greenhouse gases from the process is much larger than apparently shown by the biodiesel impact indicator alone.

6. *Demand for environmental support (emergy intensity):* The transformity of seeds is about 1.04E+05 seJ/J, of Brassica oil is 2.68E+05 seJ/J, and of biodiesel is 3.02E+05 seJ/J.* Considering that the average transformity of fossil diesel is 1.0E+05 seJ/J, results indicate that the biodiesel is three times more demanding in terms of global biosphere support (environmental inputs, land, water, indirect factors, etc.) than fossil fuels. Thus biodiesel is only 14% renewable (see Table 3.9), in spite of claims of biodiesel being green and environmental friendly. Last but not least, the EYR (Emergy Yield Ratio) is 1.17 for seeds and 1.14 for biodiesel. The EYR (the ratio of emergy globally exploited to emergy invested from outside) is an indicator of the ability of the process to exploit the local resources, and 1.30 is really too low. The EYR of fossil fuels is between 7 and 15, depending on the extraction site and reservoir typology. This means that fossil fuels provide to the economy a net contribution of resources (past solar energy accumulated) that is much higher than that supplied by bioenergy processes.

A society that developed its infrastructures and lifestyles on ready availability of cheap oil cannot run business-as-usual activities on biomass energy (low EROI and low EYR). Huge changes would be needed in the global societal organization and size for this to happen (think of the claims that prosperity would greatly decline by such authors as Nicolas Georgescu-Roegen, Howard Odum, Ivan Illich, and Serge Latouche, among others). Although it is not the goal of this chapter to discuss the complex relations among energy, population, and standard of living in present and future societies, clearly energy sources characterized by low net yield, EROI, and EYR are unlikely to support large portions of the world population at high standards of living.

---

* A transformity is defined (and calculated) as the emergy input per unit of available energy of the output. This value accounts for the cumulative work performed by the geo-biosphere to support directly and indirectly the generation of 1 J of energy output (Odum, 1996).

3.5   Results and discussion for different scenarios

We now turn to bioenergy production and use scenarios, based on production performances calculated earlier. The energy content of the biomass or the biodiesel does not indicate the net energy that is actually available to the larger economic system. In fact, as already pointed out in Section 3.3, there is an investment of energy embodied in the production factors (fertilizers, machinery, fuel) that needs to be subtracted in order to calculate the net energy delivered; in other words, a fraction of the energy yield must be reinvested back to the upstream production steps to make them independent of fossil fuels. In so doing, the actual net energy delivered decreases. To design a reliable bioenergy scenario at larger scale, the net energy actually delivered to society must be calculated.

We define *usable energy* (UE) as the energy potentially available from the process, after the heat losses due to moisture and other combustion losses are calculated and subtracted. This is more or less 50% of the theoretical energy content of dry biomass (think of the fact that biomass still has, at the time of harvesting, 20–30% of moisture and therefore its energy content is lower; moreover, it takes energy to evaporate the water content).

We next define *usable energy delivered* (UED) as the UE minus additional energy expenses for harvesting and transportation from the field to the industrial plant. Transportation energy expenses are not mandatory for the process to occur and could be largely avoided if the distance among the different steps were minimized.

The UED is not yet the *net energy* (NE), however, defined as the UED minus the energy previously invested in the process (fertilizers, machinery, fuels for production). The UED refers to how much energy is actually *available* to the user, after all energy losses and transportation expenses are subtracted, while NE refers to how much energy is actually *gained* by the society from the process, after all energy expenses for production are subtracted. In our scenario, the thermal energies of agricultural residues (30% left on field) and press cake (unfit as feedstock in the present case study) are also included as an additional energy source and added to the biodiesel energy.

Two options—low-input and high-input cropping—are explored. The results per hectare, on which scenarios are based, are shown in Table 3.11. The LI pattern has a total UED of 3.88E+10 J/ha/yr, while the HI system would provide a higher UED of 5.18E+10 J/ha/yr. However, the *net energy delivered* (NED) in the case of LI cropping is only 2.15E+10 J/ha/yr, while the HI cropping case's NED is slightly higher at 2.61E+10 J/ha/yr.

We show in Table 3.12 data on:

- The total polluted (nonirrigated) land that is theoretically available in Campania for energy cropping

Table 3.11 Usable Energy Delivered to the User by All the Coproducts of the Different Phases of the Biodiesel Production Chain (Data per Hectare)

		Energy content (J/ha/yr)	Usable Energy[a] (J/ha/yr)	Transportation of residues[b] (J/ha/yr)	Usable energy delivered (J/ha/yr)	Net energy delivered (J/ha/yr)
Low Input	Invested for process	1.73E+10				
	Biodiesel	1.46E+10	1.46E+10	7.30E+08	1.39E+10	
	Cake meal	1.17E+10	5.85E+09	5.85E+08	5.27E+09	
	Straw	4.36E+10	2.18E+10	2.18E+09	1.96E+10	
	Delivered					2.15E+10
High Input	Invested for process	2.57E+10				
	Biodiesel	2.06E+10	2.06E+10	1.03E+09	1.96E+10	
	Cake meal	1.66E+10	8.30E+09	8.30E+08	7.47E+09	
	Straw	5.51E+10	2.76E+10	2.76E+09	2.48E+10	
	Delivered					2.61E+10

[a] The usable energy (UE) is defined as the energy potentially available from the process, after the heat losses due to moisture and other combustion losses are calculated and subtracted. Residues (straw and cake meal) are only credited 50% usable energy content, due to moisture and other combustion losses. Biodiesel is credited 100% usable energy.

[b] Roughly estimated as 5% of energy content.

Table 3.12 Polluted (Nonirrigated) Land and Energy Consumption in Campania

	Amount	Unit	Source
Polluted, nonirrigated areas in Campania	43,182.86	ha	Calculated from EEA, 2000
Total energy used in Campania, all sectors (2005)	6,488 2.72E+17	ktep/yr J/yr	Campania, 2009
Total energy used in regional agriculture (2005)	200 8.37E+15	ktep/yr J/yr	Campania, 2009

- The total energy used for 2005 (based on Campania, 2009)
- The total energy used in the agricultural sector

These data are needed to design a scenario at the regional level.

Assuming that we can crop *Brassica carinata* in all the polluted, non-irrigated land of the region and that all the products and coproducts (biodiesel, straw, and press cake) are used for energy purposes, we can use the data in Tables 3.14 and 3.15 to calculate how much of agricultural and regional energy consumption can in principle be covered by means of such bioenergy (see Table 3.13).

The following assumptions are made for this scenario to be possible:

1. Energy costs of harvest of agricultural residues are already included in the energy costs of the agricultural phase because the harvesting machinery harvests and separates grains and straw at the same time.
2. Biodiesel can be used for agricultural machinery or can be sold to nonagricultural vehicles (urban buses and trucks).
3. Straw and press cakes are burned in small cogeneration boilers for production of electricity and heat, within local food manufacturing industries (mainly tomato- and vegetable-processing industries) that have huge demand for electricity and hot water and are most often equipped with their own generators fueled by natural gas.

Table 3.13 Net Energy Delivered by the Biodiesel Production Chain and Fractions of Agricultural and Regional Energy Consumption Potentially Replaced

	Low input (J/yr)	High input (J/yr)
Total net energy delivered (NED) from polluted areas	9.26E+14	1.13E+15
% of energy budget of Campania Region	0.34%	0.42%
% of energy budget of regional agricultural sector	11.07%	13.48%

4. Gradual conversion from natural gas use to biomass use is
assumed to have negligible energy and material cost per unit of
product, considering the lifetime of such machinery and based
on experience gained with previous application of the SUMMA
approach to biomass-based electricity cogeneration plants
(Buonocore et al., 2010).

The results from Table 3.13 are quite surprising. First, the very
small NED covers only very low percentages of the energy budget of
Campania and its agricultural sector, within ranges that are respectively
11.07–13.48% and 0.34–0.42%. The second surprising finding is that the
HI mode of cropping, although providing a higher UE in absolute terms,
delivers more or less the same NE as the LI mode due to the need to sub-
tract much higher input energy costs. This means that bioenergy from
marginal land supplies only a negligible amount of the energy demand
of the region and that these results cannot be improved by increasing the
cropping intensity (adding machinery or fertilizers). We did not test the
option of irrigating because the land is marginal (steep and generally far
from water sources), so this would require huge electricity costs to support irrigation.

Table 3.14 summarizes the economic investments that would be
required in our scenario, with and without the energy feedback (reinvest-
ment of a fraction of energy generated back to the process). Part of the
economic investment is direct labor, and part is indirect labor (services).
It is worth noting that the investment of a fraction of the energy generated back into the process, in order to make it independent of fossil fuel
inputs, decreases the NED available to other sectors of the economy, but
also decreases the demand for economic investment, that is, the services
associated with the fossil fuels replaced. The item "Saved cost for replaced
fuels" in Table 3.14 represents the money saved thanks to the avoided fossil energy in each phase.

One could consider that a small amount of NE from marginal land is
still a net gain and better than nothing. If there were an economic advantage (large savings on fuel purchase) and new jobs generated, it would
be good business in spite of its low NED. Therefore, we accounted for all
the labor costs required directly (labor inside the process) and indirectly
(labor to provide goods and production tools) in the process and added
these costs into the total. Table 3.14 lists all these costs in each phase for
both cropping systems (LI and HI). The costs of the agricultural phase
dominate, accounting for more than 90% of the total. This indicates that
the search for any economic improvement must address the agricultural
step, while the other steps are negligible.

The economic value of the energy delivered in the form of biodiesel
and residue biomass heat is calculated as the value of the oil (equivalent)

Table 3.14 Economic Costs for Each Phase of the Biodiesel Production Chain

		Labor (€/ha/yr)	Services (€/ha/yr)
Low input	Agriculture	476.00	269.09
	Oil extraction	4.29	59.23
	Transesterification	6.43	38.57
	Transportation	18.00	2.94
	Total costs without energy feedback	*504.72*	*369.83*
	Saved costs for replaced fuels		166.10
	Other saved costs for indirect energy replaced		44.8
	Total cost with energy feedback	*504.72*	*158.93*
High input	Agriculture	476.00	416.36
	Oil extraction	6.20	83.91
	Transesterification	9.30	54.36
	Transportation	20.60	4.16
	Total costs without energy feedback	*512.10*	*558.78*
	Saved costs for replaced fuels		190.13
	Other saved costs for indirect energy replaced		90.1
	Total cost with energy feedback	*512.10*	*278.55*

replaced (using an average oil cost of €61.25/barrel or €0.45/kg). The total NED from Table 3.13 is equivalent to 22,133 t of oil equivalent (1 t oil equivalent = 4.186E+10 J) in the case of LI cropping, and 26,961 t oil equivalent for HI cropping.

The economic investment for the production chain (from data in Table 3.14) is indicated in Table 3.15. This does not include the cost of fuels used or other costs for indirect fossil energy saved, because it is assumed that a fraction of the energy delivered was feedback to the production process in order to make it independent of fossil energy input (a precondition for process sustainability). The comparison is therefore between the economic investment for all production factors except energy costs and

Table 3.15 Economic Balance between the Total Economic Investment for the Biodiesel and Heat Production Chain and the Saving of Economic Investment Associated with Such Bioenergy in Marginal Land of Campania

	Low input (€/yr)	High input (€/yr)
Total economic investment (a)	2.87E+07	3.41E+07
Value of delivered energy (b)	1.00E+07	1.22E+07
Net economic (b–a)	−1.86E+07	−2.19E+07
Ratio a/b	2.85	2.79

the economic value of net energy delivered (i.e., after all energy costs for production are subtracted).

The HI cropping pattern, if implemented in all marginal land of the region, would require about €34 million of economic investment (within the local process and in the whole life cycle assessment chain). Again, this amount does not include the cost of energy used because it was assumed that energy costs would be covered by a fraction of the energy delivered. The economic investment required is therefore 2.79 times higher than the economic savings associated with the oil saved thanks to the NED in the form of biodiesel and heat. The LI cropping system would require 16% less investment, and its investment/savings ratio would be 2.85, suggesting that HI cropping is not the most profitable option, in spite of its higher yield per hectare and total yield.

3.6   Conclusions

The energy and environmental performance of Brassica biodiesel is hardly suitable for energy requirements in Campania as a whole or for regional agriculture alone. If straw and cake meals are also considered for energy generation in addition to the biodiesel, and if inputs are allocated to all coproducts, then the performance is slightly higher from an energy point of view, but the environmental performance (most of all the total emissions per hectare) and the economic performance make the process unacceptable. When all factors are accounted together, the option of cropping for energy on marginal land becomes much less attractive than claimed by some. Results show that the net profit is very small in energy terms and that there is no profit at all in economic terms. The economic balance would be better for much higher oil prices, however, or if labor costs could be decreased, which is not very likely to happen.

In conclusion, we can say:

1. *Brassica carinata* (and other crops with similar characteristics in terms of yield and required cultural practices) is not suitable to be used in a bioenergy production chain because of its low productivity. We had to include all the by-products to reach an acceptable performance in energy terms.

2. Fossil fuels have at present a relatively low cost. Biofuels could be competitive only if the cost of a barrel reaches €160–200, assuming that all other costs remain the same (again, unlikely). Reducing costs for labor and services by 50% would make the process profitable at an oil price around €130/barrel.

3. The marginal land is marginal because of several negative characteristics (water not easily available, poor soil organic matter content,

distance from transportation routes, excess slope, pollution due to previous use by industrial activities, etc.) that make it unsuitable for food cropping. Having realized that it is also unsuitable for energy cropping, planting a forest for better landscape and higher biodiversity might be the best solution. This forest might in the future provide wood for energy and other industrial uses.

4. Decreasing the region's dependence on fossil fuels is unlikely to be addressed by means of biomass energy. This is especially due to the high energy consumption of productive sectors (including agriculture) and the region as a whole. A high energy-intensity society cannot be supported by a diluted energy source such as biomass.

References

Bargigli, S. 2002. Analisi del ciclo di vita e valutazione di impatto ambientale della produzione ed uso di idrogeno combustibile. Tesi di dottorato (Ph.D. thesis, University of Siena).

Bargigli, S., Raugei, M., and Ulgiati, S. 2004. Mass flow analysis and mass-based indicators. In *Handbook of ecological indicators for assessment of ecosystem health*, ed. Jorgensen, S. E., Xu, Fu-Liu, Costanza, R., 353–378. Boca Raton, FL: CRC Press.

Bargigli, S., and Ulgiati, S. 2003. Emergy and life-cycle assessment of steel production in Europe. In *Emergy synthesis: Theory and applications of emergy methodology*, Vol. 2, ed. Brown, M. T., Odum, H. T., Tilley, D., and Ulgiati, S., 141–156. Gainesville: Center for Environmental Policy, University of Florida.

Berger, G., Kaechele, H., and Pfeffer, H. 2006. The greening of the European Common Agricultural Policy by linking the European-wide obligation of set-aside with voluntary agri-environmental measures on a regional scale. *Environmental Science & Policy*, 9(6): 509–524.

Biondi, P., Panaro, V., and Pellizzi, G. 1989. *Le richieste di energia del sistema agricolo italiano.* Progetto Finalizzato Energetica, Sottoprogetto Biomasse ed Agricoltura, Report LB-20. Rome: Consiglio Nazionale delle Ricerche.

Boustead, I., and Hancock, G. F. 1979. *Handbook of industrial energy analysis.* New York: John Wiley & Sons.

Brown, M. T., and Herendeen, R. 1996. Embodied energy analysis and emergy analysis: A comparative view. *Ecological Economics*, 19:219–236.

Brown, M. T., and Ulgiati, S. 2004. Energy quality, emergy, and transformity: H. T. Odum's contributions to quantifying and understanding systems. *Ecological Modelling*, 178(1–2): 201–213.

Buenfil, A. 2000. Sustainable use of potable water in Florida: An emergy analysis of water supply and treatment alternatives. In *Emergy synthesis: Theory and applications of emergy methodology*, Vol. 1, ed. Brown, M. T., Odum, H. T., Tilley, D., and Ulgiati, S., 107–116. Gainesville: Center for Environmental Policy, University of Florida.

Buonocore, E., Franzese, P. P., and Ulgiati, S. 2010. Emergy evaluation of an integrated bioenergy production system: The case study of Enköping (Sweden). In *Emergy synthesis: Theory and applications of the emergy methodology*, Vol. 6, ed. Brown, M. T., Campbell, D., Comar, V., Huang, S. L., Rydberg, T., Tilley, D., and Ulgiati, S. Gainesville: H. T. Odum Center for Environmental Policy, University of Florida, in press.

Campania. 2009. Piano energetico ambientale regionale, Regione Campania (Environmental regional energy plan for Campania Region).

Cardone, M., Mazzoncini, M., Menini, S., Rocco, V., Senatore, A., Seggiani, M., and Vitolo, S. 2003. *Brassica carinata* as an alternative oil crop for the production of biodiesel in Italy: Agronomic evaluation, fuel production by transesterification and characterization. *Biomass and Bioenergy*, 25:623–636.

CEEC (Council of the European Economic Community). 1992. Council Regulation (EEC) No. 1765/92 of June 30, 1992, establishing a support system for producers of certain arable crops. *Off. J. L* 181, 0012–0020.

Cialani, C., Russi, D., and Ulgiati, S. 2005. Investigating a 20-year national economic dynamics by means of emergy-based indicators. In *Emergy synthesis: Theory and applications of emergy methodology*, Vol. 3, ed. Brown, M. T., Odum, H. T., Tilley, D., and Ulgiati, S., 401–416. Gainesville: Center for Environmental Policy, University of Florida.

Copani, V., Cammarata, M., Abbate, V., and Ruberto, G. 1999. Caratteristiche biologiche e qualità dell'olio di crocifere diverse in ambiente mediterraneo. Presentation at the 33rd Convention of the SIA (Società Italiana di Agronomia), Padua, Italy, September 20–23.

De Mastro, G., Grassano, N., Tedone, L., Verdini, L., Lazzeri, L., and D'Avino, L. 2010. Esperienze pilota a sostegno dello sviluppo delle filiere agro-energetiche in Puglia. In Le agroenergie in Puglia potenzialità e prospettive. Regione Puglia, Area Politiche per lo Sviluppo Rurale, Bari, 134–196.

De Mastro, G., Tedone, L., Rotondo, G., and Marzi, V. 1999. Valutazione bioagronomica di crocifere ad alto erucico in Basilicata. In *Proceedings Atti 33° Convegno SIA, 1999, 21–23 settembre Padova*, 237–238.

EPA (Environmental Protection Agency). 2010. *Compilation of air pollutants emission factors*. Vol. 1, *Point sources*. 5th ed. Publication AP-42. Washington, DC: EPA. http://www.epa.gov/ttnchie1/ap42/.

EEA (European Environment Agency). 2000. *Corine land cover 2000 seamless vector database*. [Online database accessed June 16, 2011.] Copenhagen: EEA. http://www.eea.europa.eu/data-and-maps/data/corine-land-cover-2000-clc2000-seamless-vector-database.

————. 2009. *EMEP/EEA air pollutant emission inventory guidebook, 2009*. Copenhagen: EEA. http://www.eea.europa.eu/publications/emep-eea-emission-inventory-guidebook-2009.

Fischer, G., Prieler, S., van Velthuizen, H., Berndes, G., Faaij, A., Londo, M., and de Wit, M. 2010. Biofuel production potentials in Europe: Sustainable use of cultivated land and pastures, Part II: Land use scenarios. *Biomass and Bioenergy*, 34(2): 173–187.

Herendeen R. 1998. Embodied energy, embodied everything ... Now what? In *Proceedings of the 1st International Workshop Advances in Energy Studies: Energy flows in ecology and economy*, 13–48. Rome: MUSIS.

Hinterberger, F., and Stiller, H. 1998. Energy and material flows. In *Proceedings of the 1st International Workshop Advances in Energy Studies: Energy flows in ecology and economy*, 275–286. Rome: MUSIS.

Lettenmeier, M., Rohn, H., Liedtke, C., and Schmidt-Bleek, F. 2009. Resource productivity in 7 steps: How to develop eco-innovative products and services and improve their material footprint. Berlin: Wuppertal Institute for Climate, Environment and Energy.

Odum, H. T. 1996. *Environmental accounting: Emergy and environmental decision making*. New York: Wiley.

Odum, H. T., Brown, M. T., and Brandt-Williams, S. 2000. *Handbook of emergy evaluation: A compendium of data for emergy computation*. Folio #1, *Introduction and global budget*. Gainesville: Center for Environmental Policy, University of Florida. http://www.enst.umd.edu/tilley/emergy/EmergyFolio1Introduction.pdf.

Rosso, F., Cerrato, M., and Meriggi, P. 1999. Esperienze di coltivazione della *Brassica carinata* nella Valle Padana meridionale ed aspetti di tecnica colturale. Presentation at the 33rd Convention of the SIA (Società Italiana di Agronomia), Padua, Italy, September 20–23.

Schmidt-Bleek, F. 1993. MIPS re-visited. *Fresenius Environmental Bulletin*, 2:407–412.

Ulgiati, S. 2001. A comprehensive energy and economic assessment of biofuels: When "green" is not enough. *Critical Reviews in Plant Sciences*, 20(1): 71–106.

Ulgiati, S., Ascione M., Bargigli, S., Cherubini, F., Federici, M., Franzese, P. P., Raugei, M., Viglia, S., and Zucaro, A. 2010. Multi-method and multi-scale analysis of energy and resource conversion and use. In *Energy options impact on regional security*, 1–36, ed. Barbir, F., and Ulgiati, S. Dordrecht, The Netherlands: Springer.

Ulgiati, S., Raugei, M., and Bargigli, S. 2006. Overcoming the inadequacy of single-criterion approaches to life cycle assessment. *Ecological Modelling*, 190:432–442.

chapter four

Can switchgrass deliver the ethanol needed to power U.S. transportation?

Tad W. Patzek

Contents

4.1   Introduction

In this chapter, I will apply the universal laws of mass and energy conservation to an annual cycle that uses switchgrass to produce ethanol and compares that cycle with the solar photovoltaic cells that deliver equivalent mechanical power. A switchgrass-ethanol cycle consists of two parts:

1. Repeated harvests of switchgrass grown on large industrial plantations, followed by drying, compacting, and baling the harvested grass and transporting the bales to a remote refinery.
2. Repeated decomposition and fermentation of the switchgrass feedstock, followed by membrane separation/distillation/azeotrope separation of a dilute ethanol broth to give anhydrous ethanol.

In the refinery, switchgrass is used as both the feedstock for "cellulosic ethanol" fuel and the sole source of heat and electricity required to run the grass decomposition, simple sugar fermentation, and ethanol distillation processes. Both parts of the cycle are repeated each year and deliver a specified volume of anhydrous ethanol continuously for decades.

The term *cellulosic ethanol* is meant to suggest that certain components of wood and green plant materials (cellulose, pectins, and hemicelluloses) are chemically separated (mostly from lignin in wood) and partially split into hexose and pentose monomers, which are then fermented to produce ethanol.

Close to three billion years of plant evolution from cyanobacteria and algae have made cellulose very stable and resistant to biochemical attacks (Taiz and Zeiger, 1998; Zeltich, 1971). Cellulose can be quickly decomposed and hydrolyzed by extreme mechanical grinding, hard nuclear radiation, or steam exploding and severe chemical attack by hot concentrated sulfuric acid or sodium hydroxide (Schweiger, 1979; Fan et al., 1980; San Martin et al., 1986; Lynd et al., 2002; Zhang and Lynd, 2003, 2004). Biochemical enzymatic attacks are orders of magnitude slower and have an inherently low efficiency (Lee and Fan, 1983a, 1983b). For example, it takes 20 hours of a cellulase enzyme attack to shift to homogeneous reaction kinetics and 90 hours to complete the attack. When strong acids or hydroxides are used to damage the crystalline structure of cellulose, reaction kinetics accelerate by two orders of magnitude, that is, the reaction time is shortened by a ratio comparable to the speed ratio of a jet flying and a human jogging.

Cellulose fibers are separated from the rest of woody biomass in the well-known "kraft process," which is fast, efficient, and highly energy intensive. The kraft process, developed by Carl Dahl in 1884, now produces 80% of paper volume. Caustic sodium hydroxide and sodium sulfide are applied to extract the lignin from the wood fiber in large pressure vessels called *digesters*. The best energy efficiency of this process is about 30 MJ/kg of paper pulp (Ayres et al., 2002), more than the higher heating value of pure ethanol, defined in Section 4.3.4. Therefore a much milder, but slow, enzymatic process must be used to obtain simple sugars from cellulose.

It has been claimed that more than a billion tons of cellulosic biomass can be extracted from U.S. territory each year (Perlack et al., 2005; Somerville, 2006) and converted into biofuels, producing, for example, 130 billion gallons of cellulosic ethanol (Khosla, 2006). Alas, the recent Moderate-resolution Imaging Spectroradiometer (MODIS) satellite-based (Heinsch et al., 2003) calculations (Patzek, 2007, 2008a) of net primary productivity of all plants growing in the United States cast doubt on the sustainability of this claim. To produce 130 billion gallons of ethanol, one would have to use each year between one-third and two-thirds of all above-ground biomass growth in the United States.

It is customary in the agricultural literature (e.g., Parrish et al., 2003; Casler and Boe, 2003; Berdahl et al., 2005; Schmer et al., 2008) to conduct a few short-term (less than five-year) field studies and assume that average yield of switchgrass anywhere in the United States can be *quantified* by a single number $x \pm$ some deviation and that that number can only grow with time. The x estimated by Schmer et al. (2008) from their field data was 7.2 Mg/ha/yr. In this chapter, I suggest that the wide temporal and spatial variations of switchgrass yields prohibit relying on a single value x. Instead, using all available data weighted by a long-term field survival function, one can propose a probability distribution function of continuous (or eternal) switchgrass yields and calculate this distribution's expected value. That value alone (calculated in this chapter to be 6.8 Mg/ha/yr) is unreliable, but the distribution provides a *qualitative* insight into what might be expected from the future switchgrass plantations.

Quantitative models of complex processes on the surface of the Earth do not work, as illustrated in detail by Pilkey and Pilkey-Jarvis (2007). On the other hand, a qualitative model when applied correctly in what-if scenarios can be useful. A brief look back at the simplistic quantitative models of the corn ethanol industry—for example, the Greenhouse Gases, Regulated Emissions, and Energy Use in Transportation Model (GREET; Wang, 2001) and its conceptual offspring, The Berkeley Energy and Resources Group's Biofuels Analysis Meta-Model (EBAMM; Farrell et al., 2006)—have been shown to be not only physically incorrect (see, e.g., Patzek, 2006b) but also incapable of predicting the odds of survival of ethanol companies, which have been completely dependent on government subsidies regardless of the market conditions.

To calculate a realistic efficiency of conversion of switchgrass to ethanol, I "reverse-engineer" an estimate of the energy efficiency of cellulosic ethanol production in an existing pilot plant, operated by Iogen in Ottawa, Canada. I then translate the Iogen plant results from wheat straw to switchgrass and calculate how much additional switchgrass must be burned to obtain anhydrous ethanol with "zero" use of fossil energy. To account for all available field data on switchgrass yields, I perform a regression (Monte Carlo) analysis. Finally, I compare the switchgrass-ethanol cycle with photovoltaic cells.

Similar mass, energy (Pimentel and Patzek, 2005; Patzek, 2006a, 2006b), and free energy or "exergy" (Patzek, 2004, 2007, 2008c; Patzek and Pimentel, 2005) balances have been applied elsewhere to various crops and biofuels. Free energy is irreversibly consumed in all steps of all biofuel production processes and should be used to determine the relative degrees of unsustainability of different biofuels. Such applications are always based on physical data, including the repeated average yields of a plant feedstock grown on large fields, and the measured energy efficiency of ethanol refineries or cogeneration plants using biomass gasification/liquid fuel synthesis and generating electricity.

4.2 Preliminary calculations

By assuming the average switchgrass yield from Schmer et al. (2008), doubling the energy efficiency of a cellulosic ethanol refinery relative to that given in Farrell et al. (2006), and disregarding the high chemical and environmental costs of large-scale switchgrass agriculture, one does *not* obtain a viable macrosystem of automotive fuel supply. One should therefore consider other more efficient technologies of converting sunlight into motive power. The simple calculation below demonstrates this fact.

Assuming average continuous ("eternal") yield of switchgrass of 7.2 tons per acre per year on a dry mass basis (dmb)—corresponding to the estimate by Schmer et al. (2008)—a biorefinery that might one day process 2,000 tons dmb of switchgrass per day (Epplin et al., 2007) will need 101,000 acres of grassland to supply it with raw material. (Epplin et al. [2007] assume 5.5 tons dmb/acre as their average switchgrass yield.) Assume that the supply area is roughly a circle with the biorefinery at the center. The mean square radius that scales the field area encompassed by this circle is 8 km. If the dry switchgrass is compacted to high-density bales and transported by large trucks, then roughly 400 truck trips, 16 km long on average, will be required each day. Note that the real distance driven by the trucks will necessarily be longer, because one cannot fill 100,000 acres of land with a contiguous sea of switchgrass. Meandering roads, buildings, ponds, creeks, rivers, and perhaps even trees will require additional area. At 2.4 km/liter, these trucks will use about 980,000 liters of diesel fuel per year. As discussed later in this chapter, the energy requirements of a switchgrass refinery might be satisfied by burning more switchgrass, but—because of the complex transportation logistics and the doubling of field area involved—coal or natural gas use will be more likely.

Let us set aside the different estimates of process energy requirements of the nonexistent switchgrass refinery (29 MJ/liter in EBAMM; see Farrell et al., 2006). Instead, let us assume that this refinery is 43% energy efficient, requiring only 15 MJ of primary energy per liter of ethanol, just like the modern corn ethanol refineries compared in Figure 2 in Patzek (2006a). We will further assume the ethanol production efficiency from EBAMM of 0.38 liters/kg switchgrass (Patzek, 2006b, 2006c). With these assumptions, our biorefinery will produce 3,000 barrels or 493,000 liters of gasoline equivalent per day and will need 360 tons of coal equivalents per day. At 57 liters (15 gallons) of gasoline equivalent per tank (equal to 46.7/29.6×57 or about 90 liters of ethanol per tank), and assuming that a car drives for one week on a tank of fuel, our switchgrass biorefinery will be able to fill up tanks of 59,400 small and midsize cars each year (cars that use on average 11 liters of gasoline equivalent per 100 km, or 22 miles per gallon), provided that the switchgrass yield remains perfectly constant. To put it differently, one will need 1.7 acres of switchgrass and 2.2 tons of coal equivalent per year to run a midsize car in perpetuity.

If, instead, one covered these 1.7 acres of land with 10%-efficient solar photovoltaic cells (also an impossible task today), one would generate primary energy sufficient to run 115 cars. Thus, one obtains a factor of roughly 100 when comparing the future energy efficiencies of car systems powered by solar cells or switchgrass ethanol.

The simplified analysis here has neglected the energy costs required for the production of switchgrass; harvesting, compacting, and baling the switchgrass; producing and maintaining photovoltaic cells; and maintaining the transportation/distribution infrastructure. It turns out (Patzek, 2007) that after these costs are included, the photovoltaic car systems are 100 to 1,000 times more efficient than the alternative biofuel car systems.

In the metropolitan San Francisco Bay Area, there are 4 million vehicles (see Purvis, 2011, Table A-5). Thus, with the very optimistic assumptions above, roughly 7 million acres of switchgrass and 8 million tons of coal equivalent per year (roughly 1% of U.S. coal production) would be required to fuel the Bay Area's vehicles. Seven million acres is equal to 9% of the total area of corn agriculture, by far the largest agricultural crop in the nation, delivering more biomass than all other crops combined. Seven million acres is also 2% of all grassland area in the United States (Patzek, 2008a, Table 2.3) and is 80% of the 9 million acres of prime irrigated agricultural land in California (Thompson, 2009).

Note that if one went back to EBAMM for the refinery efficiency, one would need 16 million tons of coal equivalent per year to power the vehicles in a single major metropolitan area in the United States. If, in addition, one recognizes that the EBAMM ethanol yield of 0.38 liters/kg has no factual justification (see Section 4.3.4) and that 0.23 liters/kg is more appropriate, these requirements will grow to 11.6 million acres and exceed 16 million tons of coal equivalent (around 2% of U.S. coal production in 2008), respectively.

A vast acreage of switchgrass would encompass highly varied climatic conditions and soils, undoubtedly lowering the current average yield estimates. For example, corn yields in Minnesota are usually 50% higher than those in Missouri. The same general principle would apply to switchgrass grown in different regions of the United States. Two numbers for average yield of switchgrass and refinery efficiency, presented in Schmer et al. (2008) and Farrell et al. (2006), are inadequate to describe such a complex distributed system. Here I attempt to fix this deficiency by providing *contexts* of the means and standard deviations of switchgrass yields and refinery efficiencies.

Figure 4.1 is reproduced from the supporting online materials of Scharlemann and Laurance (2008). This figure summarizes the environmental impacts of 29 major biofuel schemes. The bottom line is this: The large-scale biofuel systems have very large negative impacts on the entire planet that go well beyond those of natural gas and crude oil. In addition, our planet can never produce enough of the raw biomass for these systems for a sufficiently long time (see, e.g., Patzek, 2008a), regardless of the claims by Perlack et al. (2005).

4.3   Input data

4.3.1   Switchgrass yields

Here, the published switchgrass yields are compared and analyzed. In particular, a distinction is made between

- The finite-time switchgrass plantations that take 1–3 years to establish and are commercially viable for 5–20 years at variable yields, and

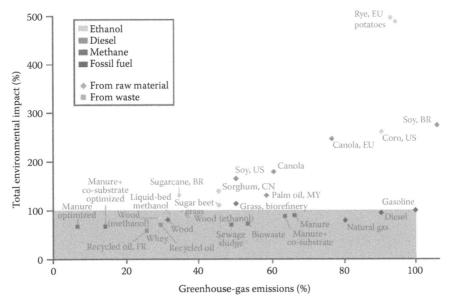

Figure 4.1 Greenhouse gas emissions plotted against overall environmental impacts of 29 transport fuels, scaled relative to gasoline. The origin of bio-fuels produced outside Switzerland is indicated by country codes: Brazil (BR), China (CN), European Union (EU), France (FR), and Malaysia (MY). Fuels in the shaded area are considered advantageous in both their overall environmental impacts and greenhouse gas emissions. (Adapted from Zah, R., Böni, H., Gauch, M., Hischier, R., Lehmann, M., and Wäger, P., *Ökobilanz von Energieprodukten: Ökologische Bewertung von Biotreibstoffen,* St. Gallen, Switzerland: Empa, 2007.)

- The "eternal" average yield of switchgrass needed for the robust, multidecade fuel supply estimates.

All plant mass is reported on a water-free basis. Switchgrass (*Panicum virgatum*) is a warm-season C$_4$ grass and a close cousin of corn (maize) (Lawrence and Walbot, 2007). It is one of the dominant species of the central North American tallgrass prairie. Switchgrass is a hardy perennial that reproduces by rhizomes, shoots (tillers), and seeds. In my calculations, I assume that switchgrass begins growth in late spring, takes only one or two years to grow sufficiently to establish a harvestable field, and may last for another 8 to as much as 19 years.

On tiny, 1.6 × 3 m and 1.6 × 1.8 m plots fertilized similarly to cornfields, switchgrass can yield from 4 to 18 Mg/ha/yr for four years (Casler and Boe, 2003). Larger switchgrass stands have been investigated for at least 10 years in one case (McLaughlin and Adams Kszos, 2006), but seldom above five years (Fike et al., 2006a). The field studies in Parrish et al. (2003),

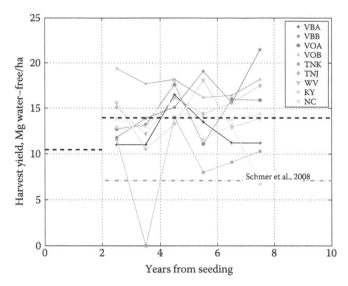

Figure 4.2 Single crops of switchgrass on large fields. Fields shown are Virginia, Blacksburg, Sites A and B (VBA, VBB); Virginia, Orange, Sites A and B (VOA, VOB); Tennessee, Knox (TNK); Tennessee, Jack (TNJ); West Virginia (WV); Kentucky (KY); and North Carolina (NC). The upper broken line represents the mean of the data; the lower one is the continuous average with a two-year delay of annual harvests. (From Parrish, D. J., Wolf, D. D., Fike, J. H., and Daniels, W. L., *Switchgrass as a biofuels crop for the Upper Southeast: Variety trials and cultural improvements: Final report for 1997 to 2001*, Report ORNL/SUB-03-19XSY163/01, Oak Ridge, TN: Oak Ridge National Laboratory, 2003.)

Casler and Boe (2003), Berdahl et al. (2005), McLaughlin and Adams Kszos (2006), and Schmer et al. (2008) are summarized in Figures 4.2 through 4.6. The reported mean harvests vary from 4 to 25 Mg/ha/yr and the stand establishment periods vary from 1 to 3 years.

4.3.1.1 Harvest delays

What matters from the point of view of transportation fuel supply is the continuous (year-after-year) mean supply of biomass and the variance of this supply. Therefore, taking an arithmetic mean of 3–4 years of harvest would be justified only if switchgrass grew "forever" with the commercially viable fluctuating yields, after it has been established for two years on large fields. Table 4 in the supporting information to Schmer et al. (2008) gives a comprehensive overview of the actual biomass yields from 10 established (2 years after planting) large switchgrass fields in the mid-continental United States. These data are plotted in Figure 4.7. It appears that all but one of the measured switchgrass yields extrapolate to zero after 4.5–11 years from seeding.

If switchgrass is not harvested for two years and the harvest data are reported only for three years, then the actual yields should be discounted by taking the mean of

$$[0, 0, \text{Yield}(3), \text{Yield}(4), \text{Yield}(5), \text{Predicted Yield}(6), \ldots] \qquad (1)$$

For the data in Figure 4.7 (Schmer et al., 2008, Table 4), this procedure discounts the yields effectively by a factor from $\frac{3}{5}$ up to $\frac{5}{7}$. In contrast, the switchgrass field studied by McLaughlin and Adam Kszos (2006) survived at least 10 harvests with the yields higher than those in all other studies considered in this chapter (Parrish et al., 2003; Casler and Boe, 2003; Berdahl et al., 2005; Fike et al., 2006a, 2006b; Schmer et al., 2008). In each of Figures 4.2 through 4.7, the mean yields of the harvests and mean continuous yields with a two-year delay are denoted by the horizontal broken lines.

My choice of a two-year delay of annual harvests for the low-yield fields and a one-year delay for the high-yield ones, is based on the following reports:

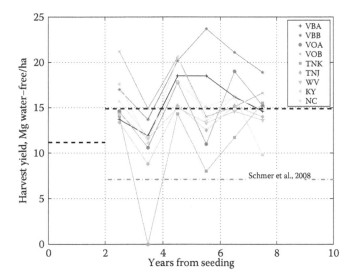

Figure 4.3 Double crops of switchgrass on large fields. Fields shown are Virginia, Blacksburg, Sites A and B; Virginia, Orange, Sites A and B; Tennessee, Knox; and Tennessee, Jack. The upper broken line represents the mean of the data; the lower one is the continuous average with a two-year delay of annual harvests. (From Parrish, D. J., Wolf, D. D., Fike, J. H., and Daniels, W. L., *Switchgrass as a biofuels crop for the Upper Southeast: Variety trials and cultural improvements: Final report for 1997 to 2001*, Report ORNL/SUB-03-19XSY163/01, Oak Ridge, TN: Oak Ridge National Laboratory, 2003.)

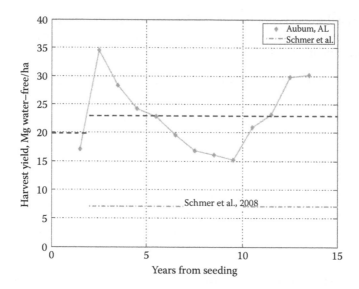

Figure 4.4 A 10-year study of switchgrass growth in Auburn, Alabama. The upper broken line represents the mean of the data; the lower one is the continuous average with a two-year delay of annual harvests. (From McLaughlin, S. B., and Adam Kszos, L., 2006, *Biomass Bioenergy* 28(6): 515–535.)

1. Schmer et al. (2008, Table 4) contains a spreadsheet entitled "Biomass yields from established (2 years after planting) switchgrass fields in the midcontinental U.S."
2. In their paper, Schmer et al. (2006, p. 161) state the following: "At the field scale in the northern Great Plains, second year biomass yields are limited by establishment stands only if initial stands are less than 40%. If establishment year switchgrass stands on a field have threshold frequency levels of 40% or more, post-establishment biomass yields and post-establishment switchgrass stands are likely influenced more by site and environmental variation than by initial stand frequency. Failure to obtain a fully successful switchgrass stand the establishment year (stand frequency of 40% or greater) can limit biomass yield in post-establishment years resulting in decreased revenue" (2006, p. 161). Since 4 out of 10 sites were below 40% frequency and 7 out of 10 were below 50% frequency, I took this finding as a statistically valid restriction to harvesting switchgrass in the first two years after establishment.
3. The Bransby et al. (2003) study states that, although switchgrass yields were low in the establishment year (2.44 Mg/ha), yields increased over the next four years showing relatively

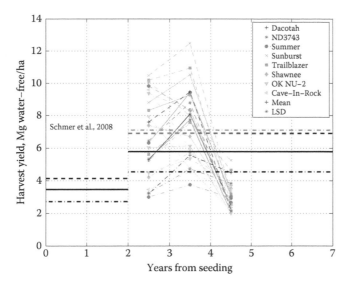

Figure 4.5 Measured yields of eight switchgrass cultivars at three North Dakota sites. Cultivars are indicated by diamonds. The continuous lines refer to the Mandan Site 1, the broken lines to the Mandan Site 2, and the dash-dot lines to the Dickinson site. Those sites were small, roughly 2×6 m. The upper horizontal lines represent the means of the data and the lower ones are the continuous average with a two-year delay of annual harvests. Note that these yields extrapolate to zero in up to seven years. (From Berdahl, J. D., Frank, A. B., Krupinsky, J. M., Carr, P. M., Hanson, J. D., and Johnson, H. A., 2005, *Agron. J.* 97:549–555.)

little response to precipitation. I took their finding as a valid restriction to harvesting switchgrass in the first two years after establishment.

4. McLaughlin and Adam Kszos (2006, p. 517–518) state: "One of the most persistent issues in producing switchgrass as an energy crop has been delineation of management regimes that will enable growers to rapidly and consistently establish strong stands of switchgrass. As a small-seeded species that initially allocates a large amount of energy to developing a strong root system, switchgrass will typically attain only 33–66% of its maximum production capacity during the *initial* and *second* years before reaching its full capacity during the *third* year after planting. Switchgrass is most susceptible to weed competition as well as the dangers of 'assumed failure' during the critical first season" (emphasis added). I took their finding as a valid restriction to harvesting switchgrass in the first two years after establishment.

5. Fike et al. (2006a, p. 198) state: "Limited information is available regarding biomass production potential of long-term (>5-yr-old)

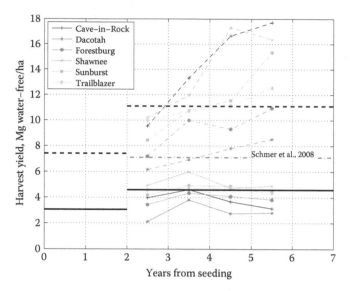

Figure 4.6 Measured yields of six switchgrass cultivars at Brookings, SD, and Arlington, WI, sites. Cultivars are indicated by diamonds. Plot sizes were *tiny*: 1.6 × 3.0 m at Brookings (continuous lines) and 1.6 × 1.8 m at Arlington (broken lines). The upper horizontal lines represent the means of the data and the lower ones are the continuous averages with a two-year delay of annual harvests. (From Casler, M. D., and Boe, A. R., 2003, *Crop Sci.* 43(6): 2226–2233.)

switchgrass.... Yields at Site B (19.1Mgha⁻¹) were about 35% greater than those at Site A (14.1Mgha⁻¹), although the sites were only 200m apart." I take this statement to mean that, except for the Auburn study, zero evidence has been gathered thus far about the long-term viability of switchgrass monocultures and that there is a great variability of switchgrass yields. Fike et al. continue: "Plots were evaluated in fall 1996 after having been established in 1992 (or 1993 in Kentucky)" (2006a, p. 204 [footnote of Table 2]; p. 205 [title of] Table 3). This means a waiting period of two to three years.

6. Fike et al. (2006b, p. 208) state: "During 1992, four switchgrass cultivars were planted at eight sites across five states.... Sites were chosen to bound broad geographic, edaphic, and climatological differences within the upper southeastern U.S.A.... Cutting managements were first imposed in the year after establishment. Biomass yields reported here are based on harvests from 1994 to 1996, with the exception of the Kentucky site (1995 and 1996 only due to later establishment)." This remark, again, is a clear indication of a two- to three-year delay between regular cutting of switchgrass and stand establishment.

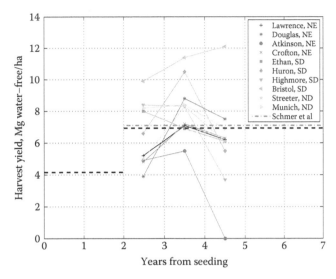

Figure 4.7 Measured yields of switchgrass established for two years on large fields. Fields shown are in North Dakota (ND), South Dakota (SD), and Nebraska (NE). The dash-dot thick line is the continuous yield prediction from Schmer et al. (2008). The upper broken line represents the mean of the data; the lower one is the continuous average with a two-year delay of annual harvests. With one exception, these yields extrapolate to zero from 4.5 to 11 years. (From Schmer, M. R., Vogel, K. P., Mitchell, R. B., and Perrin, R. K, 2008, *PNAS* 105(2): 464–469.)

4.3.2   Inputs to switchgrass agriculture

In this section, the average annual mass inputs of macronutrients (NPK and Ca) and field chemicals such as pesticides, herbicides, and fungicides, as well as the corresponding energy inputs to switchgrass agriculture, are estimated and discussed.

Switchgrass is in many ways similar to lawn grass. If one bags grass clippings from a lawn, one immediately notices that the lawn starts to become yellow, and the "yield" (the number of times one has to mow) decreases. This declining lawn productivity is caused by the depletion of macronutrients (N, P, and K) and micronutrients. Each time one removes biomass debris from an environment, one removes nutrients, and future yields suffer (Patzek, 2007). Switchgrass is exactly the same: If one harvests switchgrass for biomass, fertilizer must be applied each year at levels similar to those applied to cornfields.

The mass fluxes in industrial switchgrass agriculture and the pertinent references are listed in Table 4.1. The corresponding energy fluxes are listed in Table 4.2 and are shown in Figure 4.8. Note that by accounting for the machinery, repairs, transportation, potash, and lime costs, these energy fluxes are about 47% higher that those listed in Schmer et al. (2008).

4.3.3 Switchgrass versus corn

Average mass and energy fluxes in U.S. corn agriculture are listed in Table 4.3. Table 4.4 compares the mass balances of cellulosic and corn ethanol. Schmer et al. (2008, Tables 2 and 6) imply that a typical switchgrass plantation requires 112 kg/ha/yr N, 56 kg/ha/yr P_2O_5, and over 8 kg/ha/yr

Table 4.1 Estimates of Mass Fluxes Involved in the Standard Management of Switchgrass

Input	Mass flux (English)		Mass flux (metric)	
Switchgrass[a]	6,088	lbm/acre	6,830	kg/ha-yr
Nitrogen	100	lb N/acre	112	kg N/ha-yr
Phosphorus[b]	13	lb P_2O_5/acre	14	kg P_2O_5/ha-yr
Potassium[c]	28	lb K_2O/acre	31	kg K_2O/ha-yr
Lime[d]	545	lb CaO/acre	611	kg CaO/ha-yr
Gasoline	0.0	gal/acre	0	L/ha-yr
Diesel[e]	2.8	gal/acre	26	L/ha-yr
Liquid petroleum gas (LPG)	0	gal/acre	0	L/ha-yr
Natural gas	0	scf/acre	0	sm^3/ha-yr
Pesticides[f]	0.00	lb/acre	0.00	kg/ha-yr
Herbicides[f]	7.1	lb/acre	8.0	kg/ha-yr
Irrigation	0	inch	0	cm/yr
Seeds[g]	2.7	lbm/acre	6.7	kg/ha-yr
Field machinery[h]	25	lb/acre	28	kg/ha-yr
Transportation[i]	3,928	lb/acre	4,406	kg/ha-yr

[a] Dry mass basis, the mean of the distribution in Figure 4.2.

[b] P_2O_5 applied at 60 kg/ha-yr in the first year and the replacement rate for four more years. Based on Duffy, M., and Nanhou, V. Y., *Costs of producing switchgrass for biomass in southern Iowa*, University Extension Report PM 1866 (Ames: Iowa State University, 2001), amortized over five years.

[c] K_2O application at the replacement rate from Duffy and Nanhou (2001).

[d] Lime application from Duffy and Nanhou (2001) amortized over five years.

[e] Average diesel fuel use to operate farm equipment (Schmer, M. R., Vogel, K. P., Mitchell, R. B., and Perrin, R. K., 2008, Net energy of cellulosic ethanol from switchgrass, *PNAS* 105(2): 464–469).

[f] Average pesticide and herbicide use in switchgrass farming (Schmer et al., 2008).

[g] From Schmer et al., 2008.

[h] Machinery consists of a tractor, tandem disk, roller harrow, seed drill, fertilizer cart, sprayer, row-crop cultivator, combine with grain header, and grass baler (Schmer et al., 2008). Conservatively, I use Smil et al.'s (Smil, V., Nachman, P., and Long, T. V., II, *Energy analysis and agriculture: An application to U.S. corn production* [Boulder, CO: Westview Press, 1983]) estimate of machinery mass for a 175-acre corn farm amortized over 12 years.

[i] Transport of grass with 10% of moisture out, and fertilizers and field chemicals in.

Table 4.2 Estimates of Energy Fluxes Involved in the Standard Management of Switchgrass

Input	Specific energy (MJ/kg)	Energy flux (GJ/ha-yr)
Switchgrass	18.10	123.62
N[a]	48.00	5.38
P_2O_5[b]	6.80	0.10
K_2O[c]	6.80	0.20
CaO[d]	1.75	1.07
Gasoline[e]	46.70	0.00
Diesel[e]	45.90	1.01
LPG[e]	50.00	0.00
NG[e]	55.50	0.00
Electricity[f]	10.29	0.41
Pesticides[g]	268.40	0.00
Herbicides[h]	322.30	2.58
Irrigation[i]	131.00	0.00
Seeds[j]	45.00	0.30
Field machinery[k]	85.00	2.38
Transportation[l]	Variable	0.49
Repair & maintenance[m]	Variable[m]	0.20
Total		*14.10*

[a] Current European urea (Patzek, T. W., 2004, Thermodynamics of the corn-ethanol biofuel cycle, *Crit. Rev. Plant Sci.* 23[6]: 519–567, Section 3.1).

[b] Patzek (2004, Section 3.1.3).

[c] Patzek (2004, Section 3.1.4).

[d] Patzek (2004, Section 3.1.5).

[e] Patzek (2004, Table 12).

[f] kWh/Mg switchgrass from EBAMM and GREET 1.6 as quoted in Schmer, Vogel, K. P., Mitchell, R. B., and Perrin, R. K., 2008, Net energy of cellulosic ethanol from switchgrass, *PNAS* 105[2]: 464–469), translated to MJ of primary energy by multiplying kWh by 3.6/0.35.

[g] Patzek (2004, Section 3.1.6).

[h] Schmer et al. (2008).

[i] MJ/cm-ha (Patzek, 2004, Section 3.7). No irrigation is used in the current model, but some must be used in the drier regions.

[j] Schmer et al. (2008)

[k] Average value from Smil, V., Nachman, P., and Long, T. V., II, *Energy analysis and agriculture: An application to U.S. corn production* (Boulder, CO: Westview Press, 1983), Table 5.6.

[l] From Wang, M., *Development and use of GREET 1.6 fuel-cycle model for transporation fuels and vehicle technologies*, Technical Report ANL/ESD/TM-163 (Argonne, IL: Argonne National Laboratory, Center for Transportation Research, 2001), plus 288 MJ/ha-yr for personal commute; see Patzek (2004, Section 3.8 and Table 17).

[m] Mostly 1/12th (8%) of the energy in machinery.

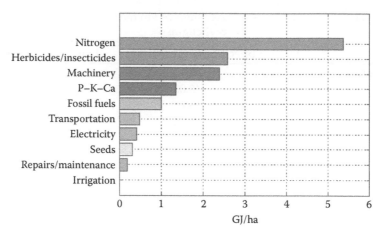

Figure 4.8 Energy fluxes in industrial switchgrass agriculture. Mostly, fossil energy fluxes in industrial switchgrass agriculture amount to about 14 GJ/ha/yr, or about 50% of those in corn agriculture (Patzek, 2004). With a one- to two-year delay of initial harvest, average fertilizer use will be lower than shown here.

of field chemicals (herbicides, pesticides, and fungicides) (see Table 4.2). As shown in Table 4.3, these fertilizer application rates are comparable with those in U.S. corn agriculture: 150 kg/ha/yr N and 64 kg/ha/yr P_2O_5 (Conway, 2005); however, only 3 kg/ha/yr of field chemicals are applied to corn on the average, or almost three times less.

It also turns out that an average cornfield can produce about two times more ethanol today than a switchgrass field might in an unspecified future using an industrial process that still does not exist. Ethanol is already being distilled twice as efficiently from corn as it might be from switchgrass one day (37% vs. 20%).

Note that the ratio of biomass energy output to the fossil energy inputs in an average switchgrass field is 124/14, or about 9, compared with the analogous ratio of 134/10.8, or more than 12, for an average cornfield, based on the admittedly low inputs from Table 4.3 (see also Patzek, 2004).

On average, corn agriculture is more efficient in delivering primary energy than switchgrass agriculture is. However, annual corn agriculture causes higher erosion losses than multiyear switchgrass agriculture.

Despite its low efficiency and environmental harmfulness (Patzek, 2004, 2006a; Pimentel et al., 2006), corn ethanol seems to be less harmful than switchgrass ethanol. The key reasons for this conclusion are the following:

1. In terms of biomass energy output/fossil energy inputs, corn agriculture is more efficient than switchgrass agriculture (the respective ratios are 12 and 9).

Table 4.3 Mass and Free Energy (Exergy) Fluxes of Direct Inputs to U.S. Corn Agriculture

Input	Quantity[a] (English units)[b]	Quantity (SI units)[b]	Specific exergy[c] (MJ/kg)[b]	Exergy (GJ/ha-yr)
Average yield	139.3 bu/acre	7,438.4 kg dmb/ha[d]	18.0	133.89
Seed	28,739.0 #/acre	23.6 kg/ha	18.0	0.42
Nitrogen, N	133.5 lbm/acre	149.8 kg/ha	24.1	3.61
Potash, K_2O	88.2 lbm/acre	98.9 kg/ha	2.7	0.27
Phosphate, P_2O_5	56.8 lbm/acre	63.7 kg/ha	4.4	0.28
Lime, CaO	15.7 lbm/acre	17.6 kg/ha	2.9	0.05
Diesel fuel	6.9 gal/acre	54.2 kg/ha	44.4	2.41
Gasoline	3.4 gal/acre	24.8 kg/ha	48.1	1.19
LPG	3.4 gal/acre	15.9 kg/ha	48.9	0.78
Electricity	33.6 kWh/acre	298.9 MJ/ha	n/a	0.30
Natural gas	246.0 scf/acre	14.5 kg/ha	46.4	0.67
Chemicals	2.7 lbm/acre	3.0 kg/ha	261.0	0.78
Total				*10.76*

[a] Conway, R., The net energy balance of corn ethanol, Presentation at the NCGA Renewable Fuels Forum, Washington, DC, August 23, 2005.
[b] The common time unit, yr^{-1}, is not listed.
[c] Exergy is free energy referenced to the average conditions of the environment. It can be regarded as an equivalent of electricity. Free energy is consumed in prodigious quantities in corn and corn-ethanol production processes (Patzek, T. W., 2004, Thermodynamics of the corn-ethanol biofuel cycle, *Crit. Rev. Plant Sci.* 23[6]: 519–567), but the use of specific exergy—as opposed to specific energy—results in values that are 10–30% lower (Szargut, J., Morris, D. R., and Steward, F. R., *Exergy analysis of thermal and metallurgical processes* [New York: Hemisphere, 1988]; http://www.exergoecology.com) (Exergy Calculator).
[d] Assuming 15% moisture by mass.

2. Corn may require more energy to grow (Patzek, 2004), but it more than makes up for this requirement with a higher yield of an easy-to-process ethanol feedstock, starch.
3. The rates of fertilization of switchgrass and corn are comparable, but switchgrass requires three times more field chemicals, potentially contributing to serious environmental problems.
4. The overall energy efficiency of a prototype switchgrass ethanol refinery, 20% (see the discussion in the following section), is half that of an average existing corn ethanol refinery (Patzek, 2004, 2006a).

Based on these observations, here is my warning for the record: An average switchgrass refinery will be at least two times less profitable than a corn refinery.

Table 4.4 Comparison of Water-Free Mass Fluxes in U.S. Switchgrass and Corn
Agriculture

Input	Continuous crop yield (kg/ha/yr)	Theoretical fuel mass yield[a] (kg ethanol/kg)	Ethanol mass flux[a] (kg/ha/yr)	Ethanol volume flux[a] (L/ha/yr)
Switchgrass	6,830[b]	0.180[c]	1,230	1,560
Corn	7,440[d]	0.345[e]	2,560	3,260

[a] Based solely on the mass balance, *not* on the energy requirements of broth distillation. Corn ethanol refineries are fueled with natural gas or coal to generate heat. The authors of Wang (2001), Farrell et al. (2006), and Schmer et al. (2008) want to burn switchgrass, instead. Note that the energy efficiency of a cellulosic ethanol refinery, 0.21, is close to 0.18. However, this coincidence may be misleading because (a) extra heat is needed to distill the dilute broth; (b) ethanol's HHV is higher than that of switchgrass by a factor of 1.6, thus lowering the cycle efficiency if switchgrass is used as refinery fuel; and (c) the switchgrass lignin is burned to generate steam and electricity, thus increasing energy efficiency of the cycle. In the end, all these factors almost balance out, and the effective ethanol fluxes based on energy are similar to those calculated here.
[b] See the discussion of mass conservation, Section 4.3.1.
[c] See Table 4.5.
[d] Conway (2005).
[e] 95% of theoretical efficiency (Patzek, 2006, Letter to the editor, *Science* 312:1747).

4.3.4   Energy efficiency of cellulosic ethanol refinery

4.3.4.1   Basic assumptions

As to the average energy costs of the future cellulosic refineries, the following assumptions are made:

1. The process energy fluxes (heat and electricity needed to separate cellulose and hemicelluloses, ferment them, and distill anhydrous ethanol) dwarf all other fluxes.
2. Switchgrass is used as the sole ethanol fuel feedstock and the source of process energy.
3. All other energy fluxes (production of enzymes, sulfuric acid, steam-exploding of biomass, water exclusion from azeotrope, and refinery hardware) are neglected for the time being. This is a very generous assumption in favor of cellulosic ethanol.
4. The average energy efficiency of the refinery is given by Equation (2) below.
5. This efficiency can only go down as the energy costs in #3 are incorporated.

4.3.4.2   Yield of ethanol from switchgrass

A commercial refinery that could produce ethanol from switchgrass did not exist as of April 2010. Nevertheless, one can analyze the fragmentary data from a cellulosic ethanol pilot plant in Ottawa, Canada, and sift through some of the published laboratory data on cellulosic ethanol yields.

The theoretical yield of ethanol from switchgrass is about 0.43 L/kg dmb, based on the conversion calculator developed by the U.S. National Renewable Energy Laboratory. The common yield assumed in GREET (Wang, 2001) and EBAMM (Farrell et al., 2006)—0.382 L/kg dmb—appears to be about 90% of the theoretical value and has no experimental validation in the context of mild enzymatic reaction conditions. For example, a recent laboratory-scale yield of ethanol from corn stover is reported to be 0.24 L EtOH/kg dmb of corn stover, at 4% of ethanol by mass (Lau and Dale, 2009). The expensive ammonia-pretreatment AFEX process used performs infinitesimally better than the industrial pilot by Iogen modeled here, but still delivers only 0.24 L/kg, or 56% of the theoretical yield.

4.3.4.3   Use the lower or higher heating value?

With regard to choosing the lower heating value (LHV) or higher heating value (HHV) in calculations, a thorough discussion has been provided by Bossel (2003), who established conclusively that only the HHV can be used to compare different fuels, especially those with different oxygen contents.

4.3.4.4   Performance of the Iogen Ottawa plant

In the process used at the Iogen plant in Ottawa, wheat, oat, and barley straw are first pretreated with sulfuric acid and steam. Iogen's patented enzymes then break the cellulose and hemicelluloses down into six- and five-carbon sugars, which are later fermented and distilled into ethanol. Standard yeast does not ferment the five-carbon sugars, so delicate, genetically modified and patented yeast strains are used. Iogen's plant has nameplate capacity of 1 million gallons of ethanol per year.

The only publicly presented history of cellulosic ethanol production is shown in Figure 4.9. From Figure 4.9, Passmore (2006), and Hladik (2006), the following can be deduced:

- 600,000 liters/year = 158,000 gallons/year of anhydrous ethanol, or 10 bbl/day = 6.7 bbl of equivalent gasoline/day, were actually produced.
- There exists 2 × 52,000 = 104,000 gallons of fermentation tank volume.
- The ratio of the annual volume of ethanol production and the tank volume is 1.5 gallons of ethanol per gallon of fermenter.

I have assumed seven-day batches with two-day cleanups. Given the reported ethanol production and the assumed batch times, there

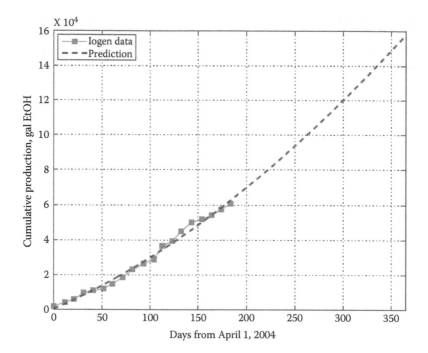

Figure 4.9 Ethanol production in Iogen's Ottawa plant. Extrapolation to one year yields 158,000 gallons. The data points are evenly spaced, as they should be for regularly scheduled batches. (Data from Passmore, 2006.)

is approximately 4% of alcohol in a batch of industrial wheat-straw broth, in contrast to 12–16% ethanol in corn-ethanol refinery broths. Shorter batch times lead to an estimation of even less favorable process parameters.

Since wheat is the largest grain crop in Canada, I use its straw as a reference (barley and oats straws are similar). On a water-free basis, wheat straw contains 33% cellulose, 23% hemicelluloses, and 17% total lignin (USDOE, 2007a). Other sources report 38%, 29%, and 15%, respectively; see Lee et al. (2007) for a data compilation. These differences are not surprising, given experimental uncertainties and variable biomass composition. To calculate ethanol yield, I use the more favorable, second set of data. The respective conversion efficiencies, assumed after Badger (2002), are listed in Table 4.5.

The calculated ethanol yield, 0.18 kg per kg straw, or 0.23 L/kg, is somewhat less than a recently reported maximum ethanol yield of 0.24 kg/kg (Saha et al., 2005) achieved in 500 ml vessels, starting from 48.6% cellulose. Simultaneous saccharification and fermentation yielded 0.17 kg/kg (see Saha et al., 2005, Table 5).

Table 4.5 Yields of Ethanol from Cellulose and Hemicellulose

Step	Cellulose	Hemicellulose
Dry straw	1 kg	1 kg
Mass fraction	×0.38	×0.29
Enzymatic conversion efficiency	×0.76	×0.90
Ethanol stoichiometric yield	×0.51	×0.51
Fermentation efficiency	×0.75	×0.50
Ethanol yield, kg	0.111	0.067

Source: Badger, P. C., Ethanol from cellulose: A general review, in *Trends in new crops and new uses*, ed. Janick, J., and Whipkey, A., 17–21 (Alexandria, VA: ASHS Press, 2002).

Because enzymatic decomposition of cellulose and hemicelluloses is inefficient, the resulting dilute broth requires 2.4 times more steam energy to distill than the average 15 MJ/L in an average ethanol refinery (Patzek, 2004, 2006a); see Figure 4.10.

One could argue that Iogen's Ottawa facility is for demonstration purposes only and that the saccharification and fermentation batches were not regularly scheduled. But, *independent* of Iogen's data, an alternative calculation yields the same result: At about 0.2 to 0.25 kg of straw per liter,

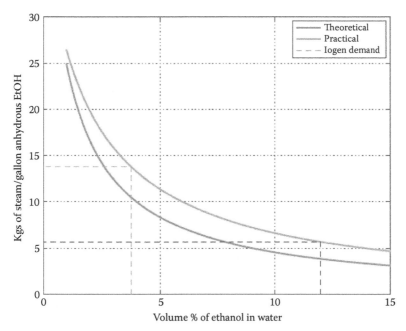

Figure 4.10 Steam requirement in ethanol broth distillation. According to Jacques et al., 2003, a 7% broth requires 2.4 times more steam than a 12% broth.

the mash is barely pumpable. With Badger's (2002) yield of 0.18 kg/kg, the highest ethanol yield is 3.5–4.4% of ethanol in water.

The HHV of ethanol is 29.6 MJ/kg (Patzek, 2004), while the HHV of wheat straw is 18.1 MJ/kg (Schmidt et al., 1993) and that of lignin is 21.2 MJ/kg (Domalski et al., 1987). With these inputs, the first-law (energy) efficiency of Iogen's facility is

$$
\eta = \frac{\text{Ethanol energy out}}{\text{Net energy in}}
$$

$$
= \frac{\text{HHV of EtOH/kg grass}}{\text{HHV of grass + Steam energy/kg grass} - \text{Lignin HHV/kg grass}} \tag{2}
$$

$$
= \frac{0.18 \times 29.6}{1 \times 18.1 + 0.18 \times 2.4 \times 15/0.787 - 0.15 \times 21.2} \approx 23\%
$$

where the density of ethanol is 0.787 kg/L. The entire HHV of lignin is credited to offset distillation fuel—another optimistic assumption for the wet-separated lignin. Given expected moisture content, probably half the HHV of lignin should be used. Equation (2) also disregards the energy costs of steam treatments of the straw at 120°C or 140°C and the separated solids at 190°C, sulfuric acid and sodium hydroxide production, molecular sieves to reject the last 4 wt % of water from the azeotrope, and so forth.

The complex enzyme production processes also use plenty of energy. Since these processes are proprietary, only enzyme prices can be used as proxies for production complexity. The enzymes necessary to split cellulose fibers and chop them into small pieces are complex proteins (Bayer et al., 1998) that need to be replicated on a mass scale. Many tons of enzymes would have to be produced each year at a dose cost commensurate with ethanol, which costs roughly $1.00 per kilogram. These enzymes biodegrade, stick to the plant mash, and are washed away and thus must be replaced after each batch.

Compare the low-cost requirement for the cellulose-splitting enzymes with an enzyme most commonly used in polymerase chain reactions. A DNA polymerase is an enzyme that assists in DNA replication. Such enzymes catalyze the polymerization of deoxyribonucleotides alongside a DNA strand, which they "read" and use as a template. The newly polymerized molecule is complementary to the template strand and identical to the template's partner strand. A common type of this enzyme is Taq DNA Polymerase from *Thermus aquaticus*, and its April 2010 price was between $300,000 and $1.2 million per kilogram (Sigma-Aldrich, 2010). According to Genentech, the pharmaceutical proteins produced in bioreactors identical to those that might be used to produce the cellulose decomposition enzymes sell for up to $12 million per kilogram (L. Leveen, pers. comm., 2007).

4.4 Calculation methodology

4.4.1 Monte Carlo simulations

Monte Carlo methods are a class of computational algorithms that rely on repeated random sampling of inputs to compute outcomes that are often represented in terms of their probability distribution functions and cumulative probability distributions. Monte Carlo simulations are useful in modeling systems with significant uncertainty in inputs, as in predicting expected ethanol yield from the geographically distributed switchgrass plantations. When Monte Carlo simulations were applied in space exploration and oil exploration, actual observations of failures, cost overruns, and schedule delays were routinely better predicted by the simulations than by human intuition or alternative "soft" methods (Hubbard, 2009).

As Casler and Boe (2003, p. 2232) conclude:

> Biomass yield of switchgrass is unstable, varying by harvest date, site, year, and cultivar. Interactions among these factors cause biomass yield to be relatively unpredictable, particularly with respect to harvest date. For single harvest of switchgrass, aimed at bioenergy feedstock production, the optimal harvest date was in late summer or early autumn, when soil and air temperatures are sufficiently low to minimize the potential for regrowth. In the short term, an earlier harvest date could increase biomass yields, but this would have detrimental long-term effects on stands. In the long term, plant mortality is apparently reduced by delayed harvest and preservation of carbohydrate reserves.

As noted earlier, Fike et al. (2006a) state: "Limited information is available regarding biomass production potential of long-term (>5-yr-old) switchgrass.... Yields at Site B (19.1Mgha⁻¹) were about 35% greater than those at Site A (14.1Mgha⁻¹), although the sites were only 200m apart."

Clearly, there is a need to capture the temporal and spatial variability of switchgrass yields in a statistical manner. This is done with the Monte Carlo procedure described next.

4.4.2 Distribution of switchgrass yields

We are interested in the most probable *continuous* biomass yield from any field, growing any switchgrass cultivar, geographically located in any state, harvested during any calendar year, and harvested after any number of years from seeding. Therefore all available field data are thrown into a single "data pool." These data are then used to construct an empirical probability distribution function (PDF) and its integral, the cumulative distribution function of switchgrass yields.

The purpose of pooling the switchgrass yield data was made clear in the discussion of switchgrass area required to power cars in a single metropolitan area. Since potentially tens of millions of acres of giant switchgrass monocultures will be needed, switchgrass will have to be grown in different geographical locations, perhaps even outside of the United States, and on the contiguous fields similar in size to the giant sugarcane plantations in Brazil in the state of São Paulo.

The switchgrass yields were made dimensionless by dividing the data by 1 Mg/ha/yr. The same procedure was applied in the derivation of all probability distribution functions and cumulative distribution functions in this chapter.

The logarithm of the switchgrass yields shown in Figures 4.2 through 4.7 is normally distributed,

$$Y_{log} = \frac{1}{\sqrt{2\pi \times 0.63}} \exp\left[-\frac{(\ln(\text{Mass flux}) - 2.19)^2}{2 \times 0.63}\right] \qquad (3)$$

with the mean $\mu_1 = 2.19$ and the standard deviation $\sigma_1 = 0.63$ (see Figure 4.11). The lognormal mean is $\mu_2 = 10.9$ Mg/ha/yr, above the Schmer et al. (2008) mean of 7.2 Mg/ha/yr, and the standard deviation is $\sigma_2 = 7.58$ Mg/ha/yr.

To perform the Monte Carlo simulations of the switchgrass-ethanol cycle, 2^{13}, or 8,192, random values are sampled from distribution (3). The result is a random variable $\{Y_{log}\}$ with a certain probability distribution function.

As discussed in Section 4.3.1, the measured switchgrass yields often decline after two to five years, and none have been reported beyond 10 years. To translate from the reported annual switchgrass yields to the equivalent continuous yields, the following *optimistic* procedure is used. Let Y_{min} denote the minimum of $\{Y_{log}\}$, and i_{max} its maximum. The weight function

$$\{W_{log}\} = \frac{3}{5} + \left(\frac{19}{20} - \frac{3}{5}\right)\frac{\{Y_{log}\} - Y_{min}}{Y_{max} - Y_{min}} \qquad (4)$$

accounts for the fact that higher yields are always associated with lower plant mortality (Berdahl et al., 2005). The maximum age of surviving commercial switchgrass plants is assumed to be 20 years and the minimum

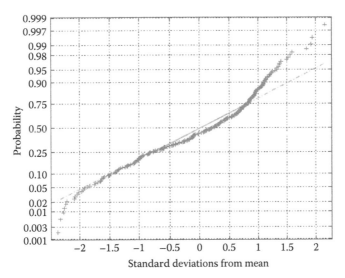

Figure 4.11 The logarithm of the switchgrass yield data. The logarithm of the switchgrass yield data in Parrish et al. (2003), Casler and Boe (2003), Berdahl et al. (2005), McLaughlin and Adams Kszos (2006), and Schmer et al. (2008) is approximately normally distributed. The underlying lognormal distribution has the mean $\mu_2 = 10.87$ Mg/ha/yr and the standard deviation is $\sigma_2 = 7.58$ Mg/ha/yr.

age five years, and there is a two-year grass establishment period with no harvest for the low-yield fields and a one-year period for the high-yield ones. The continuous switchgrass yield is then given by the following random variable:

$$\{Y_c\} = \{W_{\log}\} \otimes \{\exp(Y_{\log})\} \quad \text{Mg ha}^{-1} \text{ y}^{-1} \tag{5}$$

where $\otimes$ denotes the element-by-element multiplication of the two random variables. The mean of the "continuous yield" distribution (5) is $\mu_3 = 6.83$ Mg/ha/yr, slightly lower than that of Schmer et al. (2008), and its standard deviation is $\sigma_3 = 4.52$ Mg/ha/yr. Its probability distribution function is shown in Figure 4.12. A reviewer should not defend the analysis by Schmer et al. while at the same time criticizing the analysis here: Both means are comparable, but this analysis gives the full context of the mean.

4.4.3   A prototype of the switchgrass-based ethanol refinery

4.4.3.1   Mass balance

The composition of switchgrass in the south-central United States has been evaluated by Cassida et al. (2005). The cellulose content varies from

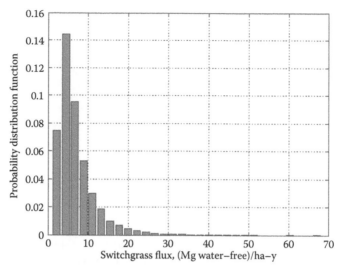

Figure 4.12 The probability distribution function of the continuous mass flux of switchgrass ("continuous yield"). The mean is μ_3 = 6.83 Mg/ha/yr and the standard deviation is σ_3 = 4.52 Mg/ha/yr.

34% to 46% by mass, with a mean of 39%. The lignin content varies from 7% to 12% with a mean of 9%. The hemicellulose content is not reported. Lee et al. (2007) list 37% cellulose, 29% hemicelluloses, and 19% lignin on average. I use the latter, more favorable estimate in my calculations, and the yield of switchgrass ethanol is almost identical to that calculated in Table 4.5 for wheat straw.

The cellulose weight fraction is assumed to be normally distributed (after Patzek, 2006d). The mean, maximum, and minimum measured values are taken from Cassida et al. (2005). The mean is μ_c = 0.39, and the standard deviation is σ_c = 0. 02. The normally distributed random mass fractions of cellulose are $\{C\}$. It is assumed that the mass fraction of hemicelluloses varies in proportion to that of cellulose:

$$\{HC\} = \frac{29}{39}\{C\}$$

(6)

and, from the mass balance, the lignin mass fraction is

$$\{L\} = 0.75 - \{HC\} - \{C\}.$$

(7)

4.4.3.2   *Refinery efficiency*
The most often quoted yield of ethanol from switchgrass, 0.38 L/kg (90% of the theoretical yield), is based on EBAMM (Farrell et al., 2006) and has no justification whatsoever in the context of chemistry relevant to this

analysis. In Section 4.3.4, I used the published fragments of industrial data (Passmore, 2006; Hladik, 2006) to arrive at a realistic efficiency of a "cellulosic ethanol" refinery of 0.23 L/kg. The ethanol yield, 0.18 kg EtOH per kg switchgrass, calculated here, is equivalent to 0.18/0.787 = 0.23 L kg of switchgrass, only slightly lower than the 0.24 L/kg reported by Lau and Dale (2009), but significantly lower than the 0.28 L/kg claimed in McLaughlin et al. (1996) or the 0.38 L/kg asserted by EBAMM.

For a lignin content between 9% and 19%, the range of energy efficiencies of a switchgrass cellulosic ethanol refinery is 20–24% if the burned lignin is bone dry and no other energy costs are incurred. With one significant digit, we get about 20% from Equation (2). At this efficiency, roughly 5 units of heat from switchgrass are necessary to obtain 1 unit of heat from 100% ethanol. The higher heating value of switchgrass is 18.1 MJ/kg (Wilen et al., 1996). Therefore, it takes 5×29.6/18.1 = 8.2 kg of switchgrass to obtain 1 kg of anhydrous ethanol, or 6.5 kg of switchgrass to obtain 1 liter of ethanol.

If the entire process of ethanol production is driven by burning lignin from fermented switchgrass, as well as burning additional bone-dry switchgrass, the average yield of ethanol is between 4,100/6.5 or about 630 L/ha/yr and 11,000/6.5 or about 1,700 L/ha/yr for the data in Figures 4.2 through 4.7.

The ethanol yield used in the Monte Carlo simulations here is assumed after Badger (2002; see Table 4.5), with the switchgrass composition calculated from Equations (6) and (7). The resulting PDF is shown in Figure 4.13. The energy efficiency of a switchgrass refinery, η (energy out as anhydrous ethanol divided by net energy in from switchgrass), is calculated from Equation (2), with the random lignin mass fraction given by Equation (7) and the random ethanol yield sampled from the distribution in Figure 4.13. The resulting PDF is shown in Figure 4.14.

4.5   Results

4.5.1   Effective volumetric flux of ethanol

The effective volumetric flux of ethanol from the switchgrass field/switchgrass-powered ethanol refinery cycle ("ethanol yield") is calculated as

$$\{Mult_1\} = \frac{0.787 \times 29.6}{18.1 \times \{\eta\}}$$

$$\{Ethanol\ Flux\} = \frac{\{Y_c\} \times 1000}{\{Mult_1\}} \quad L\ ha^{-1}\ y^{-1}, \tag{8}$$

and the resulting PDF is shown in Figure 4.15. Note that the lognormal mean μ_6 of 1,140 L/ha/yr and its standard deviation σ_6 of 750 L/ha/yr are

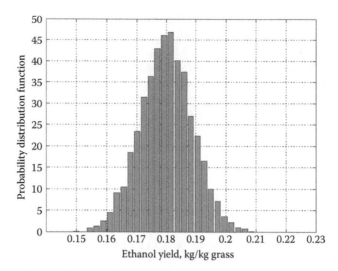

Figure 4.13 The probability distribution function of ethanol yield from switch-grass. The mean is $\mu_4 = 0.18$ kg/kg and the standard deviation is $\sigma_4 = 0.009$ kg/kg.

in the range estimated for the data in Figures 4.2 through 4.7, 650–1,700 L/ha/yr. The most probable values of ethanol yield are clustered about the lognormal mean in the bin 817–1,270 L/ha/yr. When the lognormal mean of the continuous harvest of switchgrass, 6.83 Mg/ha/yr, is multiplied by the mean refinery yield, 0.23 L/kg, the result is 1,570 L/ha/yr.

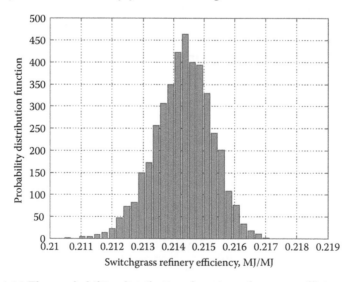

Figure 4.14 The probability distribution function of energy efficiency η of a switchgrass ethanol refinery. The mean is $\mu_5 = 0.214$ MJ/MJ and the standard deviation is $ma_5 = 0.001$ MJ/MJ.

It should be stressed that a simple multiplication of means of different probability distributions, a standard procedure in most biofuel papers, does *not* yield the most probable value of the switchgrass ethanol yield.

4.5.2   Probability of exceeding a given flux

By integrating the PDFs in Figures 4.12 and 4.15, one may estimate the probabilities of exceeding a given value of flux. The probability of achieving a continuous mass flux of switchgrass larger than the abscissa is plotted in Figure 4.16. The probability of achieving a net ethanol fuel yield (after satisfying the refinery energy needs) larger than the abscissa is plotted in Figure 4.17.

4.5.3   Summary of results thus far

The continuous mean switchgrass yield based on Figures 4.2 through 4.7 has the expected value of 6.8 ± 4.5 Mg/ha/yr (see Figure 4.12), close to the 7.2 Mg/ha/yr estimated in Schmer et al. (2008), but significantly lower than the 25.8 Mg/ha/yr claimed elsewhere (Ehrenshaft, 1998; Kumar and Sokhansanj, 2007). If switchgrass is used as the refinery fuel, the most probable ethanol yield is about 1,100 L/ha/yr. The probability of ethanol yields in the range 1,200–1,700 L/ha/yr is much less than 50%.

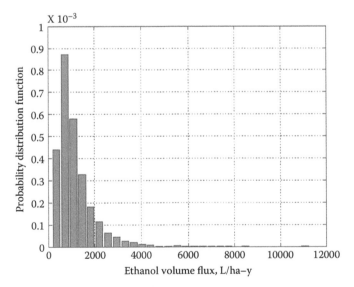

Figure 4.15 The probability distribution function of the effective volumetric ethanol flux. The lognormal mean is $\mu_6 = 1{,}140$ L/ha/yr. The standard deviation, $\sigma_6 = 750$ L/ha/yr, is comparable to the mean, indicating a long tail of the highly improbable large fluxes.

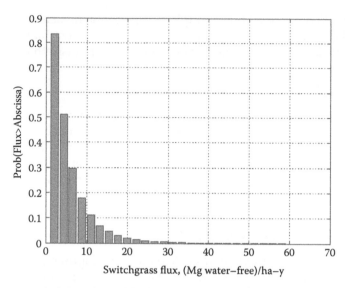

Figure 4.16 Probability that a continuous switchgrass mass flux ("continuous yield") exceeds a given value. If the abscissa is x_0, the ordinate is Prob($x \geq x_0$) derived from the yield data in Parrish et al. (2003), Casler and Boe (2003), Berdahl et al. (2005), McLaughlin and Adams Kszos (2006), Fike et al. (2006a, 2006b), or Schmer et al. (2008), weighted by the switchgrass survivability function defined in Equation (4).

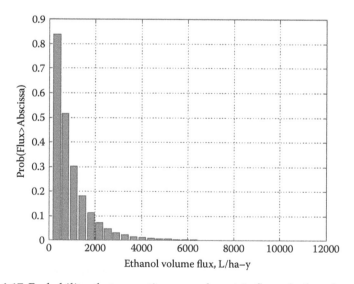

Figure 4.17 Probability that a continuous volumetric flux of ethanol exceeds a given value. If the abscissa is x_0, the ordinate is Prob($x \geq x_0$) derived from Figure 4.16 and the refinery efficiency calculated in Section 4.3.4.

4.5.4 Continuous electrical power from switchgrass ethanol

Photovoltaic (PV) solar cells generate electricity that can be converted to power a rotating shaft with almost 100% efficiency. The volumetric flux of ethanol from the switchgrass-ethanol cycle discussed here can be converted to continuous electrical power using the following formula:

$$\{Mult_2\} = \frac{18.1 \times 10^6 \times \{\eta\} \times \eta_e}{10000 \times 3600 \times 24 \times 365}$$

$$\{\text{Electric power flux from ethanol}\} = \{Y_c\} \times 1000 \times \{Mult_2\} \text{ Watts electrical m}^{-2},$$

$$(9)$$

where η_e is the average efficiency of converting ethanol to electricity. Here, $\eta_e = 0.35$.

The PDF and the exceedance probability function of continuous electrical power generated from the switchgrass-ethanol cycle are shown in Figures 4.18 and 4.19.

The probability of generating more than 0.03 W_e/m^2 continuously is less than 50%. Note that only the electrical power inputs are compared here. The substantial continuous energy inputs to switchgrass and switchgrass ethanol, as well as initial energy inputs to PV cells, are not

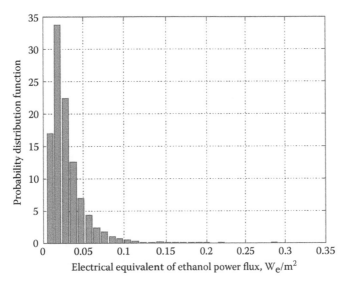

Figure 4.18 The probability distribution function of the continuous electrical power flux generated from the switchgrass-ethanol cycle.

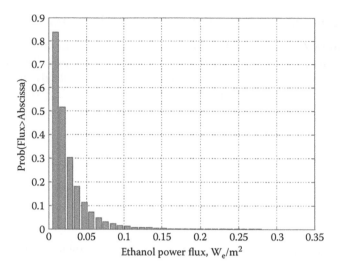

Figure 4.19 Probability that a continuous electrical power flux of ethanol generated from switchgrass exceeds a given value. If the abscissa is x_0, the ordinate is Prob($x \geq x_0$) derived from Figure 4.18.

discussed here. For more detailed approaches, see Patzek and Pimentel (2005), Patzek (2007), and Baum et al. (2009).

On average, switchgrass ethanol delivers 0.03 W_e per m^2 of field surface (1 W_e is one watt of electrical power). Also on average, a mediocre (see Figure 4.20), 10%-efficient PV cell that uses twice the area of panels

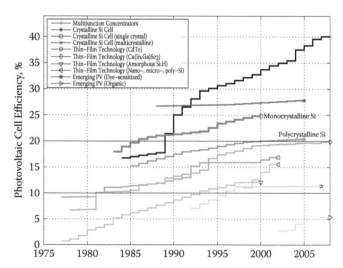

Figure 4.20 Solar light conversion efficiencies of best research photovoltaic cells. (Adapted from Kazmerski, L. L., NREL conference paper CP-520-37553, 2005.)

Figure 4.21 Efficient photovoltaic cell. 30 m² of panel and road area are necessary to generate the same electrical power from the inefficient PV cells (dark gray) as from 1 ha of average switchgrass field dedicated to producing ethanol (light gray). The barely visible 15 m² of a current efficient PV panel are in white.

for access roads and so on, delivers 0.1×200×0.5 = 10 W_e/m² continuously when it operates anywhere in the United States (Patzek and Pimentel, 2005; Patzek, 2007). A better PV panel could deliver 40 W_e/m² continuously (see Kazmerski, 2005). Thus, our present mediocre PV panel is 335 times more efficient than the switchgrass-ethanol cycle in delivering continuous power of a rotating shaft. In other words, on the average, 1 hectare (10,000 m², or about 2.5 acres) of switchgrass field is equivalent to 30 m² of spread-out and inefficient PV cells. An efficient—currently available—PV cell takes a 4-meter-square area (see Figure 4.21) and costs about $16,000 (as of March 2008) to install on a house roof. It is designed to run for 30 years with almost no maintenance.

4.6   *Equivalent CO₂ emissions*

The procedure of calculating equivalent CO_2 emissions is explained in Patzek (2004, Section 5.1, Table 19). The CO_2 emissions from the oxidation of soil humus are described in Patzek (2006a, p. 267). Only updates or parameter changes are discussed here.

4.6.1 N_2O emissions from agriculture

As stated by Crutzen et al. (2007, p. 11194):

> An evaluation of hundreds of field measurements has shown that N fertilization causes a release of N_2O in agricultural fields that is highly variable but averages close to 1% of the fixed nitrogen input from mineral fertilizer or biologically fixed N (Bouwman et al., 2002; Stehfest and Bouwman, 2006), and a value of 1% for such direct emissions has recently been adopted by Eggleston et al. (2006). There is an additional emission from agricultural soils of 1 kg N_2O–N ha^{-1} y^{-1}, which does not appear to be directly related to recent fixed N-input. The in-situ fertilizer-related contribution from agricultural fields to the N_2O flux is thus 3–5 times smaller than our adopted global average N_2O yield of 4±1% of the fixed N input. The large difference between the low yield of N_2O in agricultural fields, compared to the much larger average value derived from the global N_2O budget, implies considerable "background" N_2O production occurring beyond agricultural fields, but, nevertheless, related to fertilizer use, from sources such as rivers, estuaries and coastal zones, animal husbandry and the atmospheric deposition of ammonia and NO_x.

My old calculation (Patzek, 2004) of the total emissions from ammonium nitrate was 4.4%, significantly larger than the Intergovernmental Panel on Climate Change (IPCC) estimates. At the time, I was criticized for this calculation as exaggerated and unrealistic. Here I continue to use my old estimate, now verified independently by Crutzen et al. The smaller greenhouse gas (GHG) emissions for urea production are accounted for in the current analysis.

4.6.2 CO_2 emissions from lime

In Patzek (2004), I used an emission factor of 0.7 for calcinated lime. Here, I use the IPCC (De Klein et al., 2006) emission factor for crushed dolomite, 0.13, which is five times lower.

4.6.3   Soil erosion rate

The average soil erosion rate for switchgrass is assumed to be 3 Mg/ha/yr, or 25% of the roughly 13 Mg/ha/yr erosion rate typical of corn agriculture in recent times (United States Department of Agriculture, National Resources Conservation Service. The reported erosion rate is the sum of water and wind erosion rates. See chapter 8 in Lal et al. (2004). See also figures 9–11 in Patzek (2008c)).

4.6.4   Emissions from the refinery

As switchgrass is burned in the refinery to provide process heat and electricity, only emissions from the ethanol and denaturant transportation are included, as well as emissions from wastewater cleanup. The calculation results are shown in Figure 4.22 for the ethanol yield equal to the lognormal mean $\mu_6 = 1,140$ L/ha/yr. The total emissions are dominated by the emissions from nitrogen fertilizer production and agricultural emissions from its application.

Using this lognormal mean of ethanol yield, $\mu_6 = 1,140$ L/ha/yr, cumulative GHG emissions are 106 g CO_2 equiv. per MJ in the anhydrous ethanol (1 MJ = 10^6 joules). The GHG emissions from switchgrass ethanol are 35% higher than those from producing and burning automotive gasoline outright (see Figure 4.23) and two times higher than those from producing and burning compressed natural gas. The GHG emissions from switchgrass ethanol are generated only by the nonrenewable resources consumed in its production, and by the N_2O/NO_x emissions from

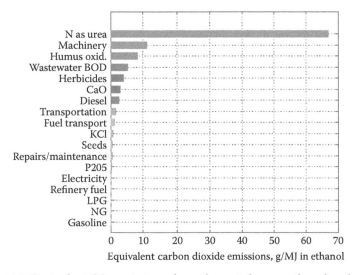

Figure 4.22 Equivalent CO_2 emissions from the switchgrass-ethanol cycle.

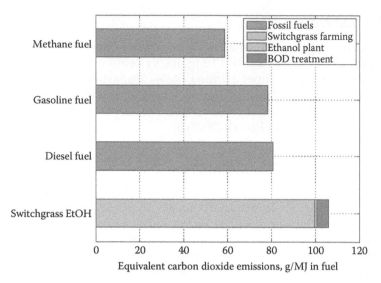

Figure 4.23 Specific greenhouse gas emissions from compressed methane, gasoline, diesel fuel, and switchgrass ethanol. The nonrenewable resources consumed to produce the ethanol generate 35% more emissions than gasoline and twice those from methane.

switchgrass agriculture. If one were to replace 10% of the current gasoline consumption in the United States with switchgrass ethanol, about 55 million tons of equivalent CO_2 would be generated each year over and above the displaced gasoline emissions. Compressed natural gas (CNG) is by far the most environmentally friendly automotive fuel, and its expansion should be considered urgently (Patzek, 2008b). Corn ethanol agriculture generates GHG emissions of about 120 g/MJ in anhydrous ethanol (see Patzek, 2006a, Figure 8). Therefore, replacing cornfields with switchgrass fields would result on the average in a 14/120, or about 11%, reduction in net GHG emissions. In both cases, no land-use changes were included in the calculations.

A word of caution is in order. Net GHG emissions from the switchgrass-ethanol cycle depend very strongly on the cycle's yield. If, for example, 1,600 L/ha/yr were produced, the net emissions of switchgrass ethanol would be zero relative to those of the displaced gasoline; nevertheless, CNG would still generate 25% fewer emissions.

4.7 Summary and conclusions

In mid-2010, all published analyses of the switchgrass-ethanol cycle were works in progress, because of the still insufficient knowledge of the system. The main strength of the approach presented in this chapter is in showing the *context* of the various measures of switchgrass field productivity and ethanol yields. The probability of achieving a continuous switchgrass yield of 8–10 Mg/ha/yr is well under 50% (see Figure 4.16). The probability of achieving a continuous ethanol yield of 1,200–1,600 L/ha/yr is also much less than 50% (see Figure 4.17). Achieving the 3,000–5,000 L/ha/yr yields asserted in the literature (McLaughlin et al., 1996; Schmer et al., 2008) is possible, but with a probability of less than 5%—in the noise of the current model.

Suppose that one would like to replace 18% of the current 20 EJ/yr the United States uses as automotive gasoline (Patzek, 2007). If the switchgrass-ethanol cycle described here were used to achieve this goal with close to 50% probability, one would need at 140 million hectares of switchgrass using the mean of the energy efficiency distribution in Figure 4.17. According to the U.S. Department of Agriculture (USDA, 2004), this is equivalent to the total U.S. cropland area (harvested, summer fallow, and failed): 140 million hectares as of 2005 (another 40 million hectares was devoted to pastures and idle cropland).

With the existing fermentation processes and technology, one obtains 0.23 L EtOH per kilogram of switchgrass, rather than the 0.38 L/kg asserted by EBAMM. The relative difference is 37%. Because the overall energy efficiency of a plausible switchgrass-ethanol refinery is only 20%, 6.5 kilograms of switchgrass would need to be processed and/or burned to obtain 1 liter of ethanol. This requirement translates into an optimistic most probable continuous yield of ethanol of about 1,100 L/ha/yr, if switchgrass were used to power the refineries.

The law of energy conservation requires that a switchgrass ethanol refinery have a highly negative difference of output energy minus input energy, or net-energy value (NEV), as shown previously in Pimentel and Patzek (2005). This follows directly from the observation that the switchgrass-ethanol process has a low energy efficiency and requires large external inputs of energy-intensive chemicals, heat, and electricity. Whether or not nonrenewable energy is used for biorefinery energy needs, NEV will be less than zero.

The industrial switchgrass plantations considered here are sun-driven, man-made "machines," whose ultimate output is *shaft work* used for generation of electricity or rotation of car wheels. These vast and complex machines should be compared against two other, much simpler devices that also convert solar energy into shaft work: solar photovoltaic cells (whenever panel areas measured in km² become commercially available) and wind turbines (as well as electricity from thermal solar). PV cells convert solar energy *directly* into electricity, the most valuable form of free energy, which can be further converted into mechanical work with small losses. Wind turbines, which produce electricity from the kinetic energy of the sun-driven wind, are not discussed here. All biofuel-producing systems should be judged on their ability to generate shaft work, not merely a biofuel.

PV cell, thermal solar, and battery research and development, as well as a large-scale implementation of already-existing PV cell manufacturing technologies, could have a much larger impact on both near- and long-term energy security of the United States and Europe than biofuels.

The major arguments advanced against the approach in this work are discussed in the appendix below.

Acknowledgments

This work was carefully reviewed and critiqued by Drs. Aaron Baum, Ignacio Chapela, David Pimentel, John Prausnitz, and Dmitriy Silin, as well as my son, Lucas, now a PhD candidate in sustainable agriculture. I am very grateful to all reviewers for their valuable suggestions, thoroughness, directness, and dry sense of humor. Special thanks are due to Drs. Prausnitz and Silin, whose inspired critiques focused the work and made it much better. Dr. Chris Somerville and his USDA helper's criticisms have resulted in several improvements to the probabilistic model presented here. Finally, I would like to thank the anonymous reviewers for their thoughtful and useful remarks and suggestions.

References

Ayres, R. U., Ayres, L. W., and Warr, B. 2002. *Exergy, power and work in the U.S. economy, 1900–1998*. El Paso: Center for the Management of Environmental Resources, University of Texas. http://www.iea.org/work/2004/eewp/Ayres-paper3.pdf.

Badger, P. C. 2002. Ethanol from cellulose: A general review. In *Trends in new crops and new uses*, ed. Janick, J., and Whipkey, A., 17–21. Alexandria, VA: ASHS Press.

Baum, A. W., Patzek, T. W., Bender, M., Renich, S., and Jackson, W. 2009. The visible, sustainable farm: A comprehensive energy analysis of a Midwestern farm. *Crit. Rev. Plant Sci.* 28:218–239.

Bayer, E., Chanzy, H., Lamed, R., and Shoham, Y. 1998. Cellulose, cellulases and cellulosomes. *Curr. Opin. Struct. Biol.* 8:548–557.

Berdahl, J. D., Frank, A. B., Krupinsky, J. M., Carr, P. M., Hanson, J. D., and Johnson, H. A. 2005. Biomass yield, phenology, and survival of diverse switchgrass cultivars and experimental strains in western North Dakota. *Agron. J.* 97:549–555.

Bossel, U. 2003. *Well-to-wheel studies, heating values, and the energy conservation principle*. Report E10. Oberrohrdorf, Switzerland: European Fuel Cell Forum. http://www.efcf.com/reports/E10.pdf.

Bouwman, A. F., Boumans, L. J. M., and Batjes, N. H. 2002. Emissions of N_2O and NO from fertilized fields: Summary of available measurement data. *Global Biogeochem. Cycles* 16(4): 6-1–6-13.

Bransby, D. I., Sladden, S. E., and Kee, D. D. 2003. *Selection and improvement of herbaceous energy crops for the southeastern USA: Final report in a field and laboratory research program for the period March 15, 1985 to March 19, 1990*. Report ORNL/Sub/85-27409/5. Oak Ridge, TN: Oak Ridge National Laboratory.

Casler, M. D., and Boe, A. R. 2003. Cultivar × environment interactions in switch-grass. *Crop Sci.* 43(6): 2226–2233.

Cassida, K. A., Muir, J. P., Hussey, M. A., Read, J. C., Venuto, B. C., and Ocumpaugh, W. R. 2005. Biofuel component concentrations and yields of switchgrass in south central U.S. environments, *Crop Sci.* 45:682–692.

Conway, R. 2005. The net energy balance of corn ethanol. Presentation at the NCGA Renewable Fuels Forum, Washington, DC, August 23.

Crutzen, P. J., Mosier, A. R., Smith, K. A., and Winiwarter, W. 2007. N_2O release from agro-biofuel production negates global warming reduction by replacing fossil fuels. *Atmos. Chem. Phys. Discuss.* 7:11191–11205. http://www.atmos-chem-phys-discuss.net/7/11191/2007/.

De Klein, C., Novoa, R. S. A., Ogle, S., Smith, K. A., Rochette, P., Wirth, T. C., McConkey, B. G., Mosier, A., and Rypdal, K. 2006. N_2O emissions from managed soils and CO_2 emissions from lime and urea application. In *2006 IPCC guidelines for national greenhouse gas inventories*, ed. Eggleston, H. S., Buendia, L., Miwa, K., Ngara, T., and Tanabe, K., Vol. 4, 11.1–11.54. Hayama, Japan: IGES. http://www.ipcc-nggip.iges.or.jp/public/2006gl/pdf/4_Volume4/V4_11_Ch11_N2O&CO2.pdf.

Domalski, E. S., Jobe, T. L., Jr., and Milne, T. A. 1987. *Thermodynamic data for biomass materials and waste components.* New York: American Society of Mechanical Engineers.

Duffy, M., and Nanhou, V. Y. 2001. *Costs of producing switchgrass for biomass in southern Iowa.* University Extension Report PM 1866. Ames: Iowa State University. http://www.extension.iastate.edu/Publications/PM1866.pdf.

Ehrenshaft, A. 1998. *Biofuels from switchgrass: Greener energy pastures.* Oak Ridge, TN: Oak Ridge National Laboratory. http://www.p2pays.org/ref/17/16279.pdf.

Epplin, F. M., Clark, C. D., Roberts, R. K., and Hwang, S. 2007. Challenges to the development of a dedicated energy crop. *Amer. J. Agr. Econ.* 89(5): 1296–1302.

Fan, L. T., Lee, Y.-H., and Beardmore, D. H. 1980. Mechanism of the enzymatic hydrolysis of cellulose: Effects of major structural features of cellulose on enzymatic hydrolysis. *Biotechnol. Bioeng.* 22:177–199.

Farrell, A. E., Plevin, R. J., Turner, B. T., Jones, A. D., O'Hare, M., and Kammen, D. M. 2006. Ethanol can contribute to energy and environmental goals. *Science* 311:506–508, with supporting online material at http://www.sciencemag.org/cgi/content/full/311/5760/506/DC1.

Fike, J. H., Parrish, D. J., Wolf, D. D., Balasko, J. A., Green, J. T., Rasnake, M., and Reynolds, J. H. 2006a. Long-term yield potential of switchgrass-for-biofuel systems. *Biomass Bioenergy* 30(3): 198–206.

———. 2006b. Switchgrass production for the upper southeastern USA: Influence of cultivar and cutting frequency on biomass yields. *Biomass Bioenergy* 30(3): 207–213.

Heinsch, F. A., et al. 2003. *User's guide GPP and NPP (MOD17A2/A3) products NASA MODIS land algorithm.* Washington, DC: NASA. http://www.ntsg.umt.edu/modis/MOD17UsersGuide.pdf.

Hladik, M. 2006. Cellulose ethanol is ready to go. Presentation at the Emerging Energies Conference, University of California, Santa Barbara, February 10–11. http://www.oilcrisis.com/ethanol/CelluloseReadyIOGen.pdf.

Hubbard, D. 2009. *The failure of risk management: Why it's broken and how to fix it.* New York: John Wiley & Sons.

Jacques, K. A., Lyons, T. P., and Kelsall, D. R. 2003. *The alcohol textbook.* 4th ed. Nottingham, England: Nottingham University Press.

Jagschies, G. 2008. Where is biopharmaceutical manufacturing heading? *BioPharm International* 21(10).

Kazmerski, L. L. 2005. Photovoltaics R&D: At the tipping point. Conference Paper NREL/CP-520-37553. http://www.nrel.gov/docs/fy05osti/37553.pdf.

Kazmerski, L., Gwinner, D., and Hicks, A. 2007. Best research-cell efficiencies. Golden, CO: National Renewable Energy Laboratory. http://www.nrel.gov/pv/thin_film/docs/kaz_best_research_cells.ppt.

Khosla, V. 2006. *Biofuels: Think outside the barrel.* http://www.khoslaventures.com/presentations/Biofuels.Apr2006.ppt.

King, C. J. 1980. *Separation processes.* 2nd ed. New York: McGraw-Hill.

Kumar, A., and Sokhansanj, S. 2007. Switchgrass (*Panicum vigratum, L.*) delivery to a biorefinery using integrated biomass supply analysis and logistics (IBSAL) model. *Bioresour. Technol.* 98(5): 1033–1044.

Ladisch, M. R., and Dyck, K. 1979. Dehydration of ethanol: New approach gives positive energy balance. *Science* 205:898–900.

Lal, R., Sobecki, T. M., Livari, T., and Kimble, J. M. 2004. *Soil Degradation in the United States: Extent, severity, and trends.* 110–130. Boca Raton, FL: Lewis Publishers.

Lau, M. W., and Dale, B. E. 2009. Cellulosic ethanol production from AFEX-treated corn stover using *Saccharomyces cerevisiae* 424A(LNH-ST). *PNAS* 106(5): 1368–1373.

Lawrence, C., and Walbot, V. 2007. Translational genomics for bioenergy production from fuelstock grasses: Maize as the model species. *Plant Cell* 19:2091–2094.

Lee, D.-K., Owens, V. N., Boe, A., and Jeranyama, P. 2007. *Composition of herbaceous biomass feedstocks.* Report SGINC1-07. Brookings: South Dakota State University, Plant Science Department.

Lee, Y.-H., and Fan, L. T. 1983a. Kinetic studies of enzymatic hydrolysis of insoluble cellulose: Derivation of a mechanistic kinetic model. *Biotechnol. Bioeng.* 25:2707–2733.

———. 1983b. Kinetic Studies of enzymatic hydrolysis of insoluble cellulose: Part 2, Analysis of extended hydrolysis times. *Biotechnol. Bioeng.* 25:939–966.

Lynd, L. R., Weimer, P. J., van Zyl, W., and Pretorius, I. 2002. Microbial cellulose utilization: Fundamentals and biotechnology. *Microbiol. Mol. Biol. Rev.* 66:506–577.

McCarthy, S. 2010. Canada's ethanol boom, minus the boom. *Globe and Mail,* March 2. http://www.theglobeandmail.com/report-on-business/canadas-ethanol-boom-minus-the-boom/article1486331/.

McLaughlin, S. B., and Adams Kszos, L. 2006. Development of switchgrass (*Panicum virgatum*) as a bioenergy feedstock in the United States. *Biomass Bioenergy* 28(6): 515–535.

McLaughlin, S. B., Samson, R., Bransby, D., and Wiselogel, A. 1996. Evaluating physical, chemical, and energetic properties of perennial grasses as biofuels. In *Proceedings of BioEnergy '96, the seventh national bioenergy conference: Partnerships to develop and apply biomass technologies.* Nashville, TN: Oak Ridge National Laboratory. http://bioenergy.ornl.gov/papers/bioen96/mclaugh.html.

Parrish, D. J., Wolf, D. D., Fike, J. H., and Daniels, W. L. 2003. *Switchgrass as a biofuels crop for the upper southeast: Variety trials and cultural improvements: Final report for 1997 to 2001.* Report ORNL/SUB-03-19XSY163/01. Oak

Ridge, TN: Oak Ridge National Laboratory, http://www.ornl.gov/info/reports/2003/3445605367634.pdf.

Passmore, J. 2006. Prepared statement. Hearing before the Committee on Agriculture, House of Representatives, Serial No. 109-34, pp. 53–71. http://agriculture.house.gov/hearings/109/10934.pdf.

Patzek, T. W. 2004. Thermodynamics of the corn-ethanol biofuel cycle. *Crit. Rev. Plant Sci.* 23(6): 519–567.

———. 2006a. A first-law thermodynamic analysis of the corn-ethanol cycle. *Nat. Resour. Res.* 15(4): 255–270.

———. 2006b. Letter to the editor. *Science* 312:1747.

———. 2006c. *The real biofuels cycles.* Online supporting material for *Science* letter. http://www.oilcrisis.com/ethanol/RealBiofuelCycles.pdf.

———. 2006d. A statistical analysis of the theoretical yield of ethanol from corn-starch. *Nat. Resour. Res.* 15(3): 205–212.

———. 2007. How can we outlive our way of life? In *20th Round Table on Sustainable Development of Biofuels: Is the cure worse than the disease?* Paris: OECD. http://www.oecd.org/dataoecd/2/61/40225820.pdf.

———. 2008a. Can the Earth deliver the biomass-for-fuel we demand? In *Biofuels, solar and wind as renewable energy systems: Benefits and risks,* ed. Pimentel, D., 19–58. Berlin: Springer.

———. 2008b. Exponential growth, energetic Hubbert cycles, and the advance-ment of technology. *Arch. Min. Sci.* 53(2): 131–159.

———. 2008c. Thermodynamics of agricultural sustainability: The case of US maize agriculture. *Crit. Rev. Plant Sci.* 27(4): 272–293.

Patzek, T. W., and Pimentel, D. 2005. Thermodynamics of energy production from biomass. *Crit. Rev. Plant Sci.* 24(5–6): 329–364.

Perlack, R. D., Wright, L. L., Turhollow, A. F., Graham, R. L., Stokes, B. J., and Erbach, D. C. 2005. *Biomass as feedstock for a bioenergy and bioproducts industry: The technical feasibility of a billion-ton annual supply.* Joint report, prepared by U.S. Department of Energy and U.S. Department of Agriculture, Environmental Sciences Division. Oak Ridge, TN: Oak Ridge National Laboratory.

Pilkey, O. H., and Pilkey-Jarvis, L. 2007. *Useless arithmetic: Why environmental scien-tists can't predict the future.* New York: Columbia University Press.

Pimentel, D., and Patzek, T. W. 2005. Ethanol production using corn, switchgrass, and wood; biodiesel production using soybean and sunflower. *Natural Resource Research* 14(1): 67–76.

Pimentel, D., Patzek, T. W., and Gerald, C. 2006. Ethanol production: Energy, eco-nomic, and environmental losses. *Rev. Environ. Contam. Toxicol.* 189:25–41.

Purvis, C. L. 2011. Auto ownership in the San Francisco Bay Area, 1930–2010. Oakland, CA: Metropolitan Transportation Commission. http://www.mtc.ca.gov/maps_and_data/datamart/forecast/ao/aopaper.htm.

Saha, B. C., Iten, L. B., Cotta, M. A., and Wu, Y. V. 2005. Dilute acid pretreatment, enzymatic saccharification and fermentation of wheat straw to ethanol. *Process Biochem.* 40:3693–3700.

San Martin, R., Blanch, H. W., Wilke, C. R., and Sciamanna, A. F. 1986. Production of cellulase enzymes and hydrolysis of steam-exploded wood. *Biotechnol. Bioeng.* 28:564–569.

Schmer, M. R., Vogel, K. P., Mitchell, R. B., Moser, L. E., Eskridge, K. M., and Perrin, R. K. 2006. Establishment Stand Thresholds for Switchgrass Grown as a Bioenergy Crop. *Crop Science*, 46:157–161.

Scharlemann, J. P. W., and Laurance, W. F. 2008. How green are biofuels? *Science* 319:43–44.

Schmer, M. R., Vogel, K. P., Mitchell, R. B., and Perrin, R. K. 2008. Net energy of cellulosic ethanol from switchgrass. *PNAS* 105(2): 464–469.

Schmidt, A., Zschetzsche, A., and Hantsch-Linhart, W. 1993. *Analyse von biogenen Brennstoffen*. Vienna: TU Wien, Institut für Verfahrens-, Brennstoff- und Umwelttechnik. http://www.vt.tuwien.ac.at/Biobib/fuel98.html.

Schweiger, R. G. 1979. New cellulose sulfate derivatives and applications. *Carbohydrate Research* 70(7): 185–198.

Sigma-Aldrich. 2010. Taq DNA Polymerase from *Thermus aquaticus*. Accessed April 19, 2010. http://www.sigmaaldrich.com/catalog/search/ProductDetail/SIGMA/D1806.

Smil, V., Nachman, P., and Long, T. V., II. 1983. *Energy analysis and agriculture: An application to U.S. corn production*. Boulder, CO: Westview Press.

Somerville, C. 2006. The billion-ton biofuels vision. Editorial. *Science* 312:1277.

Stehfest, E., and Bouwman, L. 2006. N_2O and NO emission from agricultural fields and soils under natural vegetation: Summarizing available measurement data and modeling of global annual emissions. *Nutr. Cycling Agroecosyst.* 74(3): 207–228.

Szargut, J., Morris, D. R., and Steward, F. R. 1988. *Exergy analysis of thermal and metallurgical processes*. New York: Hemisphere.

Taiz, L., and Zeiger, E. 1998. *Plant physiology*. 2nd ed. Sunderland, MA: Sinauer.

Thompson, E., Jr. 2009. California agricultural land loss and conservation: The basic facts. Washington, DC: American Farmland Trust. http://www.farmland.org/documents/AFT-CA-Agricultural-Land-Loss-Basic-Facts_11-23-09.pdf.

USDA. 2004. Estimating U.S. cropland area. *Amber Waves*, November. Washington, DC: USDA Economic Research Service. http://www.ers.usda.gov/AmberWaves/July06SpecialIssue/pdf/BehindDataJuly06.pdf.

USDOE (U.S. Department of Energy). 2007a. Biomass program. *Biomass feedstock composition and property database*. [Online database accessed July 25, 2007.] http://www.eere.energy.gov/biomass/progs/search1.cgi.

———. 2007b. DOE announces up to $200 million in funding for biorefineries: Small- and full-scale projects total up to $585 million to advance President Bush's Twenty in Ten Initiative. Press release, May 1.

Wang, M. 2001. *Development and use of GREET 1.6 fuel-cycle model for transporation fuels and vehicle technologies*. Technical Report ANL/ESD/TM-163. Argonne, IL: Argonne National Laboratory, Center for Transportation Research.

Wilen, C., Moilanen, A., and Kurkela, E. 1996. *Biomass feedstock analyses*. VTT Publications 282. Espoo, Finland: Technical Research Centre of Finland. http://www.vt.tuwien.ac.at/Biobib/fuel207.html.

Zah, R., Böni, H., Gauch, M., Hischier, R., Lehmann, M., and Wäger, P. 2007. *Ökobilanz von Energieprodukten: Ökologische Bewertung von Biotreibstoffen*. St. Gallen, Switzerland: Empa.

Zeltich, I. 1971. *Photosynthesis, photorespiration, and plant productivity*. New York: Academic Press.

Zhang, Y., and Lynd, L. R. 2003. Cellodextrin preparation by mixed-acid hydrolysis and chromatographic separation. *Anal. Biochemistry* 322:225–232.

———. 2004. Towards an aggregated understanding of enzymatic hydrolysis of cellulose: Noncomplexed cellulase systems. *Biotechnology and Bioengineering* 88(7): 797–824.

Appendix: The standard arguments against the probabilistic model of the switchgrass-ethanol cycle

This work has raised strong opposing arguments. Here I comment on the key assertions made by the proponents of "cellulosic" ethanol. As a result, clarity can be gained and arguments sharpened. What follows is a summary of the actual written questions and assertions (Q) by a critic and my answers (A).

4.A.1   General questions

Q: This paper presents an attempt to quantify net energy yield of a hypothetical *industrial process.*

A: The process I am modeling is based only on the available field- and industrial-scale data and is not hypothetical.

Q: Since one may envision quite a wide range of variables that can be speculatively evaluated by the kind of analysis presented here, one doubts that even the policy journals would welcome this kind of speculation.

A: How can a quantitative probability distribution of outcomes—based solely on the available hard data—be speculation?

Q: Additionally, this kind of paper is highly susceptible to superficial reviewing, since it requires as much work on the part of reviewers to assess the paper as was required to produce the paper. Thus, aside from the significant investment of time required to review a paper, I am concerned that there will be further proliferation of essentially nonreviewed papers that can be used to support incorrect claims.

A: According to this reasoning, anything that requires a little effort to understand is unworthy of publishing in scientific literature because it is difficult to review. I think that it is far more important to have a full probability distribution of switchgrass and ethanol yields (a difficult and uninteresting result, according to this assertion), rather than two average numbers contained in an often-quoted paper by Schmer et al. (2008) that apparently is easy and interesting to read.

Q: Whatever the case, this paper does not represent leading-edge science or engineering.

A: I always welcome arguments against my thinking that are based on hard science and engineering, as such arguments will allow me to improve the current model of the switchgrass-ethanol cycle.

4.A.2   Data sources

One of the perennial sources of disagreement is which data are relevant, and which are not. Have all the relevant data been included? Here is another Q&A session, devoted to the input data.

Q: By selectively choosing values for component processes, the author is essentially just composing an opinion that has no value as far as I am concerned. I think this is why his papers are generally ignored by the scientific and engineering community who work in this field. By contrast, the EBAMM that was published in Science *a few years ago has been an important contribution because it provides a model that anyone can use to plug in numbers to evaluate the effect of variables on the outcomes.*

A: It is with great trepidation that I am forced to resurrect the ghost of the EBAMM paper in *Science* (Farrell et al., 2006), since this model has been demonstrated beyond any reasonable doubt to violate the mass balances of the biofuel cycles it attempts to represent and, as a consequence, also the energy balances of these cycles (Patzek, 2006a, 2006b, 2006c, 2006d). In particular, the cellulosic refinery efficiency of 0.382 L of EtOH per kg dmb of switchgrass, listed in the EBAMM spreadsheet, has no justification other than quoting another spreadsheet, GREET, by Wang (2001), with no source of *that* information given. (I have an electronic copy of the originally published EBAMM and the student author's comments are quite interesting to read.) The EBAMM number corresponds to approximately 90% of the theoretical efficiency of conversion of cellulose to ethanol (more on this subject later). The EBAMM also assumes that biomass is the sole source of fuel for the refinery (as I do here) and assumes that 94,492 BTU will be burned per gallon of ethanol, or 29.5 MJ/L. An existing average corn ethanol refinery is twice as efficient, at 15 MJ per liter of ethanol (Patzek, 2006a).

According to EBAMM, 2.618 kg dmb of switchgrass is needed to obtain 1 liter of ethanol. This switchgrass contains 15% by weight of lignin, whose higher heating value (HHV) is 21.2 MJ/kg. The bone-dry lignin in the switchgrass processed for ethanol might deliver 8.2 MJ of process heat* per liter of ethanol. The remaining 21.3 MJ/L (29.5 MJ/L – 8.2 MJ/L) of the process heat would have to come from burning more switchgrass; alternatively, a combination of switchgrass lignin (4–8 MJ/L) and fossil fuels (21–25 MJ/L) could be used. One liter of ethanol has the HHV of 23.2 MJ/L. Thus, according to EBAMM, a switchgrass refinery would use about the same amount of energy from burning additional switchgrass or fossil fuels as the HHV of the product. This does not stop the EBAMM authors from claiming a 26 MJ/kg energy "credit" from recycling switchgrass lignin and a net energy value of 8.4, which made such a splash in the U.S. media. Unfortunately, in the context of a *closed* energy balance, this energy credit must originate from a *perpetuum mobile*. In other words, the *magic* energy credit asserted in EBAMM is physically impossible, and more switchgrass must be burned, lowering the overall yield of ethanol from switchgrass.

---

* 50%-wet lignin would deliver half this quantity of heat.

Apparently, all this utterly unscientific manipulation of unjustified numbers does not discourage some scientists from regarding EBAMM as "a model that anyone can use to plug in numbers to evaluate the effect of variables on the outcomes." The purpose of my work has been to evaluate and validate the very numbers on which most of the published biofuels papers are based. Perhaps this approach is one of the reasons why my papers have had such a strong impact worldwide and have helped to change biofuel policies of several EU countries and the U.S. attitude toward corn ethanol. But the critic's statement is correct: My most widely read[*] paper ever (Patzek, 2004) has yet to be cited once by a small and tightly knit clan of political *biofuelistas* in the United States, who ignore it completely.

Q: Except for the errors in the paper (outlined below), and lack of sensitivity analysis, I consider some aspects of this paper to represent the kind of analysis that should be provided by consulting engineers to companies who are contemplating construction of a bioprocessing facility rather than something that is of scientific interest. I assume that the more than 20 companies that have built or are building such facilities have been prudent enough to carry out such analyses to their own satisfaction.

A: I fully agree with this assumption. These companies are certainly self-satisfied, because—based on the unrealistic or physically impossible proposals—they have been awarded large sums of taxpayer money. Figure 4.A1 presents at *face value* the stated energy efficiencies[†] of the six proposed cellulosic ethanol plants awarded $385 million by the U.S. Department of Energy (USDOE, 2007b). For details and comparison with other liquid transportation fuel systems, see Patzek (2007, Section 6.2).

In fact, at the time of the last review of this chapter in July 2011, Range Fuels, which offered plans for the highest efficiency biorefinery, was already bankrupt despite ample help from USDOE and Vinod Khosla. One could say that I predicted this and other pending failures in my 2007 paper.

Note that the Iogen proposal for a future cellulosic refinery strives to reach the current average energy efficiency of corn ethanol refineries. More on this subject later.

As to the lack of sensitivity analysis, I am puzzled. The critic seems to be suggesting that, based on a few short-term field studies that are confounded by a wild variation of switchgrass yields, the average yield of switchgrass everywhere across the United States can be quantified by a single number x, with some deviation, and that that number can only grow with time. The x he chooses is 7 Mg/ha/yr, and it is cast in stone. I

---

[*] For the last six years, it has appeared on Google as the second most popular link with my name after my homepage, out of some 50,000–80,000 links.

[†] The HHV of ethanol out, divided by the HHV of biomass in. No fossil-fuel inputs into the plants or raw materials they use are accounted for.

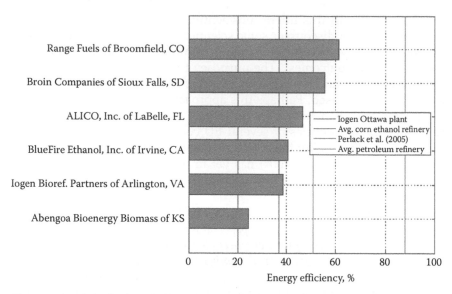

Figure 4.A1 Stated energy efficiencies of the six future cellulosic ethanol refineries awarded $385 million in DOE grants in 2007. The calculated energy efficiency (left line) of an existing cellulosic ethanol refinery in Ottawa serves to calibrate the rather inflated efficiency claims of five of the six grant recipients. Energy efficiencies of an average ethanol refinery and petroleum refinery (second and last line from the left; data from Patzek,2006a) are also shown. Subsequently, the Broin Companies became Poet.

am trying to say that we do not know enough to specify a single value x. But based on all available switchgrass yield data, weighted by a long-term field survival function, I can propose a probability distribution function of continuous (or eternal) switchgrass yields and calculate the distribution's expected value. That value is unreliable, but the distribution provides a *qualitative* insight into what might be expected from the future switchgrass plantations. Quantitative models of complex processes on the surface of the Earth do not work. On the other hand, qualitative models, when applied correctly in what-if scenarios, can be useful. The simplistic quantitative models of the corn ethanol industry (GREET, EBAMM, etc.) have already been shown by the events on the ground to be *non gratum anus rodentum.*[*]

4.A.3   Fertilization rates

Q: The author's assumptions on inputs are also incorrect. For example, in the large-scale Schmer study, no fertilizers other than N were applied.

---

[*] This poor-Latin term was coined by the "Tunnel Rats," an elite band of volunteer soldiers in Vietnam, selected both for their bravery and, above all, their small stature.

A: The Oak Ridge National Laboratory report states in its Executive Summary in Parrish et al. (2003, p. 1):

> Management studies at eight locations showed that lowland varieties when harvested once at the end of the growing season produce high sustainable yields if N fertilizer is properly managed. Upland varieties yield slightly more if harvested twice, but the increased yield is likely not economically significant. Nutrient element removal is generally greater than the amount applied.... The soil should be tested using a 0 to 10-cm sample depth. Apply 50 kg ha^{-1} of P when the soil-P test is low and 100 kg ha^{-1} of K when the soil-K test is low-plus to medium (considering the typical soil test basis for agronomic crop recommendations). No limestone would be needed if the soil pH is above 5.... While we do not have direct experimental evidence for such, our observations lead us to believe many soils may not need any N for several years after planting when using a one-cut (November) harvest management, since much of the N located in the herbage is translocated to the crown/root system at the end of the growing season. Thus, no more than 50 kg ha^{-1} N may be needed for sustained yields. For a two-cut management, 50 kg ha^{-1} N in the spring and 50 kg ha^{-1} N after the first harvest should be adequate. (Executive Summary)

Other studies report: "Nitrogen, P, and K were applied at 50, 24, and 46 kg/ha, respectively, in early May 1999, 2000, and 2001.... After four years of growth, all selections had excellent stands as judged by tiller numbers" (Parrish et al., 2003, p. 28). "Applications of N, P, and K were made in the spring of 1997 and 1998 to an established stand of Cave-in-Rock switchgrass" (Parrish et al., 2003, pp. 31–35). Other strategies were tried, too.

Fike et al. (2006a, p. 198) state:

> Switchgrass plots under lower-input management received 50 kgNha^{-1}yr^{-1} and were harvested once, at the end of the season. Plots under higher-input management received 100 kgNha^{-1}[yr^{-1}] (in two applications) and were harvested twice, in midsummer and at the end of the season.... Total P and K applications over the course of the 5-yr study ranged from 40 to 100 kg P ha^{-1} and 0 to 315 kg K ha^{-1}.

Schmer et al. (2008, Table 6) lists 5,488 MJ/ha/yr and 128 MJ/ha of P amortized over five years. Their sources are GREET (Wang, 2001) and EBAMM, which in turn relies on GREET for most of its inputs.

4.A.4   Ethanol refinery efficiency

Q: The author formulates an estimate of ethanol yield from switchgrass. He concludes from an arbitrary choice of values that the theoretical ethanol yield is 0.18 kg ethanol/kg of switchgrass and notes this is substantially lower than the 0.38 L/kg used in the widely used EBAMM published in Science.

A: The choice of ethanol yield of 0.18 kg ethanol/kg switchgrass = 0.23 L/kg is not arbitrary. It is based on the work of Badger (2002), translated to switchgrass composition. The choice of 0.38 L/kg in EBAMM *is* completely arbitrary.

Q: I cannot see any basis for such an assumption. The theoretical yield of ethanol from switchgrass is about 0.43 L/kg based on the conversion calculator developed by the National Renewable Energy Laboratory. I think it safe to assume that the process of converting sugars to ethanol can be made 90% efficient, which would be 0.387 L/kg.

A: This is a perfect example of an arbitrary subjective estimate of ethanol yield, based on an optimistic guess that has *no* factual confirmation. When one reads the biopharmaceutical literature (e.g., Jagschies, 2008), one finds that the state-of-the-art yields of most biopharmaceutical processes are but small fractions of their theoretical values. Similar but even more optimistic estimates were performed earlier for the corn ethanol refineries, which are now being forced en masse into bankruptcy.

Q: The author's modeled information on conversion should be compared to the paper by Lau and Dale (2009).

A: I am very happy that Dr. Dale and his student were able to obtain the yield of 0.191/0.787 = 0.24 L EtOH/kg of corn stover at 4% of ethanol by mass. Their laboratory-scale process performs infinitesimally better than the industrial pilot by Iogen modeled in my paper (0.23 L/kg), but it delivers only 0.24/0.387 = 63% of the yield guessed by the critic, who is so busy criticizing my numbers that he fails to see the comical side of his comments. Four years ago, the phrase "based on outdated information and misused data" was routinely applied to me by the early corn ethanol biofuelistas. Where are they today? Lau and Dale also suffer from an unacceptably high solids loading, as pointed out in my paper, and a kinetic limitation of xylose consumption that is well known for the unstable, delicate, genetically engineered yeast strains. Either of these two limitations will prevent extending the Lau and Dale approach to an industrial scale.

Q: The author then calculates net energy efficiency using 15 MJ/kg as the cost of dehydration. However, according to a detailed study of ethanol dehydration published in Science *(Ladisch and Dyck, 1979), the energy required to dehydrate*

12% ethanol is about 2.26 MJ/L. Taking the value of solids loading used by the author (25%) and assuming 0.38 L ethanol per kg of switchgrass suggests that the solution would be about 9.5% ethanol. Adjusting the energy requirement for a 9% versus 12% solution indicates the energy required to be about 3.15 MJ/L ethanol rather than 15. Of course, with moderately good reuse of heat, the actual net energy input would be much less. For the sake of argument, I assume 50% recovery of heat, so that the energy required for dehydration would be about 1.58 MJ/L.

A: The switchgrass refinery efficiency has been calculated by reverse engineering of data published from the *only* such refinery in existence. This heavily subsidized, small demonstration refinery is run by Iogen,[*] which is also behind at least two of the proposals shown in Figure 4.A1. One can demonstrate in two independent ways that the ethanol broth in the Iogen refinery has about 4% of ethanol, not 9% as asserted by the critic, or 12–16% in cornstarch refineries. In fact, 4% of ethanol by mass corresponds to a 40 g/L yield, which is significantly above the state-of-the-art product titers of 1–5 g/L and occasionally above 10 g/L, achieved by the pharmaceutical industry in their bioreactors.[†]

Because the energy cost of distillation of ethanol-water mixtures increases hyperbolically as ethanol concentration decreases, the value I *calculated* follows directly from the dilute nature of the cellulosic ethanol broth. More information on the energy costs of industrial ethanol distillation can be found in the standard industry reference *The Alcohol Textbook* (Jacques et al., 2003). In addition, Berkeley's Dr. King (1980) has written a widely used textbook on liquid mixture separations. Classically, recovery of ethanol from the fermentation broth is at least a three-step process:

1. Distillation of dilute aqueous alcohol to the azeotrope concentration (96% of ethanol by weight)
2. Distillation with a third component, benzene or a strong salt solution, to remove the remaining 4% of water
3. Distillation of water from the third component so that it can be recycled

Recently, molecular sieves have been used to bypass steps 2 and 3 and save energy.[‡] Either way, the distillation process consumes most of the process heat in ethanol refineries. I remind the reader that, according to Iogen in Figure 4.A1, the new design of its cellulosic ethanol refinery will

---

[*] Bruce Orr, a former Ontario government biofuels adviser, said Iogen and other cellulosic ethanol makers still face tremendous hurdles in scaling up from small demonstration plants because their enzymes simply don't break down the cellulose into ethanol efficiently enough. "There has been a lot of hyperbole about cellulosic ethanol, but there are still a lot of doubts [the Iogen project] will ever happen" (McCarthy, 2010).

[†] Jagschies, 2008.

[‡] But the zeolite must be oven-dried to be reused.

reach only the average efficiency of existing ethanol refineries. Most of the energy in the distillation process is used to concentrate ethanol from 90% to 100%.

As to the Ladisch and Dyck (1979) paper in *Science*, it is somewhat inappropriate in the context of producing 20 billion gallons of ethanol per year, which appears to have been the critic's goal. Instead of carrying out the expensive distillation in the 90–100% ethanol region, one could use a dehydrating agent, such as CaO, NaOH, cornstarch, sucrose, corn, cellulose from various sources, or corn residue. The process demonstration at the scale of 10^1 ml *once*, carried out by Ladisch and Dyck, apparently has not yet scaled up to the 10^9–10^{10} ml *per year* for decades required in a real refinery.

Eighty years of distillation research has not solved this scale-up issue. It turns out that use of expensive heat exchangers to recover some of the process heat in biorefineries has not been economical, regardless of the price of natural gas. Thus, modern corn ethanol refineries still use close to 50% of the anhydrous product energy to obtain that product. For dilute broths, as in the case of switchgrass fermentation, close to 100% of the product energy is used, as properly calculated by GREET and copied into EBAMM. EBAMM then introduces a *deus ex machina* energy credit of almost 100% of the process heat.

chapter five

Biofuels, climate change, and human population

Edwin Kessler

Contents

5.1   Introduction

Fossil liquid fuels and other resources are being extracted on finite planet Earth and utilized to the point that they are becoming more costly and some are expected to become scarce within the next decade. There are tight interrelationships and overlapping effects among proposed use of biofuels, climate change, and overall consumption by still-growing human populations. Countries with a preponderance of advanced technologies consume disproportionately large quantities of fossil fuels, with the accompanying production of carbon dioxide, a greenhouse gas whose increasing concentration in Earth's atmosphere is warming the planet and producing mostly negative consequences for coastal cities, ocean life, and agriculture.

According to the International Energy Agency (IEA) and the Organization for Economic Cooperation and Development (OECD), North America and the OECD member states together in 2007 consumed about 44% of the world's production of petroleum, even though together they comprised only about 23% of the world's population. The United States, with about 4.4% of the global population, produced 20% of total global emissions of carbon dioxide.[*]

A political system that produces poorly thought-out responses has led to inappropriately large increases in use of fuels with a current biological base—that is, biofuels. Analyses of the production/consumption cycle of biofuels often show that associated carbon dioxide releases involve little more than the carbon that was temporarily sequestered during growth, but analyses often neglect negative impacts on land, water, and wildlife habitat. Moreover, impacts of biofuel production on the cost and supply of food are downplayed or even ignored. And biofuels cannot become abundant enough to replace more than a small fraction of present U.S. usage of about 20 million barrels of oil daily.

5.2   Present societal conditions

Present conditions, including the imminent decline of the petroleum resource, have been considerably anticipated for more than a hundred years. For example, in *Infinite Resources: The Ultimate Strategy* (Goeller and Zucker, 1984), concern is expressed about resource depletion, but it is concluded as unlikely that the world would run short of any resource before about 2050. Considerable time would therefore be available to develop saving technologies. However, societies have proved refractory to change, global population and demand have surged beyond the authors'

---

[*] China, with four times the U.S. population, has recently overtaken the United States in carbon dioxide emissions.

predictions, and too little ameliorative action has been done to date. Remedies present great cultural and political challenges, but are urgently required.*

Today, we are assaulted by a new multitude of urgent issues. Most prominent is the number of human beings, increased by more than a factor of three in only the last 80 years, with enormously expanded and still increasing consumption and with a multiplication of associated issues that demand attention. The present global condition has been vividly, though partially, described by Kathryn Dodgson (2010, p. 1) as including:

> the extremes of deeply entrenched belief and unremitting religious (or ethnic) strife; the mindboggling daily expense of ongoing wars over commodities fought in the name of justice or democracy juxtaposed with the tragic costs of worldwide poverty and the injustice of human degradation; the excessive swings of financial markets; the frighteningly rapid rate and extent of ecological and environmental damage wreaked by practices that, in the best light, can be laid at the feet of human ignorance or shortsightedness, and at worst are the results of an excess of blatant human greed.

All of the above in their extreme manifestations can be viewed as unintended consequences of science and technology, which have brought unprecedented material well-being to many in the developed world, but have had no noticeable impact on human qualities. Relatively luxurious lifestyles in the developed world depend on abundance of energy in forms of electricity and of liquid fuels for transportation. Seeing this, the less developed world struggles to achieve similar living standards. Strengthened desires are portentous of further increased consumption of limited resources and aggravation of the ills listed above.

Why are political and social conditions important in the present context? There is a simple answer: While science and technology are a source of the present conditions, the response to the latest findings and applications of science and technology depends on political and social conditions and the perceived causation of these conditions.

The influence of major corporations on legislation and on social conditions in the United States has been debated since the founding of this

---

* Growing scarcity of some commodities is indicated by their rate of cost increase. At this writing, the U.S. consumer price index has increased about 20% since 2002, but a metals index has increased by a factor of three, and the average global price of food, according to the Food and Agriculture Organization, has increased by 70% in the same period.

republic, with a major concept of corporate personhood evolving from *Santa Clara County v. Southern Pacific Railroad*, 118 U.S. 394 (1886), a case in which the railroad argued for corporate rights asserted in the First, Fifth, and 14th amendments to the U.S. Constitution. The Internet notwithstanding, there has been some recent decline of democracy in the United States as a consequence of powerful influences by wealthy corporations and financial centers. Further decline may ensue as a result of a 5–4 decision of the U.S. Supreme Court in *Citizens United v. Federal Election Commission* on January 21, 2010. Concerning this latest case, Ronald Dworkin (2010, p. 39), in the *New York Review of Books*, notes that "five right-wing Supreme Court justices have now guaranteed that big corporations can spend unlimited funds on political advertising in any political election.... The court overruled established precedents and declared dozens of national and state statutes unconstitutional." He observes that this "appalling decision" identifies corporations with some of the same attributes and capabilities of individuals.

A different view presented by Jan Witold Baran (2010), writing in the *New York Times*, notes the importance of much that was left in place by the Supreme Court decision. Thus, corporations and unions are still unable to give money directly to candidates or political parties, may spend money only in the form of "independent expenditures," and must continue to file reports with the Federal Election Commission within 24 hours of the spending.

Omitted from Baran's analysis is discussion of an inability of many citizens, even in the presence of the Internet, to become knowledgeable of late filings. And there is also an absence of some significant information in media largely owned by corporate entities with the same interests as those contributing large sums of money to political campaigns. This condition may be partially ameliorated by an emphasis on strengthened disclosure laws, possibly enabled by the Court's essentially unanimous affirmation of disclosure as a suitable adjunct to *Citizens United*. For example, the public interest group Public Citizen, in testimony before Congress (Holman, 2010), called for passage of a strengthened disclosure law, H.R. 5175, by the U.S. Congress. On the other hand, there are groups proclaiming that H.R. 5175 does not go far enough and that it merely represents a way to conceal business-as-usual. In the presence of such angry debate, no significant disclosure legislation was passed by the Congress during 2010 nor to the date of this writing in June 2011.

Short-term financial concerns of wealthy interests often (though not always!) dominate candidates' and their supporters' thinking and actions. In some cases, special interests are seen as seeking and obtaining the best seats on the *Titanic* as it sinks with all on board. And the nationwide elections of November 2010 saw money in unprecedented amounts donated from largely unknown sources, with a sharp swing to the right both in state offices and in Washington, D.C.

Added public concern has been related to suggestions that Supreme Court justices Antonin Scalia and Clarence Thomas, who voted with the majority in the *Citizens United* case, were inappropriately indebted to the Koch brothers, who preside over very wealthy business interests that contribute to right-wing causes. Furthermore, Justice Thomas was shown to have omitted disclosure of the six-figure income of his spouse. At this writing, there has been no resolution of these apparent ethical violations, which are being especially pursued by the national organization Common Cause.

5.3   Human population

5.3.1   Overview

Human population growth has become the major global issue on planet Earth because of the positive relationship between population and consumption and because Earth's resources are finite. For millennia, human population was static or increased relatively slowly, but science and industrialization dramatically reduced death rates during the 20th century and thereby spurred population growth, since birthrates respond at lag and, depending on social context, sometimes hardly at all. At the turn of the 20th century, with a global population of about 1.6 billion, a majority in the West still deemed unrealistic, even foolish, the thought that humankind could actually deplete global resources of freshwater and petroleum, for example, or alter the climate through carbon burning. It is painful in 2011, with a global human population about 6.9 billion, to accept such beliefs, but they are a new reality, though hardly accepted as a basis for actions by society as a whole, including much of the leadership. Leadership notwithstanding, the impact of humans on the environment has become enormous, and in recognition of this, a new word, Anthropocene, labels the new epoch of human influence.[*]

Richard Hilbert (1976) has dealt with the problem presented by a lag of development of social norms that are appropriately responsive to technological developments that come rapidly with modernity. Hilbert focuses on a classic statement of the theory of anomie and deviance by Robert K. Merton (1938), but in a discussion of its origins calls attention to works of Emile Durkheim (1951, 1964). *Anomie* is disrespect for the goals of a society and/or the norms that prescribe legitimate means for their attainment. *Deviance* is any behavior, such as several forms of criminality, that depart from societal norms, even hypocrisy. We are concerned with the way in which societies of the American type—modern societies in

---

[*] Coined a decade ago by Nobel Prize winner Paul Crutzen, the Anthropocene, with numerous extinctions, postdates the Holocene, which began at the end of the last ice age, about 12,000 years ago.

general—generate anomie and thus the deviance that follows from it. In effect, the theory of anomie and deviance deals with an important cost of modernity.

In the early discussions of anomie by Durkheim, there is a suggestion that the problem of anomie appeared in Western societies when they became industrialized, when certain changes in the normative system became necessary in order to meet needs associated with new roles of work. For example, a greater emphasis developed on competition and on common success goals in order to motivate people to participate. Such motivation increases performance, but it also creates problems when large segments or classes of the population lack the access—or in some cases may be purposely denied access—to the means by which these goals can be attained, notably education and investment capital. Such disjunction is one of the more important sources of anomie. Appropriate solutions to such problems may develop as societies come to understand them. But, of course, development of understanding takes time, and there is no assurance that rational application of understanding will prevail. Hilbert and I are more pessimistic on this subject than was Durkheim, mainly because structural features of the American Way—for instance, capitalism—that are responsible for anomie are among the more valued features of its normative system.*

Hilbert's discussion of changes that might be expected to solve the problem of anomie calls attention to the costs involved. For example, if aspirations are stratified by class so that no one would ever want more than what was attainable by available means, anomie would be reduced, but so would average performance. Clearly, in the United States today this kind of change is unacceptable, which helps to explain why anomie and associated deviances are a persistent problem. In any event, failure to develop respected new norms in response to changes that are both large and very much at the foundation of societal functioning associates with a maladaptive and poorly functioning society and brings grave portents for future human welfare.

The Food and Agriculture Organization recently reported that about one-sixth of the world population is malnourished (FAO, 2010). Furthermore, in a progress report titled *Nutrition for Health and Development*, issued in 2000, the World Health Organization (WHO, 2000, p. 9) states:

> Hunger and malnutrition remain among the most devastating problems facing the majority of the world's poor and needy, and continue to dominate the health of the world's poorest nations. Nearly 30%

---

* The "too big to fail" doctrine produces a form of socialism (government subsidy) for the wealthy, while capitalism (competition) is promoted for middle and lower economic classes.

of humanity—infants, children, adolescents, adults and older persons in the developing world—are currently suffering from one or more of the multiple forms of malnutrition. This remains a continuing travesty.... The tragic consequences of malnutrition include death, disability, stunted mental and physical growth and as a result, retarded national socio-economic development. Some *49% of the 10.7 million deaths among under-five children each year in the developing world* are associated with malnutrition. *Iodine deficiency* is the greatest single preventable cause of brain-damage and mental retardation worldwide. *Vitamin A deficiency* remains the single greatest preventable cause of needless childhood blindness.[*]

We are reminded of a letter by P. L. Abplanap (1999) to *Technology Review*, in which he wrote, "Any kind of agricultural 'green revolution' which is not accompanied by effective population control merely resets the limiting parameters at higher levels and enables countries with a large proportion of starving citizens to increase the absolute numbers of starving people." Adam and Eve were instructed: "Be fruitful and multiply, and replenish the earth, and subdue it; and have dominion over the fish of the sea, and over the fowl of the air, and over every living thing that creepeth upon the earth" (Genesis 1:28). Virtually all of these, save replenishment, have now been done to excess. We have driven and are driving other species to extinction, and in many places we have even fouled our own nests. So, when will humans accomplish their mandate for replenishment of the Earth?

An important perspective on human population growth was provided in 1798 by Thomas Robert Malthus. Malthus was a graduate of Jesus College at Cambridge University, excelled there in mathematics, and from 1793 to 1804 was a fellow of that college. He was an economist and sociologist and, from 1795 to 1804, curate of the Anglican church of Asbury, Surrey. Malthus wrote several books and pamphlets on history and political economy, and his principal work was edited in six subsequent editions, the first being in 1803 (Malthus, 1803). A main point of Malthus's most famous work was that, under accommodating conditions, the rate of increase of human populations exceeds the rate of increase of the food supply, with ominous implications for future human welfare.

---

[*] Emphasis in original. The cited iodine deficiencies are of particular concern in relation to the 2011 nuclear disaster in Japan. Persons with iodine deficiency would be more apt to suffer from accumulation of radioactive iodine in the thyroid gland. Some radioactive iodine may be deposited globally as a result of the disaster at the Fukushima nuclear power plant, which has been rated at the same maximal level as the Chernobyl event.

According to the *Encyclopedia Britannica* (1910–1911, p. 515), Malthus's work was "very welcome to the higher ranks of society, that they tended to relieve the rich and powerful of responsibility for the condition of the working classes, by showing that the latter had principally themselves to blame." This is a reminder of how little political conditions have changed, but Malthus himself was not unconcerned with the condition of the poor. He is further described as

> one of the most amiable, candid and cultured of men. In all his private relations he was not only without reproach, but distinguished for the beauty of his character. He bore popular abuse and misrepresentation without the slightest murmur or sourness of temper. The aim of his inquiries was to promote the happiness of mankind, which could be better accomplished by pointing out the real possibilities of progress than by engaging in vague dreams of perfectibility apart from the actual facts which condition human life. (*Encyclopedia Britannica*, 1910–1911, p. 515)

Malthus was correct about rates of population increase, but his concept was 200 years premature, owing to the phenomenal growth of science and technology with abundant resources, especially in applications to agriculture. Malthus was not alone in underestimating the role of burgeoning science and technology. For example, *The Population Bomb* (Ehrlich, 1968) could hardly be more urgent in its appeal for fast action to reverse population growth, but rapid developments in science and technology have delayed onset of some predicted consequences of population increase. The recent book *The Dominant Animal* (Ehrlich and Ehrlich, 2008) calls again and properly for fast action from a more studied background. Other important analyses of resource and population problems have been provided by Garrett Hardin (1964, 1968), William R. Catton Jr. (1980), Al Gore (1993), Martin Rees (2003) with a cosmological perspective, Donella Meadows et al. (2004), and Jared Diamond (2005), among others.

Some polygamy continues, illustrative of extreme cultural differences. Where men have multiple wives, as allowed in some Islamic countries,* some men who would otherwise marry confront a shortage of possible wives. A fundamentalist offshoot of the Mormon faith in the United States exists in the state of Arizona, where a local leader has had as many as 80 wives in violation of U.S. laws. According to Scott Anderson (2010), most

---

* Under the condition, I am told, that the husband can provide reasonably for the material needs of all wives.

accepting women who are nurtured in this culture usually have little knowledge of the outside world, and where physical abuse is absent, they continue to be accepting. Other important cultural phenomena include forms of population control that are anathema in advanced societies but remain elsewhere. One situation of this kind in Africa has been documented by Neil Shea (2010). Another is the well-sustained culture on the Pacific island of Tipokia documented by Diamond (2005).

The historical development of attitudes on contraceptive birth control and abortion in the Western world is embodied in the 123 commentaries and articles assembled by Hardin (1964). Further development of attitudes culminated in the *Roe v. Wade* decision of 1973 by the U.S. Supreme Court, which declared a Texas law unconstitutional and caused suspension of many other state laws as well. The Supreme Court decision rests primarily on a right to privacy and also distinguishes permissible actions on bases of the viability of the fetus and protects the life of the mother. Regrettably, it has not quelled controversy in the United States.

Wide cultural differences exist globally and flow from the mists of past time. In the presence of such differences, advances in technology, especially in agriculture and medicine, have lowered the death rate and generally started to enable rapid physical growth in Third World countries. An important part of the ongoing discussion relates to negative resource impacts from newly developing nations, several of which have very large populations and rapid population growth, owing in large part to relatively large proportions of young people. For example, vivid illustrations in Paul and Anne Ehrlichs' *The Dominant Animal* (2008, p. 151) show that 45% of Nigeria's population is under 15 years of age, while only 20% of the U.S. population is. The developed world, with smaller populations and smaller projected growth, has consumed much more natural resources and continues to consume disproportionately more. Thus, there are several substantial forces in opposing directions.

Aspects of traditional religion that promote population growth are probably rooted in challenges to population maintenance in ancient times and are reflected in the instruction to Adam and Eve.[*] They find expression today, for example, by unqualified opposition of some religious bodies and nations to abortion[†] and even to contraception. Less publicized are urgings for restraint of population growth and calls even for

---

[*] See also White (1967) and extensive subsequent discussion.

[†] In the United States, fundamentalist Christians are especially outspoken in opposition to abortion, although Jesus is not reported as speaking specifically to this. Diamond (2005, p. 512) writes: "My best friends in the Third World, with families of 4 to 8 children, lament that they have heard of the benign forms of contraception widespread in the First World, and they want those measures desperately for themselves, but they can't afford or obtain them, due in part to the refusal of the U.S. government to fund family planning in its foreign aid programs." Perhaps this U.S. policy will change.

population decline from thoughtful people, including numerous authors, who observe depletion of resources in the land, water, and the biosphere. In short, there is grave loss of biodiversity and a sense of foreboding among a substantial minority of Western populations. Increasingly feared are intensified struggles among humans for scraps. Perhaps an ultimate negative viewpoint has been presented by Cormac McCarthy in *The Road* (2006), subsequently a movie, and with an important review by Charles Stang (2011).* These same matters are discussed in depth with presentation of some alternative perspectives in several articles in the May/June 2010 issue of *Mother Jones* (Whitty, 2010).

A powerful business community, represented by Chambers of Commerce in cities and towns throughout the United States, continues to promote physical growth. With a few important exceptions, all cities and towns in America vie for new industry and new population—always of the clean, smart, good-jobs types, of course, and always to "increase the tax base," "reduce unemployment," and/or accommodate expected population increases. Nevertheless, after much development and growth in the United States for more than 200 years, significant unemployment continues, and cities and towns struggle to provide and maintain the infrastructure necessary for provision of modern services to their citizenry. Growth of population in the United States and other Western countries is slower now, a consequence of the so-called Demographic Transition, which accompanies a condition wherein children become a lesser asset and women are better educated and more politically powerful.

Promotion of economic growth with increase of human numbers has been somewhat facilitated by increased use of some expressions in spoken and written language. Thus, *smart growth, sustainable growth*, and *economic growth* have become more widely used, and these terms are frequently used in connection with a portion of activities directed toward environmental conservation. However, none of these forms is truly sustainable if their central premise is physical growth, since physical growth can't be indefinitely maintained on a finite planet. Perhaps *cultural growth* is the only truly sustainable human form.

The preceding discussion does not treat some important economic factors. As this is written, the United States and most other developed countries are experiencing substantial unemployment, have assumed massive international debt, and are experiencing large deficits in internal accounts. There is concern about the probable length and ultimate impact of the present global recession, whose causes lie in part in the human condition in interaction with political systems that remain insufficiently responsive to the effects of burgeoning science and technology, as discussed above

---

* The book ends with a turn to some optimism. It is not uncommon to find some optimism in despair, that is, some good in evil and, occasionally, the converse.

and further implied by Hilbert. Thus, developing resource scarcities with increasing prices are as much effect as cause.

Aging and shrinking populations, as in Japan, pose very difficult moral and practical problems because care of an increasing elderly population rests on a diminishing number of young workers. The situation in China is also very trying. Although enforcement of a one-child-per-family policy in China has drastically reduced the birthrate, the population of 1.3 billion reported by the 2010 census is still growing while aging. The recent rate of increase is about 5.8% per decade—only half the rate of the previous decade, but still very significant. Furthermore, abortion of fetuses identified as females in utero has led to a large sexual imbalance among new births, with about 118 boys for 100 girls; this large imbalance poses an ominous outlook for marriage as this new population ages (Hvistendahl, 2011). These matters are more extensively discussed in current magazines and books, of which an article by Jonathan Shaw is a good example (Shaw, 2010). A comprehensive discussion of the highly variable human population situation in various countries is presented in a special section of *Science* (Chin et al., 2011).

5.3.2   Moral and political dilemmas

Although moral and political dilemmas are deeply embedded in matters discussed in the previous section, some of these matters are outside the concerns of other societies. Thus, in the United States, efforts of business communities toward growth are reinforced (sometimes inadvertently) by some who act from deeply held moral beliefs. For example, in February and November 2010, inserts urging organ donation appeared in newspapers across Oklahoma.* The February inserts presented the sad story of a child, born with a defective heart, who received surgery for repairs that proved temporary, but became healthy after a heart transplant.

At least two moral issues arise concerning increasing applications of such technology, given the tendency in many cases that "if technology enables, it will be applied." One consideration involves the large dollar expenditures, surely well into six figures in this case, which otherwise could be devoted to treatment of more widespread deficiencies, such as insufficient neonatal care. A related problem that I'm told is deemed more serious in the medical community (and surely elsewhere) is the proportionately very large costs involved with relatively short life extensions of ill older people. In some cases, life-threatening conditions that inspire

---

* This paragraph and section illustrates the complexity of the issue. I contributed to a costly organ donation that has enabled a woman of middle age to lead a productive life for decades. I do not take a moral position on this issue, which is little discussed in public space but is receiving attention in the medical community.

successful remediation involve genetic defects that may increase the incidence of similar defects in future generations.

An important recent case in the United States was a prolonged legal battle over Terri Schiavo, who was hospitalized after suddenly collapsing, possibly a result of self-fasting in an effort to lose weight. Previous, somewhat similar cases in the "right to die" movement in the United States involved Karen Ann Quinlan (1975–1984) and Nancy Cruzan (1983–1990). Eventually, many entities became involved in the Schiavo matter over 15 years (1990–2005): family, the Catholic Church, political parties, the public, hospitals, hospices, state and federal courts, and even the Congress and President George W. Bush. Republicans and other right-wing factions in the United States strongly opposed the denial of life support, but Terri Schiavo technically died on March 31, 2005, 13 days after her feeding tube was removed on the order of a local court. An autopsy showed that she had been in a vegetative state. Moral issues connected with such matters are incompletely resolved today, but will become more urgent as human population increases and resources dwindle.

Suicide by elderly and seriously infirm persons in great pain does not impact global population in any important way, but such issues are related to dealings with individual humans or human populations. Suicide is not illegal, but *assisted* suicide is seriously controversial and has been legalized in only three U.S. states, although it has become legal in several European countries and in Japan. Strict safeguards against criminal behavior are universally applicable.

A controversial advocate of assisted suicide in the United States was Dr. Jack Kevorkian, a pathologist, painter, and writer, who assisted more than a hundred persons to die. Kevorkian was convicted of manslaughter and was in prison for eight years for having delivered means of death to Thomas Youk, age 52. This was different from previous cases wherein Kevorkian provided the "setup," but its manipulation was by the clients. Youk was in the terminal stages of Lou Gehrig's disease and was physically unable to deliver the means of death to himself. Dr. Kevorkian died on June 3, 2011. Was he deviant?

We note that there is a strong ethic in both general populations and medical communities toward saving lives and functions. We also note situations seen by many as contradictions. These are manifested, for example, in warlike tendencies. Subcultures involved with drugs seem to have no hesitation in embracing mass murder, religious extremists kill many outside of their own culture and justify the killing of some who are nominally within their own culture but with whom they disagree, and "suicide bombers" are seen in Western societies as showing lack of respect for human life generally.

Although the global situation today remains remarkably different from the "Reverence for Life" advocacy of Albert Schweitzer, who died at 90 years of age in 1965 at the hospital he founded in Africa, there are intense activities devoted to saving and improving lives. A major program promoted as lifesaving is the U.S. AIDS program in Africa. This program involves annual U.S. appropriations of several billion dollars and brings nearly normal lives to millions who would otherwise die. The program in Africa is backed up by intensive research in the United States and elsewhere.

Another program is the private sector marketing of snus, a tobacco substitute supposed to help people avoid cigarettes or break addiction to tobacco. It is an alternative addiction, but is credited with considerable reductions of often-fatal lung cancer. Still another is represented by Doctors without Borders, which provides free surgical and pharmaceutical assistance toward remediation of congenital birth defects and cultural abuses around the world. A further example largely from the private sector is the wide charitable distribution of the drug Mectizan by its manufacturer, Merck (Rea et al., 2010). Mectizan is a derivative of Ivermectin, widely used in U.S. cattle operations, and is very effective against river blindness, a disease that has afflicted about 40 million worldwide and blinded about 300,000 in Africa. The Carter Center, headquartered in Alabama, has recently reported striking progress toward elimination of guinea worm infections in Africa.

Regrettably, none of these welcome efforts has much impact on global population numbers, but they reflect attitudes that are central toward shaping fluctuations in global population. Human population must be controlled, but controls should be moral, and the enormous disparities in related outlooks within Western cultures are both symptoms and causes of anomie and deviance.

In the United States, in opposition to the "religious right" are such organizations as Call to Action, Catholics for Free Choice, the Guttmacher Institute, the National Abortion Rights Action League, and Planned Parenthood. Surveys have shown that a large majority of Catholics in the United States use contraceptive methods that are opposed by Church doctrine as announced from the Vatican.

Exposure and discussion of sex scandals involving priests in the Catholic Church prompted an essay in the *New York Times* by Maureen Dowd (2010), a Catholic. Dowd's column calls for modification of Church doctrine with regard to celibacy, acceptance of marriage for priests, ordination of women as priests, and greater presence and therefore influence of women in the Church. Of course, greater influence of women generally is well recognized as a desideratum in the global conferences discussed in the following section and elsewhere, even as it is stoutly resisted in some places and cultures.

Kissling (2010) has discussed Catholic doctrine on contraception with emphasis on the most recent 140 years, and she demonstrates that there are political components such as a felt need for consistency to the Church's pronouncements. At this writing, basic Church doctrine as announced remains immutable.

Diamond (2005) discusses the contrasting collapse and long-term maintenance of different societies. Ultimate outcomes are invariably related to treatment of the environment, related in turn to qualities of governing individuals, their behavior, and the nature of political and social systems. Sometimes, modern conditions bear only a remote resemblance to their portrayal by the press. For example, one of Diamond's detailed analyses shows that the slaughter of up to a million Rwandans in 1994, about 11% of the then total population, was rooted as much in overpopulation with resultant food shortages as in the well-reported ethnic rivalries.

Diamond, among many others, presents an extensive analysis of Norse settlements during 985–1500 in Greenland, with widespread agreement that change toward a cooler climate played an important role in loss of the colonies. Other factors included inability or lack of political will* to adopt Eskimo (Inuit) cultural adaptations to a spare land and conflicts with the Inuit.

5.3.3   International conferences, countercultures, and communications

Issues related to population and women's health were treated at the International Conference on Population and Development in Cairo during September 5–13, 1994. Subsequently, conclusions were discussed and endorsed at a plenary session of the United Nations November 17–18 of the same year. A principal theme was maintenance of economic growth and sustainable development, and it was recognized that stabilization of global population is a strong desideratum, though not attainable for some decades. Although the populations of many countries are moving toward both lower death and birthrates, populations of the least developed countries are growing fastest, and birthrates are highest among the most impoverished.

The 1994 Cairo Conference recognized that the treatment, status, and education of women are tightly related to birthrates, which are substantially lower for educated women than for the uneducated and impoverished. This and womens' rights and related subjects were extensively discussed a year later, at the Fourth World Conference on Women, held in Beijing, September 4–15, 1995. The 218-page report of the conference is very carefully worded, but dozens of countries and the Holy See provided extensive documentation of qualifications and exceptions. During February 28–March 11, 2005, in a follow-up to the Beijing Conference of 1995, the 49th Session of the Commission on the Status of Women of the United Nations reaffirmed the Beijing Declaration and Platform for

---

* How are these different?

Action, welcomed the progress made, and called for continuing and renewed efforts toward gender equality.

Citizens act individually and in groups. Citizen activism is represented in part by the sustainability movement. Citizens advocate for measures at various points along the political spectrum; they may urge the government to adopt measures consonant with recommendations of the scientific community or may push for the opposite, depending on intricacies of human behavior and societal order.

A growing gentle counterculture is represented by the Transition movement, which originated with Rob Hopkins in the United Kingdom. (See the web site, http://www.chelseagreen.com/content/the-transition-movement-%e2%80%93-preparing-for-a-world-after-peak-oil/.) It is a response to both excessive consumption and climate change. According to the web site, Hopkins describes a Transition initiative as "a process which acts as a catalyst within the community to get people to explore themselves, [to respond] to peak oil and climate change," helping community members to "develop a really attractive, enticing vision of how the town could be beyond its current dependence on oil and fossil fuels." The Transition movement is represented in several tens of United States cities, and in Oklahoma City by a sustainability network (http://www.oksustainability.org). The movement in Oklahoma is locally very active in several cities, although it does not have firm support or participation by the "establishment" at this writing.

Behavior opposite to aims of the Transition movement was presented at a meeting of the Edmond (Oklahoma) City Council on December 6, 2010. The meeting had been called to discuss sustainability and how that might be developed in Edmond, but the meeting was disrupted and eventually canceled after a raucous group denounced sustainability as a plot of the United Nations, which, they claimed, seeks to take over government in the United States. This idea, a manifestation of ultranationalism, also called jingoism, subsequently spread to Norman, Oklahoma, and is widely manifested elsewhere in the United States.

In the midst of enormous complexity and variety, there is growing public realization of dark implications of human population increase. As previously noted, nearly 200 years after Malthus, there were, among others, Hardin (1964, 1968, p. 1248) and Ehrlich (1968). The former concludes:

> The only way we can preserve and nurture other and more precious freedoms is by relinquishing the freedom to breed, and that very soon. "Freedom is the recognition of necessity"—and it is the role of education to reveal to all the necessity of abandoning the freedom to breed. Only so, can we put an end to this aspect of the tragedy of the commons.

Another illustration of rising awareness is the opinion piece "The Earth is Full" by Thomas Friedman (2011).

While there was no significant societal response to Ehrlich and Hardin in 1968 (indeed, global and U.S. populations have increased about 50% since then), there is now increased documentation concerning related issues. For example, Carter Dillard (2007) has offered a comprehensive and exhaustive legal analysis of procreative rights, concluding, "The right to procreate, correctly defined, is a right at least to replace oneself, and at most to procreate up to a point that optimizes the public good." And the Optimum Population Trust has numerous free downloads of authoritative papers at its website.* Population and its increase are also dealt with directly in books by Catton, Diamond, the Ehrlichs, Gore, Meadows, and Rees.

Actual means to address population numbers with preservation of human rights has been addressed only marginally in popular journals. Is population reduction feasible outside of government enforcement, as in China? In the presence of strong earthquakes, fearsome tsunamis, large fires, repressive governments, and other negatives, global population is increasing at this writing by about 80 million persons—1.2%—annually, and if sustained, such a growth rate would produce a doubling of global human population in less than 60 years. Population numbers are stationary when the average birth number is near 2.1 per female inhabitant, and averages vary widely, with some countries losing population and most gaining. Large birthrates among immigrants have actually reversed population loss in Europe and are a source of some civil unrest there.[†] Large families, still promoted in some cultures, become symbolic of a conflict for demographic advantage—note, for example, the large families characteristic of both Orthodox Jews and many opposed Palestinians.

If the disparities were much smaller, then reduction to 2.0 of the average number of children per female would have little impact on individuals, but the eventual impact on world population would be major. For example, if the global population were reduced just 1% annually, the number of people would be halved within 70 years. But population growth rates remain very refractory to direct action, with China an exception, and Hardin has noted that human populations present problems without technical solutions. It is realized by the UN conferences and elsewhere that population growth will be reduced humanely only through effectively improving human rights, especially women's rights, providing increased access to contraceptives and education about them, reducing poverty and economic disparities, and improving health.

---

* http://populationmatters.org.
[†] Immigration is also a contentious issue in the United States and is reversing population decline in some areas.

Thus a path toward some alleviation of excessive human population has been suggested, but travel along that path has made tortuous and uncertain around the world for a variety of reasons, among them societal resistance itself, included in the comprehensive discussion by Hilbert (1976) and presented in brief in Section 5.3.1. I believe that transition, including population control, will come in a mixture of force from natural causes and ready acceptance, but the timetable is uncertain.

5.4   Climate change

5.4.1   Overview

The issues with global climate change are very numerous, and the science is complex, in part because Earth's climate is both quite variable and subject to many influences. Climate change issues are deeply intertwined with issues of resource depletion, and recommendations drawn from both sets of issues are often strikingly similar.

The scientific community well understands most of the climate influences, although a few of importance are not fully understood. Some influences, such as solar variability, are not accurately predicted in detail and are fully beyond human control. Earth's climate has large societal impacts on a global scale, and we should foreshadow and act on projected changes that would be significant to human beings, including effects on the biodiversity of flora and fauna. Basically, the present situation of most concern involves large emissions of the greenhouse gas carbon dioxide, which has greatly increased due to industrial development during the past hundred years or so. The carbon dioxide content of Earth's atmosphere has increased from its preindustrial value of about 280 parts per million by volume (ppmv) to about 390 ppmv in 2010. A doubling of the preindustrial value has been calculated to produce a global temperature anomaly of about 1°C, or larger with positive feedbacks. These matters are discussed in some detail starting with Section 5.4.6 below.

Regrettably, there is weakness in our ability to respond effectively even in the presence of considerable understanding in the scientific community. Such weakness lies in the human condition and in political conditions both in the United States and around the world, partially described above and especially in Section 5.2. These conditions give rise to some intense controversy and sometimes weakness in rational support to science, even in the presence of widespread use of the fruits of science. Historically, the United States has been among leaders in science and in science education, but has recently fallen behind, as indicated by numerous surveys.

Most of this section can be further studied in numerous reports of the Intergovernmental Panel on Climate Change (IPCC), which was established in 1988 by the United Nations Environment Program and the World Meteorological Organization (WMO) to provide a scientific view of climate change and its potential consequences.[*] Thousands of scientists from around the world contribute, and it is open to all member countries of the UN and WMO. Conclusions of the IPCC are presented in reports of substantial length, but there are sections such as executive summaries and policy implications that can be read in an hour or so.[†] Major IPCC concerns are reiterated and supported by others, for example, in a statement from the Geological Society of London (Summerhayes, 2011). Brian Lovell, president of the society declared, "Climate change is a defining issue of our time" (Geological Society, 2010, p. 10). The climate situation in 2010 is presented very comprehensively in Blunden et al. (2010).

While this essay presents summaries of most of the major points developed by the IPCC, the reader should refer to the IPCC website, especially because related references are not included directly with this essay. The reader might also pursue most of the same subjects in a small book by Goodstein (2004) and more comprehensively by Blunden et al., cited above.

5.4.2   Historical record

Although Earth has been warmer than today through most of its existence, there have been ice ages. For example, in North America an ice sheet advanced beginning about 110,000 years ago, eventually covered practically all of Canada, and extended south to Long Island in New York. It receded to produce conditions somewhat like the present only 12,000 years ago. More recently, the so-called Little Ice Age lasted from approximately the start of the 14th century until late in the 19th century, including the Maunder Minimum (see Section 5.4.3).

Analysis of Earth's climate in relation to greenhouse qualities of Earth's atmosphere was anticipated as early as 1827 by Joseph Fourier, and the effect of relative opacity of carbon dioxide to some infrared radiation was elaborated by the Swedish chemist Svante Arrhenius in 1896. The greenhouse theory was first set on firm bases by Guy Stewart Callendar in the late 1930s and by Gilbert N. Plass in 1956 (Fleming, 2010). Earth's temperature is about 33°C (60°F) warmer than it would be in the absence of the several greenhouse gases.

Earth's average global temperature has increased about ¾°C (about 1.3°F) since the start of the 20th century, and most of the change has occurred during the past 25 years. A large majority of glaciers worldwide

---

[*] Numerous IPCC reports are archived at http://www.ipcc.ch/.
[†] While this chapter presents summaries of most of the major points developed by the IPCC, the reader should refer to the IPCC website, http://www.ipcc.ch/, because related references are not included directly with this chapter. The reader might also pursue most of the same subjects in a small book by Goodstein (2004).

are receding very noticeably, and sea level is rising about 2 mm annually. These trends are expected to continue, with further temperature increase during the 21st century of between 1°C and 6°C, especially in the absence of strong controls on emissions of carbon dioxide. The large uncertainty in model predictions relates to variations in model sensitivities and varied projections of future carbon dioxide emissions.

5.4.3   Astronomical influences on Earth's climate

About 100 years ago, a Serbian scientist, Milutin Milankovitch, computed the effects of the major planets Jupiter and Saturn on Earth and its orbit. He found variations of Earth's tilt on its axis at periods of 100,000 years and 40,000 years. The tilt variations amount to about 3°, which is enough to produce significant variations of differences between winter and summer. In other words, there are periods when winters are colder and summers hotter, and other periods when variations between seasons are reduced. Hotter summers and colder winters tend to produce reduced glaciation because there is less snow in winter in the colder and drier winter air and more melting in summer. Milankovitch's analysis is confirmed in a report by a former director of the Meteorological Office in Great Britain (Mason, 2010).

Another astronomical effect on climate is produced by orbital structure. Earth's orbit is an ellipse, not a perfect circle; Earth's distance from the sun varies between 94.6 million and 91.4 million miles. We are currently in a phase where Earth is nearer the sun during the Northern Hemisphere winter and further during the Northern summer. This means that seasonal variation in Earth's Northern Hemisphere is diminished in comparison with the times when our planet is further from the sun during the Northern winter and closer in summer. When the difference between seasons is least in the Northern Hemisphere, of course, the seasonal difference tends to be greatest in the Southern Hemisphere. This variation, a consequence of the so-called precession of the equinoxes, has a period of about 26,000 years and tends to influence glaciation as described in the previous paragraph. Earth's seasons and their variations are produced by the tilt of Earth on its axis from the plane of Earth's orbit around the sun (now about 23½°). The gravitational tug of the sun and moon on Earth's small equatorial bulge causes Earth to wobble like a child's top, though much more slowly.

Solar variation can also be important. About three hundred years ago, there was a cool period on Earth, known as the Maunder Minimum, with relatively cold weather recorded in Europe from about 1650 to 1710. This, coming during an otherwise cool period known as the Little Ice Age, was associated with an extended minimum of sunspots, which appear on the sun as disturbed magnetic areas that vary in number with a period of about 11 years. During the Maunder Minimum, very few sunspots were observed for about 65 years.

At this writing in Spring 2010, there is a surprising sunspot minimum, with the least sunspots since 1913. Will this be followed by the dearth last observed during the Maunder Minimum? We can't know, surely, but according to scientists at the National Center for Atmospheric Research (Henson, 2010) and the National Solar Observatory (2010), there are substantial indications that this decrease is temporary.

Hundreds of millions of years ago, Earth's atmosphere contained much more carbon dioxide than it does at present, and coal was being formed. However, the greater carbon dioxide was compensated by a fainter sun, and temperatures were not greatly unlike the past million years. The sun is predicted to continue to become larger and more radiant over forthcoming billions of years.

5.4.4   Feedbacks

Increases and decreases of glaciers are complicated by feedback. For example, if there is less ice, then there is less reflection of sunlight and more absorption by Earth of solar heat. So, once there is a trend toward reduction of glaciers and sea ice, this tends to be reinforced by decreased albedo. The opposite is true, as well, and a shift toward increased ice is self-reinforcing also. A strong push from some source or another helps to start ice to increase or decrease.

Carbon dioxide and methane are greenhouse gases, and they tend to increase air temperatures near Earth's surface, with an implied increase of the most influential greenhouse gas, water vapor, because the amount of water vapor that can be held in air increases with temperature (about 3% per °F). This is another example of positive feedback, and the associated increase of water vapor underlies an expectation that heavy rains and floods will increase. There are data indicating that such phenomena have already begun.

5.4.5   Some other natural influences on Earth's climate

If there is a volcanic period, the additional particles ejected by volcanoes are, through blockage of solar radiation, a force toward cooling. In recent times, there have been well-verified observations of cooling following volcanic eruptions in Mexico and the Philippines and of Mount St. Helens in the United States. These cooling effects have been observed to persist for a few years. However, another effect of volcanism is introduction of large amounts of carbon dioxide into the atmosphere. Increased carbon dioxide produces a force toward warming at Earth's surface, as discussed further in Sections 5.4.6 and 5.4.7 below.

Still another factor in climate variations is ocean currents such as the Gulf Stream. The Gulf Stream is driven mainly by large-scale wind currents interacting with the distribution of land and sea. It is warm water flowing rapidly northeastward along the east coast of the United States and thence toward Europe and contributes to a climate in Europe that is warm for its latitude. The British Isles, for example, enjoy a milder climate than much of Canada at same latitudes. But ocean currents are more than just the Gulf Stream and the analogous Kuroshio along the east coast of Japan. Thus, ocean waters eastward of eastern Greenland and eastern Canada sink as a result of the strong cooling that occurs with icy blasts from continental areas in winter. Sinking waters in these areas move at deep levels even to far places in the Southern Hemisphere. With a few exceptions, ocean currents by themselves are not a cause of global warming and cooling, but rather are affected by global warming or cooling.

An important exceptional variation is the El Niño phenomenon, part of El Niño/Southern Oscillation. The Southern Oscillation was discovered by Sir Gilbert Walker in the 1920s. It involves changes in the distribution of atmospheric pressure and associated winds, primarily in Australia and Indonesia. The wind changes produce El Niño, an occasional warming of equatorial Pacific waters that lasts a year or two. Sometimes warm water appears off the coasts of Peru and northern Chile, where ocean waters are normally cooler than the latitudinal average. When these waters are warm, the adjacent land areas, usually among the world's driest deserts, receive rare heavy rains. Weather globally tends to be warm during a strong El Niño, a probable principal cause of record global warmth during 1998.[*]

The corresponding La Niña phenomenon presents relatively cold water in the tropical Pacific Ocean. During La Niña years, temperatures in the southeastern United States tend to be warmer than normal, and temperatures in the Northwest tend to be cooler than normal.

Other pressure variations of special importance to seasonal climates in the United States and Europe are the Arctic Oscillation and its closely related North Atlantic Oscillation. These mainly involve pressure differences between high and low latitudes over the northeastern Atlantic Ocean. Thus, when the Icelandic Low is well established, there are strong westerly winds over the North Atlantic Ocean. However, when the Icelandic Low is weak, winters tend to be cool in the United States and Europe, as actually occurred during 2009/2010, with relative warmth in most of Canada and the Arctic.

5.4.6   Greenhouse gases

Earth's atmosphere contains several components that are called *greenhouse gases* (GHGs). The most important (i.e., the most opaque to infrared radiation) is water vapor. Next in importance is carbon dioxide. Methane

---

[*] Record global warmth was duplicated in 2005 and 2010.

(natural gas), ozone, and nitrous oxide are also significant. The GHGs are transparent to sunlight, but they tend to absorb crucial parts of Earth's outgoing infrared radiation. As noted in Section 5.4.2, this makes Earth at its surface about 33°C warmer than it would be otherwise. The GHGs help make Earth habitable by human beings and other creatures. We should be grateful for the presence of GHGs, since Earth would be substantially ice-covered without them.

Consider effects of the sun's radiation on temperatures here on Earth. The temperature of the solar surface* is about 6,000°C (11,000°F). As solar radiation arrives here, the much-weakened sunlight passes through our atmosphere and heats the Earth's surface. Because Earth is much cooler than the sun, Earth's outgoing radiant heat is at infrared wavelengths, invisible to human eyes. When there are more GHGs, there is more absorption of the outgoing infrared radiation, so Earth's surface becomes warmer. Then there is more heat transfer by convection (overturning) from Earth's surface.

A prediction of this theory is that warmer surface layers are accompanied by cooler upper layers. This is because Earth reradiates all of the heat that it receives from the sun. With more carbon dioxide in the atmosphere, Earth can radiate just as much at infrared wavelengths at lower temperatures as it previously radiated with less carbon dioxide at higher temperatures. And at heights great enough to be above most blocking of radiation-absorbing gases at still greater heights, temperatures are lower for effective loss of heat. And substantially lower temperatures in Earth's high atmosphere (stratosphere) are indeed being observed, even while there are higher temperatures at Earth's surface.

This is a rather simple explanation of greenhouse warming, based essentially on the teachings of 50 years ago. Study of atmospheric circulations nowadays reveals a somewhat more complicated picture—complicated enough to stimulate controversy. For example, Richard Lindzen (2009) emphasizes that shifts of climate comparable to the present have recurred on numerous occasions (Davies, 2010).[†]

---

* Since the sun is gaseous throughout, it does not have a solid surface. The temperature of 6,000°C applies to the photosphere, which is just above the opaque layer that carries heat by convection from the solar interior.

† Richard Lindzen is Alfred P. Sloan Professor of Meteorology at the Massachusetts Institute of Technology and is well known and appreciated for his technical contributions, especially to radiation theory. His recent views on global warming are summarized and explained via presentations delivered at the University of Massachusetts–Lowell on October 13, 2010. Among other details, Lindzen believes that effects of climate change, both now and in the future, are exaggerated. I agree in part with Lindzen. For example, on page 7 of the fall 2010 issue of *Defenders*, a publication of Defenders of Wildlife, support is given to recent studies that concluded in part that a 12% decline of certain Mexican lizards is a consequence of global warming. With global temperature rise of just ¾°C?

Major climate variations over time are associated with variations of carbon dioxide in Earth's atmosphere, which is not surprising, since carbon dioxide is a GHG, and the solubility of carbon dioxide in ocean waters declines with increasing water temperature. The historic and prehistoric records of both carbon dioxide and climate can be tracked through investigations of the constituents of ice cores, ocean sediments, and fossils. Ice cores are taken from glaciers in Greenland, Antarctica, and at high altitudes in the tropics.

The ice cores show a layered annual structure, and in some areas, nearly a million years of ice accumulation can be revealed in the cores. When snow falls, some air is trapped in the snow, and this air can be analyzed for its content of carbon dioxide. There are also indications of volcanic eruptions, revealed by dust in the ice. The ratios of two isotopes of oxygen of atomic weight 16 and 18 in sediments, fossils, air, and water are also indicators of global temperatures at the time the materials analyzed were formed.

Analyses show periods of higher and lower temperature that are nearly contemporaneous with increases and decreases of carbon dioxide in the atmosphere. Usually, temperature increases before carbon dioxide increases, and the subsequent carbon dioxide increase is attributed to outgassing from warmer oceans in which carbon dioxide solubility diminishes with increasing temperature. Occasionally, as in the present case, carbon dioxide increases first.

It is also important to trace the history of GHGs over tens and hundreds of millions of years, and Jeffrey Kiehl (2011) has presented a summary on this subject. One hundred million years ago, the content of carbon dioxide in Earth's atmosphere was about 1,000 ppmv, and, with a slightly weaker sun, Earth was substantially ice free, with an average temperature on Earth about 10°C warmer than today. Tens of millions of years passed before atmospheric carbon dioxide was reduced to 280 ppmv, yet a scenario of business as usual has the carbon dioxide content rising from its present value near 390 ppmv to 1,000 ppmv in only 100 years! Very regrettably, even if societies become more responsive to scientific implications and reduce carbon emissions dramatically and very soon, the short time span that is nevertheless implied for further carbon dioxide increase and further temperature rise carries distressing implications for diversity of life on Earth. A science of biodiversity conservation continues under development (Dawson et al., 2011).

5.4.7   Recent history of greenhouse gases and effect on oceans

Various analyses show that the carbon dioxide content of Earth's atmosphere, before the start of the industrial revolution about 200 years ago, never exceeded about 300 ppmv during the past three quarters of a million years. Now, it is above 390 ppmv, as shown in Figure 5.1a. Why is that?

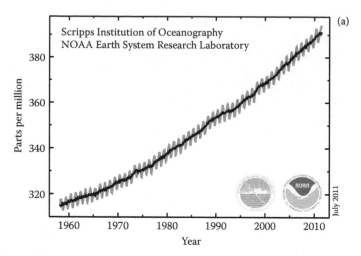

Figure 5.1a Historical record of cumulative carbon dioxide measurements at the Mauna Loa Observatory on the island of Hawaii. Notice that the curve on the long-term graph is concave upward from the start of precision measurements in 1958 through March 2011. Concavity reflects the increasing rate of global industrialization and carbon dioxide emissions.

And does it matter? And, if it matters and needs to be corrected, what can we do about it?

The current high level of carbon dioxide in Earth's atmosphere is principally a result of human burning of fossil fuels: coal, oil, and natural gas. Coal is typically about 70% carbon. Consider a 110-car coal train in the United States. Each car carries more than 100 tons of coal, and all 12,000

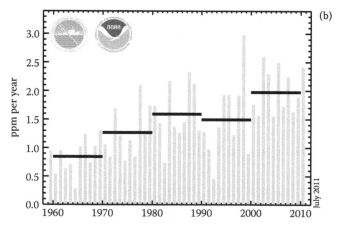

Figure 5.1b Historical record of carbon dioxide measurements at the Mauna Loa Observatory on the island of Hawaii. This graph records the increase in ppm of carbon dioxide for each year from 1958 through March of 2011.

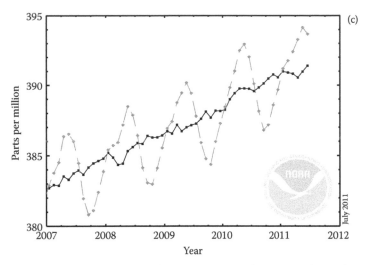

Figure 5.1c Record of carbon dioxide measurements from 2007 through mid-year 2011. The annual fluctuations are a consequence of seasonal variations of plant growth, principally in the Northern Hemisphere. Carbon dioxide is maximal in Earth's atmosphere in May of each year.

tons carried by one train are burned in a typical 1,000-megawatt power plant in a single day. Carbon dioxide production from such a power plant is thus about 30,000 tons per day.

There are also net releases of carbon dioxide as forests are converted to pastures, palm oil plantations, and other uses. Emissions are from both the carbon sequestered in trees and from newly uncovered soil. The IPCC found that such emissions amount to about 17% of the global total. In response, the REDD program (Reducing Emissions from Deforestation and Forest Degradation in developing countries) has been established (United Nations, 2008).

According to the International Energy Agency, carbon dioxide emissions globally amount to about 30 billion metric tons annually, and, rhetoric notwithstanding, the rate of carbon dioxide emissions is still increasing. About half of present emissions dissolve in surface waters of the oceans and the other half stays in the atmosphere. The lifetime of carbon dioxide in the atmosphere is uncertain and is variously estimated from a few tens of years to several centuries. The atmospheric content of carbon dioxide is increasing by 1.5 to 2 ppmv per year, about three times faster than the rate of increase recorded when measurements were started by Charles Keeling on Mauna Loa, Hawaii, in 1958, just 50 years ago.[*] This acceleration of the

---

[*] The carbon dioxide annual global maximum occurs in May (see Figure 5.1c). The 2010 maximum was nearly 3 ppmv greater than that of May 2009.

rate of increase, shown in Figure 5.1b, is believed to lie in the astonishing rate of industrialization around the world with enhanced emissions of carbon dioxide.

As noted above, carbon dioxide is dissolved in Earth's waters, and about a quarter of all this GHG produced by human activities over the past 200 years remains so dissolved. Effects of this dissolution on ocean chemistry were described in the 1970s, and there were important related studies of effects on ocean life. However, there was little public attention to this matter until a report was issued by the Royal Society in Britain in 2005, amplified four years later by the Interacademy Panel on International Issues (2009).

The rate of solution of carbon dioxide and its amount in Earth's waters is responsive to its concentration in Earth's atmosphere. The pH of ocean surface waters has already diminished by about 0.1 from preindustrial values, with present reduction of carbonate ion concentration to levels below any during the past 800,000 years.

The rate of present increase of carbon dioxide in Earth's atmosphere and waters may well exceed the rate at which some organisms in the ocean can adapt, and unless carbon dioxide in the atmosphere diminishes, it is predicted that ocean life will be significantly affected. Details are not predictable at present, but oceanographers believe that all of ocean's calcifiers and phytoplankton could be affected, resulting in alterations of food chains and whole ecosystems and significant consequences. Such changes would be irreversible for hundreds of thousands of years.

Research on these issues are discussed (with many references) in a special 2009 issue of *Oceanography* (see especially Doney et al., 2009; Feely et al., 2009). These reports include graphs that show the irregular decline of pH* since direct measurements began about 1990 as well as an associated increase of the carbon dioxide content of Pacific Ocean waters, especially in the surface layers. Future effects are expected to be more pronounced and of earlier onset in Arctic and Antarctic waters, because the solubility of carbon dioxide in water is larger at lower temperature. These subjects are further discussed in a special 2010 section of *Science* (see Smith et al., 2010).

The final sentence in the article by Feely et al. (2009, p. 46) is, "If anthropogenic CO_2 emissions are not dramatically reduced in coming decades, there is the potential for direct and profound impacts on our living marine ecosystems." These consequences are being further investigated.

The heat content of the oceans is vast, but ocean temperatures respond over decades to climate change, and the expansion of water with rising

---

* The present pH of the oceans is about 8.1, slightly alkaline, and has declined slightly more than 0.1 in the past century. The decline seems small at first glance, but entails an increase of 30% in hydrogen ion concentration.

temperature is an important contribution to rising sea levels globally. Of course, the temperature of surface waters changes more rapidly than that of water at depth, and mixing of the waters is a process that occurs over thousands of years. Average ocean temperature has increased about 1°F over the past hundred years.

Methane in Earth's atmosphere has increased greatly since preindustrial times, perhaps as a partial result of increased usage of natural gas, with pipeline leaks and incomplete combustion. Although methane on a molecule-to-molecule basis is more than 20 times more potent as a GHG than carbon dioxide, its present concentration of about 1.8 ppmv in Earth's atmosphere produces only about 10% of the effect of carbon dioxide. Methane increase in Earth's atmosphere has resumed at an annual rate of about 0.03 ppmv since 2007, after its global average content was nearly stationary during 1999–2007. Methane has become too valuable to waste, but as usage of natural gas increases, pipelines must be carefully checked and leaks sealed.[*]

Methane is released from thawing tundra. It has also been found in ocean sediments, especially on Arctic shorelines, as methane clathrate, which is an icy combination of methane with cold water. There have been some investigations of means to mine clathrates for natural gas, but a general concern is that methane may be released to Earth's atmosphere in very large volumes if sea temperatures rise sufficiently. There are occasional reports of such releases from the Arctic Ocean, especially where that ocean borders Siberia,[†] and it seems possible that such emissions may be a cause of the recent resumption of methane increase in Earth's atmosphere.

5.4.8   What is happening to the climate now?

The temperature record shows an irregular increase since the Maunder Minimum of 300 years ago. The year 1998 remains as the warmest year of recent record, but it has been very closely matched subsequently, and as this is written in 2010, the global average of March–May temperatures is reported by the U.S. National Oceanic and Atmospheric Administration and the Goddard Institute for Space Studies to have set a new global high

---

[*] New large volumes of natural gas have been discovered in shale and are being exploited with new technologies of horizontal drilling. Increased usage of natural gas for fuel in motor vehicles, for conversion to liquid fuel, and for heating and power generation are being promoted. Newly identified natural gas reserves may be adequate for a century of increased usage. Environmental effects, especially related to water quality, are controversial.

[†] There have also been analyses of methane inputs to Earth's atmosphere from livestock operations, but I have never seen numbers about this that include a subtracted portion related to marked decline of wild animal populations.

record; The Goddard Institute reported that the temperature for the whole year of 2010 slightly exceeded 1998 and 2005. The warmth of 1998 may have been a consequence of El Niño (warm Pacific Ocean waters), which was very strong that year. So, one might conclude that the subsequent temperature record in the absence of El Niño is actually an indication of continued warming and that actual global temperatures will rise again, stimulated by continued increase of carbon dioxide. Overall, global temperatures have risen about ¾°C (1.3°F) since the start of the 20th century, with most of the change having taken place since the 1970s.

Although global average temperature has hardly increased since 1998, a large majority of glaciers are continuing to melt worldwide. Average sea level is rising about 2 mm (an eighth of an inch) per year and about double that in the Bay of Bengal, with severe consequences for the Bengalis and for some island populations. There are some local exceptions to sea level rise where glacial melt and reduction of glacial weight associates with a local rise of land surfaces. Large variations have to be taken into account.

Arctic ice is continuing to recede, on average. There has been some significant recent reduction of ice shelves on northern Ellesmere Island, the far northern reach of Canada. There was an astonishing reduction of Arctic ice during the summer of 2007, and after the minimum of 2007 there was a slight rebound of Arctic ice during 2008 and 2009. A comparison of the lateral extent of Arctic ice on December 3, 2010, with its extent on the same date in 2008 shows noticeably less ice in all the bays of the Arctic except to the north of Scandinavia and Novaya Zemlya. The distribution of ice is responsive to patterns of wind and weather, as well as to temperatures. A negative trend in ice coverage also appears in the comparison of April 2011 with April 2010.

The average thickness of winter Arctic ice is about 2.5 m. The recession of ice during 2007 was accompanied by reduction of its average thickness by about 0.5 m. In order to monitor ice thickness accurately, a satellite equipped with a type of radar for that purpose was launched successfully on April 10, 2010, and in the autumn of 2010, the thickness of Arctic ice was reported to be continuing to diminish.

The waterway between western Greenland and Canada, the Parry Channel, is the entry to the Northwest Passage, which opened briefly in 2007. If the Parry Channel were to open on an annual basis, it would become a very important shortcut for international trade between Europe and Asia.

There are indications of more severe droughts in subtropical regions and increased rainfall in middle and high latitudes of the Northern Hemisphere. Some studies are finding statistical support for an increased frequency of very heavy rainfall events with disastrous floods such as occurred during spring 2010 in Arkansas, Oklahoma, and Tennessee in the United States, in southwestern France, and in southern China. These findings are supportive of model simulations, but conclusions in this area are not robust, given the very large natural variability of precipitation. More robust is a recent study by Gerald Meehl et al. (2009) that identifies a clear trend toward an increase in the ratio of occurrences of record high temperatures to record low temperatures over several decades in the United States.

The overall data indicate that increases in average warmth are being manifested around the world and that model findings are considerably supported at present. As already noted, this is attributed to increased carbon dioxide in Earth's atmosphere. The IPCC predicts that, owing to lags related to the large heat content of the oceans, warmth and sea level will probably increase for decades even if emissions of carbon dioxide are strongly controlled. Further, the IPCC predicts that there will be intensified drought in subtropical regions, hotter summers in the U.S. Midwest and elsewhere, and serious water shortages where there is dependence now on seasonal melting of snows and glaciers.* Within a hundred years, sea level may rise several feet, to the point that coastal areas of importance near or in such places as New York City, Washington, Miami, New Orleans, coastal Bangladesh, and various ocean islands will be inundated[†] unless barriers still to be constructed are effective. Predictions also include more intense storms in and out of the tropics and great negative impacts on agriculture, with severe impacts on much of civilization as we have known it recently.

5.4.9   Naysayers and skeptics

There are activists who display ignorance about the science and implications of climate research and who dispute the dire predictions of the IPCC. Some are political figures who either don't see the larger picture or are ignoring it for reasons of politics or short-term personal gain. Some are opposed to government and oppose the IPCC statements as actions of governments rather than results of research by independent scientists.

---

* Water shortages are already occurring at many places on Earth, as described at length in the April 2010 issue of *National Geographic.* Some of these shortages relate directly to human activities, such as dam building and overpopulation, and others may relate to climate change. For example, there was a severe drought in the southeastern United States (focused on Atlanta, Georgia), during 2007–2009. This was interrupted by exceptional rains. The large natural variation in rainfall prevents our knowing accurately the extent to which one or both of such flood and drought events are natural or a consequence of climate change. Many other droughts and floods have occurred around the world recently, but these have also occurred in the past. Decades from now, hindsight should yield better interpretation of present conditions.

† Total inundation of a small island in the Bay of Bengal was reported during March 2010, removing a point of contention between Bangladesh and India. The island, first identified by satellite in 1974, was named Purbasha by India and South Talpatti Island by Bangladesh. Forecasts of sea level rise vary with different model assumptions.

Some are saying that Earth's climate has always been fluctuating (mostly true) and that the present fluctuations are unrelated to human presence (quite false). I think that it is absurd for anyone, even members of the U.S. Congress, to dismiss the entire global warming phenomenon as a hoax. Although a few of thousands of statements from the IPCC are erroneous or exaggerated, the presence of an important anthropogenic contribution is properly undeniable and has been predicted by thoughtful scientists since early in the 19th century.

Global warming is occurring and carbon dioxide emissions by industrialized human kind are a critical cause—those are facts. The nature and tactics of a few very effective deniers have been explored in depth by Oreskes and Conway (2010). Their book is properly acclaimed because it clearly illustrates destructive properties of our political condition today, largely enabled by advances in communications and by the disproportionate wealth of vested interests.

Two articles in *Physics Today* illustrate that denial of "inconvenient truths" is not new, and propose means to counteract denial (Sherwood, 2011; Somerville and Hassol, 2011). Consider that Copernicus' theory that placed the sun at the center of the solar system instead of Earth, was published on his death bed, and Galileo's findings in support of Copernicus were not fully accepted by the Catholic Church for hundreds of years!

"The lingering public uncertainty about anthropogenic climate change may be rooted in an important but largely unrecognized conflict between climate science and some long-held beliefs. In many cultures, the weather and climate have historically been viewed as too vast and too grand to be directly influenced by people. Examples of higher powers asserting control over the weather and climate are found throughout ancient and modern religious texts. In these belief systems, humans may indirectly influence the climate through communication with the divine, but they cannot directly influence the climate. Skepticism about anthropogenic climate change may therefore be reasonable when viewed through the lens of religion or the lens of history. To create a lasting public understanding of anthropogenic climate change, scientists and educators need to appreciate that the very notion that humans can directly change the climate may conflict with beliefs that underpin the culture of the audience" (Donner, 2011, p. 1298–1299).

The above notwithstanding, some naysayers are simply parroting others for reasons of base politics, though a few are reasoning and present credible if limited arguments.* We have already noted the prominent scientist, Richard Lindzen, who makes a case that indications of oncoming catastrophe are exaggerated. Nigel Lawson, not a meteorologist but a prolific author and former chancellor of the exchequer and secretary of state

---

* We may call them "deniers."

for energy in the United Kingdom, has prepared one of the more lucid critiques (Lawson, 2009) with significant arguments concerning the costs of mitigation and uncertainties related to projections of future conditions. However, both Lawson's book and Lindzen's recent editorials omit consideration of intertwined resource depletion issues that call for remedies similar to those indicated by the IPCC climate findings.

Some of the support for skeptics comes from the very volume of scientific materials. Bill McKibben (2010), a popular writer and authority on environmental issues, has offered a comparison with the O. J. Simpson jury trial for murder during 1995. That trial produced an acquittal in spite of overwhelming evidence of apparent guilt, and the reason, as explained by McKibben, lay in major part with the great mass of materials, a little of which was flawed. Acquittal was obtained through defense attorneys' emphasis on the flaws. In the same way, a few flaws exist in the several thousand pages of scientific discussion on the evidence concerning global warming. Skepticism in the public mind is amplified by repeated emphasis by some deniers on these few flaws, with important financial support from individuals and entities whose short-term financial and personal interests are thereby well served.

During a panel discussion at the University of Oklahoma in 2008, Kenneth Crawford, director of the Oklahoma Climatological Survey, noted that if the IPCC forecasts of climate change and associated recommendations are disregarded even though valid, the penalty for having disregarded them would be very severe. On the other hand, even if IPCC projections turn out to be exaggerated, the penalty for having followed the IPCC recommendations would not be dire.* This, again, is partly a consequence of entanglement of issues: The resource depletion issue and absorption of carbon dioxide by the oceans demand actions strikingly similar to those recommended by the IPCC in light of climate change.

An argument similar to Crawford's is made by Ehrlich (1968) in his closing remarks, and the same is presented emphatically and convincingly by Meadows et al. (2004). There is emphasis on the existence of a large penalty for delay. Calls such as those by Ehrlich, Crawford, Hardin, and Meadows are akin to calls for application of the Precautionary Principle, which has been embodied in legislation by the European Union.†

Notwithstanding the condition of present large emissions, there are continuing strident calls in the United States and elsewhere for further

---

* A friend mentioned that if changes are made under the assumption that IPCC predictions are correct, we would then be stuck with clean water, clean air, and independence from petroleum suppliers even if the IPCC projections are false.

† The European Commission issued a Communication on the Precautionary Principle on February 2, 2000. Application of the principle is designed to provide a high level of protection in the presence of scientific uncertainty. Its application has been extended beyond the environmental policies that first prompted it.

economic growth. This can only increase the demand for electrical energy, coal, and liquid fuels, all of which usually associate with increased emissions of carbon dioxide. If physical growth and associated growth in demand continue, the ultimate demand for fuels would increase along with associated emissions. And thousands of chambers of commerce across the United States are calling for no less than resumption of economic and physical growth just as soon as the global economic recession, which continues in early 2011, relaxes. Although there are many learned calls for constraint, for reasons of both climate change and resource depletion, too few are listening, and sufficient action by the establishment is notable for its absence (see also Section 5.3.1 and the final paragraphs of Section 5.6).

In the United States, the search for a replacement for petroleum-derived liquid fuels reflects ardent wishes to preserve the present strength of the automotive economy. The search involves investigation of alternative fuels and recovery of energy resources via activities made economically feasible by the high and irregularly rising price of traditional oil and by federal subsidies.

Resistance to change is further illustrated in Oklahoma with striking examples of efforts to continue to expand highway travel while ignoring opportunities for providing improved public transportation and freight service via rail, which is much more energy efficient and, if better utilized, would significantly reduce both the threats of global warming and the dependence of the United States on imported petroleum. For example, powerful highway interests are replacing the Oklahoma City Crosstown Highway (Interstate 40) at a cost of nearly a billion dollars for less than four miles of new road in a very wrong place. The head of a consulting firm in Dallas, Texas, has produced a report that presents many reasons for saving the Union Terminal for passenger rail in Oklahoma City (Latham, 2008).[*]

5.4.10 Possible mitigation

Efforts toward mitigation of global warming involve the scientific, citizen, and political communities and governments, all interacting and overlapping. The call by a great majority of the scientific community is for drastic reduction of carbon dioxide emissions and for strong measures toward conservation of resources. Reduced emissions requires fewer coal-burning power plants, greater efficiency in travel, and perhaps less travel, as it may be somewhat replaced by improved communications. A properly comprehensive list of needed actions would be long! A changed mind-set by a majority of societies and populations is essential, but at this writing there is little evidence of sufficient tendencies toward change.

---

[*] I have discussed the awful political history of this situation in a local newspaper, the *Norman Transcript* (Kessler, 2009).

Releases of carbon dioxide to the atmosphere would be reduced if coal for power generation were replaced by more use of natural gas,[*] nuclear,[†] or wind.[‡] Strong conservation measures and control of population might allow some coal-burning plants to be phased out or not built. Another option being explored would involve use of biofuels (see below), and still another would involve separation of carbon dioxide from the gas stream emanating from fossil fuel plants and its permanent sequestration in the ground. Carbon capture and sequestration (CCS) is receiving some emphasis. Furthermore, on December 7, 2009, the U.S. Environmental Protection Agency (EPA, 2009) declared that emission of GHGs "threatens the public health and welfare of the American people." The EPA thereby assumed authority to control such emissions, but strong controls are not presently being exercised and are controversial in the U.S. Congress.

CCS is costly and is currently practiced only rarely, where carbon dioxide under high pressure produces enhanced recovery of oil (tertiary recovery). However, in hopes of identifying a sufficiently economical process, Secretary of Energy Steven Chu indicates that $3.4 billion is being invested by the U.S. Department of Energy (DOE) in CCS research and development (Chu, 2009). Chu notes also that DOE is testing the permanence of sequestration in different geologic formations and is establishing an investigatory partnership with China.

---

[*] Coal is mostly carbon, and it includes other elements, some of which humans see as pollutants. Oxidized carbon as carbon dioxide weighs more than three times the original weight of carbon. Natural gas is methane, CH_4, whose burned products are two molecules of water vapor and one molecule of carbon dioxide. With oxidation, coal produces about twice as much carbon dioxide per unit of heat as natural gas.

[†] Although nuclear power plants emit no greenhouse gases, nuclear fission has become a much more volatile issue recently because of the devastation with major releases of radioactivity at nuclear power plants in Japan caused by a tsunami on March 11, 2011. Remaining unsolved problems include waste disposal, very high costs of construction, threats of terrorism, and release of carbon dioxide attendant to manufacture of the materials needed for nuclear power plants. An international effort is under way to build an experimental power plant at Cadarache, France, to explore production of electricity by nuclear fusion, that is, fusion of heavy hydrogen to make helium in a variant of stellar processes. This process produces very low levels of residual radioactivity. An article in the April 2010 issue of *Physics Today* discusses delays, cost overruns, and the resolution of disagreements among partners. In it, Steve Cowley, head of the British fusion program, says: "We should be trying to make fusion a commercial reality as soon as possible.... What worries me is, if you push ITER [the International Thermonuclear Experimental Reactor] back, it becomes less relevant. Fusion must deliver on a time scale that is relevant for the energy debate.... [Otherwise] we might opt for decisions in energy that we regret—like the long-term burning of coal. Things that you would not do if you had fusion" (Feder, 2010, p. 20). Depending in part on the degree of success of ITER, there might be fusion energy for thousands of years of human civilizations.

[‡] Wind power is a rapidly growing industry lately. In favorable places, its continued development may match Denmark's 20% of total electrical power supply. See its advantages and limitations in Pimentel (2008, Chapter 6).

Christian Lastoskie (2010) and Ramanathan Vaidhyanathan et al. (2010) discuss promising findings that involve sequestration of carbon dioxide in amines, a derivative of ammonia. Lastoskie notes that a global switch to renewable energy sources is slow in coming and that development of techniques for CCS remain of great importance.

In 2008, the Oklahoma legislature produced Senate Bill 1765, which called for a task force to investigate carbon sequestration. I am a member of that task force. There was interesting discussion of differing approaches during meetings on November 4, 2008, and February 25, 2010. However, no new actions toward sequestration are occurring in Oklahoma at this time, inhibited by large costs in the absence of tertiary recovery of oil and sufficient government subsidies. Overall, the future of CCS remains doubtful at this writing.

A portion of the scientific community has recently suggested geoengineering, that is, directed efforts of an engineering nature to reduce temperatures.[*] One suggestion is to place sulfate particles in the high atmosphere, where they would block some sunlight and thereby lower Earth's temperature. Another is to fertilize seas with iron, for better capture of atmospheric carbon dioxide by microorganisms already in the seas.[†] Still another is to place solar reflectors in a low-Earth orbit. The societies are agreed that geoengineering should be a strategy of last resort and that none of these strategies or others should be employed soon. However, renewed discussion of this subject has been presented by Rachel Hauser and Nicole Gordon (2010). More research is clearly required to reduce uncertainties about side effects, costs, and uneven distribution of effects—with some probably detrimental—and to raise assurance that objectives of a proposed action would be met.

Representative of government moves toward remedial actions is, first, the Kyoto Protocol, adopted in Kyoto, Japan, on December 11, 1997, which entered into force on February 16, 2005. The protocol calls for a reduction of GHGs by 5.2% from the 1990 level and provides various mechanisms for accomplishment of this. Currently, the protocol has been ratified by 187 states, but the United States is not among them because of resistance in Congress.

A conference of thousands was held in Copenhagen, Denmark, during December 5–20, 2009, in a further effort to obtain agreement of governments toward mitigation of global warming. A result was a document of 12 paragraphs. This acknowledges the seriousness of the situation, notes

---

[*] The American Meteorological Society issued a statement, "Geoengineering the Climate System," on July 20, 2009, and this was adopted with some small changes by the American Geophysical Union on December 13, 2009 (Chell, 2010). The various proposals have also been discussed in *Technology Review* (Bullis, 2010).

[†] But, what about the pH of oceanic waters? See section 5.4.7.

the requirement for deep cuts in global emissions to hold increase of global temperature below 2°C, recognizes that forest practices must be compatible with reduced emissions, recognizes market mechanisms for achievement of goals, and recognizes the need for facilitative financial commitments from developed countries. It also describes needed administrative mechanisms and pathways. Writing in the *New York Times* on December 20, 2009, Andrew Revkin and John Broder (2009) note that the final accord is a statement of intention and not a binding pledge, but although this is seen as a flaw, it is nevertheless seen as an essential step forward.[*]

A successor conference, held in Cancún, Mexico, from November 29 to December 10, 2010, involved 190 countries and was geared toward acceptance, extension, and reinforcement of the Kyoto Protocol, which expires in 2012. Disagreements between the developed and undeveloped world were clearly manifested, with the former declining to accept binding limits on emissions of carbon dioxide and the latter demanding that this be done as a precursor to effective arbitration. Achievements of the conference were modest and, according to the *New York Times* on December 11, 2010, fell well short of broad changes that the IPCC says are needed to avoid dangerous climate change. However, the Cancún Agreements, a statement of intention but not legally binding, gave participating countries another year to decide whether to extend the Kyoto Protocol.

According to the *New York Times*, the agreement establishes "a new fund to help poor countries adapt to climate change, creates new mechanisms for transfer of clean energy technology, provides compensation for the preservation of tropical forests and strengthens the emissions reductions pledges" (Broder and Rudolf, 2010, p. A12) developed during the Copenhagen meetings.

Cap-and-trade legislation in the U.S. Congress is recognized in international agreements. Such legislation has passed the House of Representatives but is stalled in the Senate at this writing. The legislation provides for defined limits on emissions by individual entities and sale of allowances in the event that emission goals are overachieved. However, there has been notice that adoption of cap and trade would penalize the middle class and lower wage earners disproportionately through increased prices for petroleum and natural gas, on which they spend a larger share of their income than the wealthy. Therefore, there is a movement to substitute the concept of "cap and dividend," in which receipts from the sale

---

[*] The thrust of the meetings at Copenhagen was somewhat blunted by controversy over numerous emails illegally hacked from files at the University of East Anglia, England, a center of IPCC research. They were subsequently quoted out of context by deniers, thereby spreading doubt about the principal conclusions of the IPCC. An investigation within Britain has exonerated IPCC authors except for a finding that there had been efforts to withhold information, and greater transparency was urged. Two other investigations have fully exonerated the scientists.

of permits would be returned on a per capita basis to the people. This has been the subject of testimony before the Congress (Hansen, 2009) and an extensive report from the Political Economy Research Institute at the University of Massachusetts (Boyce and Riddle, 2007).

One of the arguments widely voiced in the U.S. Congress concerns the perception that the economy would be damaged by actions to limit U.S. dependence on still cheap hydrocarbon fuels. This doesn't respond to the finding of scientists who maintain that delay will later cause increased suffering. In response to this condition, the Citizens Climate Lobby is under development to effectively confront Congress with ordinary citizens who see the legislature as being under the sway of established special interests, such as railroads and highways.[*] The path ahead will be difficult in any event.

5.5   Biofuels

5.5.1   A general view

As already noted, Western cultures are strongly dependent on cheap energy and have been shaped by the availability of liquid fuels at low cost for transportation and principally by coal for electric power generation. Coal is supplemented with natural gas, hydropower, enriched uranium, and, more recently, wind. Coal is plentiful,[†] but its combustion produces the GHG carbon dioxide, as well as toxic elements. The liquid fuel resource (petroleum) is becoming noticeably depleted, although, at this writing, it is fully meeting demand at fluctuating prices between \$80 and \$110 per barrel.[‡]

The impending shortage of this critical source of energy has led to a vigorous search for alternatives, and biological sources (biofuels) are

---

[*] Much more information on these subjects is available on the Internet (see Further Readings).

[†] The U.S. supply of coal, burned at the current rate, would endure for several human lifetimes. However, much coal has been mined with lax federal controls and with negative consequences for the environment, especially in Appalachia. These conditions and the attendant emission of carbon dioxide and pollutants hardly warrant its depletion with continuation of business as usual.

[‡] An article, "Peak Oil Production May Already Be Here," appeared in *Science* on March 25, 2011. The article notes an abundance of unconventional oil at high cost, including oil beneath deep water and as a viscous component of some Canadian sands. The price of conventional oil received a push during spring 2011, related to new substantial unrest in the Middle East. On the other hand, Saudi Arabia has expressed concern about anticipation of peak *demand*, because of substantial investments in maintenance and even increases of production from that country. Simmons (2005) and, as reported by Ghiselin (2010), presents balancing views. The rate of extraction of petroleum may be a million times faster than its rate of creation, but even if greatly overestimated, a large imbalance is cause for concern.

among the candidates. This subject has been comprehensively documented by the America's Energy Future Panel on Alternative Liquid Transportation Fuels (NAS et al., 2009), which presents a basically optimistic perspective on the likelihood that biofuels obtained from cellulose will provide sustainably for a substantial portion of needs via applications of technologies, both presently available and developable in the near future with stimuli from government subsidies.

On the other hand, in a different approach to the sustainability issue, Jeffrey Dukes (2003) has shown that relatively inefficient ancient processes formed the fossil fuels used now. Dukes investigated and reported on the amounts of carbon lost, preserved, and finally recovered nowadays from geological deposits that underwent physical and chemical transformations as they were converted to the fossil fuels utilized by humans. Probably more than 400 times as much carbon was involved in the coal, oil, and natural gas extracted and burned annually today as is produced annually today in biota as solar-enabled net primary production (NPP).* Dukes further estimates that replacement of Earth's carboniferous resources with biofuels would require a 50% increase of humankind's relatively efficient use of NPP, with heavy impact on the environment, including competing flora and fauna.

However, there is also use of traditional fuels during the manufacture of alternative fuels from biota, and a perspective different from the America's Energy Future Panel's (NAS et al., 2009) has also been presented by Richard Heinberg (2009). In a comprehensive survey of the energy and environmental pictures in relation to alternative fuel sources,[†] Heinberg finds that the only options for preservation of Western civilizations lie in rapid realization of the limitations of alternative sources of energy, application of extreme measures of conservation that result in drastic reduction of liquid fuels for transportation, and no less than very substantial alteration of lifestyles. This may be described as *adaptation*, and there has been important support for this recently (Lempinen, 2009, 2010).

The *Progress Report of the Interagency Climate Change Adaptation Task Force* (Sutley et al., 2010) first presents a broad description of the anticipated

---

* All life forms depend on photosynthesis by plants (a small exception being primitive organisms in high sulfide waters). The conversion of solar energy during plant growth is called *primary production* and can be measured in energy units or by the mass of dry plant matter. Gross primary production includes the energy or mass of respiration, which is subtracted to obtain NPP. The solar energy received on Earth exceeds the energy used by humankind the world over by about 10,000 times, and less than one part in a thousand of solar energy, on average, is utilized in photosynthesis.

† See also Meadows et al. (2004), Millennium Ecosystem Assessment Board (2005), Brown (2008), and Murphy (2008), for example. A previous publication edited by Pimentel (2008) presents a detailed analysis of biofuel production and implications.

effects of climate change and then focuses on the administrative and coordinative actions that are necessary between the federal government and numerous other authorities that implement particular adaptive strategies.

We must bear in mind the much lower ratio of energy returned on energy invested (EROEI) with biofuels in comparison with that traditionally associated with petroleum. Biofuels present a ratio of 10 or less, while oil presented about 100 a century ago, diminished to perhaps about 20 today. Low values of EROEI have dramatic negative implications for the still-burgeoning human population.

Introducing the concept of *energy sprawl*, Douglas Fox (2010) notes the limitations of biofuels and calculates the amount of land needed for production of the same amount of energy from different sources. All of the so-called alternative energies require many times the land area required for natural gas or coal, with corn ethanol requiring about 500 times more land than needed for the extraction of fossil fuels!

Thomas Sinclair (2009) has offered a quantitative analysis of plant growth that also leads to great cautions regarding biofuel limits. He refers especially to the congressional mandate that calls for vastly increased production of ethanol, writing, "Nitrogen fertilizer [manufactured with natural gas] of annual biofuel crops will inevitably be needed." He continues that, concerning

> ethanol production from cellulosic feed stock, somewhere between 25 and 50 million hectares of new land must be brought into high and sustained agricultural production.... It would be the most extensive and rapid land conversion in U.S. history.... The amount of water transpired by these crops could be large enough to affect the hydrologic balance. (Sinclair, 2009, p. 407)*

Sinclair also disagrees with the America's Energy Future Panel (NAS et al., 2009). The panel writes with some optimism about bringing 550 million tons of dry biomass to the conversion table (for 607 million barrels of gasoline-equivalent fuels), while Sinclair writes with some pessimism about 430 million tons of dry biomass.

There are other problems intrinsic to biofuels from cellulose, because cellulosics contain only about 30% as much starch and/or sugar as corn grain, and this means that transportation costs are correspondingly higher. It should also be noted that no commercial net energy from cellulosics has been produced anywhere as of this writing. Biofuels should

---

* The harvested cropland in the United States during 1997–2007 was near 310 million acres, according to the Economic Research Service of the U.S. Department of Agriculture.

probably be regarded as transitional to developed solar, wind, and perhaps thermonuclear in the future.

We nevertheless discuss current efforts toward utilization of biofuels while realizing that their utilization can provide only a small fraction of the energy required by current practice and that their use must be carefully managed to keep environmental damage within decent limits and to prevent conflict among nations.

Of course, there is the possibility that breakthroughs in science and technology will alter the outlook radically. For instance, Joule Unlimited, a company based in Cambridge, Massachusetts, is claiming to produce diesel fuel and other fuels with genetically modified bacteria that utilize only sunlight, water, and carbon dioxide, with efficiencies up to 50 times that of biofuel production. At this writing, the company website states that it is currently targeting commercial delivery of 15,000 gallons of diesel fuel per acre per year with no impact on farmland (see also the discussion in Section 5.5.6, below).

A strongly contrasting approach to Joule Unlimited is pursued at the Land Institute of Salina, Kansas (Bontz, 2011). There ethanol production is examined as one of several possible side benefits to development of a perennial wheatgrass via crossbreeding.

5.5.2 Ethanol from corn and sugarcane

The U.S. Energy Independence and Security Act of 2007 calls for annual production of 36 billion gallons of biofuels within the United States by 2022, but at this writing, it appears this target may never be reached and, perhaps, should not be reached. Ethanol is produced from corn by enzymatic conversion of cornstarch to sugar, with subsequent fermentation and distillation. This process, coupled with the cultivation and fertilization of corn and its transport to mills, has been found by David Pimentel and Tad Patzek (2008) to require more energy than the ethanol creates during its combustion as a motor fuel and to involve net emissions of carbon dioxide, partly as a consequence of the need for human travel and transport of inputs during the growth and harvesting stages.

Although their result is controversial, others more optimistic still indicate only a slightly positive EROEI in the case of corn ethanol. Also, while ethanol is transported now by barge and truck, its more widespread use would require installation of new infrastructure in the form of a costly pipeline system, because ethanol dissolves in water, contains some water, and would corrode existing pipelines. Furthermore, ethanol production from corn and otherwise increased demand for corn has increased the cost of corn to the point that many existing ethanol-from-corn manufacturing facilities are without profits, and a few have shut down.

Some producers have shifted from other crops, including wheat and cotton, to corn due to its higher price.* Owing to such shifts, there has been a strong rise in the price of wheat and a decline of cotton production, with some loss of jobs because cotton production requires more workers than corn. This has had a continued negative impact on poorer nations, whose citizens can no longer afford the rising costs of basic food (McDonald, 2010). On February 24, 2011, former president Bill Clinton told the U.S. Agriculture Department's annual Agricultural Outlook forum that we must become energy independent, but not at the cost of food riots (Jalonick, 2011).

There are also environmental consequences of increased corn production because corn requires nitrogen fertilizers, entails abuse of the environment via soil erosion, and requires considerable water in areas where corn is irrigated. These negative aspects are omitted from a recent internet posting by Abengoa Bioenergy Corporation. Abengoa emphasizes sustainability in its posting, but use of corn for ethanol production implies increased soil erosion and the use of petroleum in the manufacture of fertilizer and herbicides for weed control.

It is clear that governmental stimuli encouraging the conversion of corn to ethanol have entailed numerous unintended consequences, some detrimental. Of course, initial legislation, including a tax of 54 cents per gallon on importation of ethanol from Brazil, was enacted under pressure from corn growers (at this writing, corn growers are protesting the existence of loopholes), which amounts to an illustration of inadequacies in the U.S. political system.

Partly for the reasons implied above, there is a strong research effort toward production of ethanol from cellulose, although there are inherent problems with cellulose, as mentioned in section 5.5.1. The America's Energy Future Panel report already cited (NAS et al., 2009) presents an optimistic outlook toward research under way in the United States and supported by hundreds of millions of dollars in federal subsidies. The principal problem involved in this process is the digestion of the lignin that provides rigidity to plants and reduces access to cellulose and hemicellulose. Whether this problem will remain unsolved or not is itself an important question.

Production of ethanol from sugar in Brazil is much more efficient than its production from corn in the United States, and where the system has incorporated advanced technologies, there is also little abuse of workers,

---

* See http://www.abengoabioenergy.com/corp/web/es/acerca_de/general/introduccion/areas_actividad/index.html. See also Meadows et al. (2004), Millennium Ecosystem Assessment Board (2005), Brown (2008), and Murphy (2008), for example. A previous publication edited by Pimentel (2008) presents a detailed analysis of biofuel production and implications.

low or even negative net emissions of carbon dioxide, and little or no impact on food prices. Practically all automobiles in Brazil can use ethanol as fuel, and this industry, with some recent discoveries of offshore oil, has enabled Brazil to become a net exporter of oil. However, this industry is not transferable to the United States because there is insufficient area in the United States for sugarcane production, and the U.S. demand for liquid fuel is more than 10 times that of Brazil.

Total Brazilian production of ethanol annually—on only 1.5% of its arable land—was about 150 million barrels in 2010, which, because the energy of ethanol is less than that of gasoline, is equivalent to about 130 million barrels of gasoline. This could supply two weeks of liquid fuels for light-duty vehicles in the United States at present usage.* While two weeks of present usage may not seem like very much at first thought, it could be quite appreciable if society were transformed to reduce demand substantially. This situation becomes an argument for a generalized "adaptation" (Lempinen, 2009, 2010).

5.5.3   Biodiesel fuel from sugar, palm oil, and residual lipids

Biodiesel is diesel fuel with a recent biological origin and can be made from sugar and from various lipids (oils and fats). Plant lipids contain glycerin, which must be removed before the oil becomes diesel fuel. This is done in a refining process known as *transesterification*, which involves an interaction of the lipids with an alcohol, water, and later heat to remove the water, along with the addition of catalysts, principally sodium hydroxide, which are recovered for reuse. The removed glycerin is used in soapmaking and has other industrial and commercial applications.

Although Brazil has a flourishing industry for conversion of sugar to ethanol, there is also great interest in Brazil in conversion of both cellulose and sugar to biodiesel in order to provide work and corresponding income to farmers in northeastern Brazil, to reduce toxic emissions from vehicles that use conventional diesel, and to reduce diesel imports. The manufacture of biodiesel from sugar is experimental, as recently discussed in *Technology Review* (Regelado, 2010) in connection with Amyris Biotechnologies, a California-based company that is investing heavily in pilot plants located in Brazil. The conversion of sugar by Amyris involves genetically modified yeast that produces a forerunner of biodiesel rather than ethanol.

---

* The NAS (2009) reports as follows on page 1: "the U.S. transportation sector consumes about 14 million barrels of oil per day, 9 million of which are used in light-duty vehicles." A breakdown of total usage is given in Table 1.1 on page 53 of the NAS report. Data from the U.S. Energy Information Agency shows U.S. gasoline demand at about 9 million barrels per day. The amount of gasoline produced by refinement of a barrel of crude oil is variable.

Oil from soybeans is presently the principal source of biodiesel. According to the U.S. National Biodiesel Board, biodiesel production in the United States, which had risen from 112 million gallons in 2005 to 691 million in 2008, declined to 545 million gallons in 2009. These totals included contributions from both soybeans and other sources of oil. The decline in 2009 was related in part to a rise in price of soybeans to the high level of nearly $10 per bushel and partly to expiration of a federal tax credit of $1 per gallon on January 8, 2010. At this writing, the tax credit has not been renewed.

According to the Ag Marketing Research Center at Iowa State University, the oil production was enabled by use of 5.7% of the soybean crop for production of 440 million gallons (10.5 million barrels) of biodiesel during the 2008–2009 production year. During 2009–2010, 259 million gallons of biodiesel fuel was produced from soybeans.

The America's Energy Future Panel (NAS et al., 2009) predicts a positive outcome of U.S. research to develop efficient methods for conversion of cellulose to biodiesel and envisions production of large amounts of biomass for this purpose. Note again that there are no commercial plants producing biofuels from cellulose at this time. Success cannot come too soon, because conversion of tropical forests to plantations of oil palms for biodiesel production is proceeding at an alarming rate, with many species of tropical fauna at risk, as angrily discussed by George Monbiot (2005).

5.5.4   Uses of wood

Research may lead to cost-effective means for conversion of woody cellulose to liquid fuel. The America's Energy Future Panel (NAS et al., 2009) projects that 124 million tons of woody biomass in the form of remainder forest products might be available for conversion to liquid biofuels (perhaps 135 million barrels) by 2020. The effectiveness of a future conversion process is also problematic, however, although the America's Energy Future Panel is optimistic.

Another use of wood is for direct heat; wood wastes now provide about 3% of thermal heat in the United States. For example, the removal and use of mesquite trees on ranch lands in Texas, which are damaging to ranching interests,* can provide an additional benefit (Associated Press, 2010).

I have similarly suggested use of perennial ragweed (*Ambrosia psilostachya*) as a fuel for the generation of direct heat. Perennial ragweed sometimes infests farms in Oklahoma, especially under conditions of

---

* The trees are spread by cattle, which eat the beans. Surviving beans then sprout readily in feces. The trees suppress growth of desired grasses.

heavy grazing by cattle, which avoid it and thereby contribute to its proliferation in overgrazed pastures. In central Oklahoma, September usually brings a secondary maximum of rainfall, and the shallow-rooted
plants are then mature and easily pulled up when the soil is moist. A
special machine might pull up the plants, which would be left on the
field to dry for a few days, then raked, baled, and transported to a power
plant for use as fuel. Field-dried perennial ragweed burns with a very
hot flame. A benefit of such power generation would be reduction of a
weedy infestation that greatly reduces desirable grasses, but I have not
yet found enabling interest toward construction of an appropriate harvesting machine.[*]

In a personal test of "green living," I have experimented with wood
burning for heating my home and have modified a traditional fireplace to
the efficiency of an airtight stove. A direct-current fan, backed up by car
batteries for use in the event of a power outage, circulates room air into a
tube that goes through the firebox and out again, and the temperature of
the exiting air can be as high as 350°F. Approximately seven cords of wood
heats 1,000 square feet to a temperature of 72°F during the cold season in
Oklahoma,[†] and the winter bill for natural gas is thereby reduced from
about $200/month (for the whole house of 1,600 square feet) to about $30
(I use natural gas to maintain a small greenhouse above freezing during
the winter season and for hot water). This experimental project is made
practical by (1) my good health and (2) an abundant supply of wood at no
cost on a farm that I manage.[‡] However, the wood must be cut, hauled
home and stacked, and taken into the house during the heating season.
All of this involves much work and some cost and a dramatic shift in lifestyle from that associated with reliance on automatic controls! Although
neighbors in the small city of Norman, Oklahoma, have no objection to
the smell of burning wood, pollution would be considerable if the practice
were widespread,[§] and without a sufficient supply of firewood from outside the city, Norman's many trees with their biostorage of carbon dioxide
would be lost within a few years.

---

[*] Furthermore, mechanical reduction of ragweed would reduce the use of herbicides that are
manufactured from petroleum. Another proposal for conservation of resources suggests
use of warm water outlets from power plants for heating of plastic-covered hoop-houses.
The hoop-houses (cold frames or greenhouses, depending on temperature maintenance
within) would grow local vegetables, especially hardy greens during the cold season,
with the expectation of ultimate savings of energy and direct costs for transportation.

[†] The average January temperature in central Oklahoma is 36°F.

[‡] Costs would balloon to more than the cost of natural gas service if the wood had to be purchased.

[§] Owing to issues of air quality, some cities have strongly controlled uses of wood for heating. For example, in Aspen, Colorado, a law passed in 1983 limits the number and types of
devices people can install. Facilities previously installed can be retained, but the number
of wood-burning stoves is gradually being reduced by wear and tear.

5.5.5   Switchgrass and other grasses and forbs

Legislation passed in Oklahoma in 2007 provided for an experimental planting of 1,000 acres of switchgrass in the Oklahoma Panhandle. Switchgrass is somewhat resilient to extremes of drought and flood, and the purpose of this planting is primarily to learn good management techniques for a sustainable partial source of cellulose to be provided to a biofuel plant being built by Abengoa Bioenergy in nearby Hugoton, Kansas.*

The Hugoton plant is presently scheduled for startup in 2012, with a proprietary technology for conversion of 2,500 dry tons of daily biomass to 15–18 million gallons (about 330 million barrels) of ethanol annually, and with 75–80 megawatts of electricity for the national grid, equivalent in energy to another 16 million gallons of ethanol (pers. comm.). The estimated production is somewhat less than that projected theoretically in the America's Energy Future Panel study (NAS et al., 2009), but may be more realistic. In any case, the projected production represents little impact on demand for liquid fuels.

Research at Oklahoma State University has improved the productivity of switchgrass by 20%; historic harvests amount to about 2.5 tons per acre, with greater amounts up to a factor of three, depending on fertilization. In Kentucky, an experiment with pelletized switchgrass is demonstrating reduced toxic emissions when switchgrass is burned with coal for power generation.

The Hugoton project received a subsidy of $76 million from DOE in 2007, and as much as $500 million in subsidies and loan guarantees for all manufacturing facilities, buildings, and outlying units is contemplated. A question for our nation relates to the quality of this investment. I will not explore this question in depth, but merely note that with the price of gasoline at $4/gallon, the annual retail cost of 30 million gallons of equivalent liquid fuel is about $120 million.

The fact remains that the technology for conversion of cellulose to ethanol or other biofuel is significantly incomplete at this writing.

5.5.6   Fuel from algae

The production of fuel from algae is discussed briefly, but somewhat optimistically, by Pimentel (2008). However, a principal venture in this field, Greenfuels Technologies, entered bankruptcy during 2009 (Bullis, 2010).

---

* Abengoa is based in Spain and has projects in Africa, Europe, and North America. An Abengoa project for electrical generation via concentrated solar power in Algeria contemplates a partnership with New Energy Algeria for transmission of up to 6,000 megawatts to Europe via undersea cables by 2020 (see Pimentel, 2008, p. 270).

At present, discussion of this subject seems somewhat muted, and it is not discussed in the extensive America's Energy Future Panel report (NAS et al., 2009).

Thomas R. Sinclair of the University of Florida and North Carolina State University has written that algae have the same requirements for light and nutrients as other plants and that freshwater algae have a greater requirement for water than land-based crops (pers. comm.). Nevertheless, possibly illustrative of the rapidity of technological advances, bioengineered algae under development by Solazyme can reportedly convert ordinary plant material to oil without sunlight (ABO, 2010). It is noted that plants already contain the energy of sunlight. Solazyme has received a federal grant of $22 million to commercialize their algae-to-fuel-oils process and to lower the price per gallon, and the company has produced fuel that has been successfully tested and used by the U.S. Navy (ABO, 2010). Final results are awaited with interest (see also the discussion in the last paragraph of Section 5.5.1).

5.5.7   Use of manure and other wastes

Waste oil from restaurants and some other residues can be used for production of biofuels, and both biodiesel and biomethane have obvious applications for transportation. The Heartland Flyer, which makes a daily round-trip between Oklahoma City and Ft. Worth by rail, started a year-long test in April 2010 with diesel fuel that is 20% from beef tallow produced in Texas.

Methane is produced at waste landfills and is also being produced from manure on a few large farms in the United States and elsewhere. It is occasionally transported by pipeline from farms to small communities. In Holland, methane produced from cow manure, grass, and food-industry residues is being used in a new thermal plant operated by Essent, a Dutch company, for heating about 1,100 homes in the northern part of that country, around Leeuwarden.

Practices such as these being implemented in Europe and North America represent decent steps in the right direction, but outputs are too small to reverse the effects anticipated from the long-term decline in availability of liquid fuels. About two and a half tons of manure is required to produce methane in an amount equivalent to one gallon of gasoline, and one may ask if the conversion of manure to methane is better than its use for soil improvement.

Generally, these and other bioprocesses are not as beneficial as advertised, because the necessary energy and environmental costs of production, collection, and delivery are not usually fully included.

5.6   Concluding remarks

Climate change, resource depletion, human population growth, and diminution of biodiversity are issues that are deeply intertwined and call for similar responses. There have been numerous warnings, but societal change has not been substantially forthcoming, and the warnings are accordingly becoming foreboding for human prosperity and even survival.

We humans have the intellect to meet the challenges, but effective application of that intellect is problematic and worse. Deficiencies in response are the subject of sociological studies that help us understand why the situation is as it is. Societies do not change readily or easily, and there is a cost for change, as discussed in Section 5.3.1 in terms of anomie and deviance. W. V. Reid et al. (2010) call for better integration of the social and other sciences in order to address desperately needed societal change on a global basis. Ignorance and corruption in political systems and, in the United States and elsewhere, the tremendous influence of financial contributions on political decision making, are holding back necessary societal change.

A compelling related example is provided by the environmentally disastrous flow of oil into the Gulf of Mexico that started on April 20, 2010, after an explosion, a fire, and complete destruction of an oil rig and well occurred in mile-deep water. President Barack Obama called for a moratorium on deepwater drilling until a study of the cause of the Gulf disaster was complete and means for guaranteeing the prevention of a similar future disaster were determined. He further called for the immense oil company BP (formerly British Petroleum) to establish a large fund for compensation of the many people injured physically and economically. These actions were opposed by some leading political figures, who voiced concern about shortages of jobs in the oil industry and scarcity of money for jobs should BP decline. This indicates, again, that there will be no simple or easy transition from the oil-based economy in the United States to a structure relatively free of the declining oil resource and its foreign suppliers. But such a transition is essential and will occur through natural forces if not enabled sooner by human agency.

Perhaps there is hope in new developments in science and technology that include promise for energy-efficient human wellness, and perhaps abundant energy will be provided by fusion of hydrogen to helium, and perhaps new communications technology will provide for improved education and replacement of some energetically costly physical transportation. And there is hope in numerous local efforts toward energy efficiency, individually small but cumulatively possibly large.

Nevertheless, there are grave concerns for the ultimate capability of any future mix of technologies to produce a sustainable supply of liquid fuels (or other fuels) for transportation, use in buildings, and industrial processes in quantities to which Western societies have become accustomed, and a very large infrastructure in developed countries will probably have to change dramatically at crippling cost. The human population and its consumption may be too large, checks to population growth too weak, and human attitudes too resistant to change to offset impacts of climate change and resource depletion and to avoid irreparable damage to the environment and irremediable conflicts between and among cultures and nations.

Development of biofuels as a response to present and looming energy shortages has tended to downplay or ignore negative impacts of this course and is grossly inadequate even in the best case. As Charles A. S. Hall and John W. Day Jr. (2009, p. 327) have written:

> The world today faces enormous problems related to population and resources.... The concept of the possibility of a huge, multifaceted failure of some substantial part of industrial civilization is so completely outside the understanding of our leaders that we are almost totally unprepared for it.... Evidence of negative impacts has historically preceded general public acceptance and policy actions by several decades.... If we are to resolve these issues, including the important one of climate change, in any meaningful way, we need to make them again central to education at all levels of our universities.... We must teach economics from a biophysical as well as a social perspective. Only then do we have any chance of understanding or solving these problems.

And, of course, principles underlying energy and its use must be a major part of curricula.

The time is very late, and the book *Limits to Growth* (Meadows et al., 2004) should be widely read. Its persuasive arguments are insufficiently heeded, and a large number of students of the subjects discussed here are like so many Cassandras, that is, prophets without listeners, as far as the political leadership is concerned:

> Waiting to introduce fundamental change reduces the options open for humanity's long-term future. Waiting longer to reduce population growth and stabilize productive capital stocks means that population is larger, more resources have been consumed, pollution levels are higher, more land has deteriorated, and the absolute flows of food,

services, and goods required to sustain the population are higher. Needs are greater, problems larger,
and capacities less. (Meadows et al., 2004, p. 250)

This message is emphasized by a report to the U.S. Congress from the
National Academies (NAS et al., 2010, p. 5): "The longer the nation waits
to begin reducing emissions, the harder and more expensive it will likely
be to reach any given emissions target." Further, Diamond (2005, p. 509)
notes, "A society's steep decline may begin only a decade or two after the
society reaches its peak numbers, wealth, and power." This is confirmed
by detailed analyses in Diamond's book and by the collapse of the Soviet
Union. Those who manipulate political systems for their personal material benefit in the short term should instead act on the realization that we
are one human family and that the challenge to *Homo sapiens sapiens* as a
whole family is both unprecedented and very severe.

There has been a spate of articles in learned journals to commemo-
rate the 50th anniversary of the start of a search for extraterrestrial intelligence. Paul Davies (2010) notes that we have hardly begun to investigate
possibilities of life elsewhere. With a related point of view, in a recent
article in *Sky and Telescope* entitled "Where Have All the Aliens Gone?"
Jacob Haqq-Misra and Seth Baum (2010) briefly explore the possibility that
intelligent life and civilizations developed elsewhere in our vast galaxy
might have visited us by now, unless they ran out of resources and col-
lapsed as we may do.[*] "If unsustainable rapid growth led to the demise of
past galactic civilizations, then perhaps the challenge rests with human
civilization to become responsible consumers and ensure our own long-
term survival" (Haqq-Misra and Baum, 2010, p. 86).

We have been well warned, but have not heeded the warnings so far.
Changes will be forced by nature whether or not we humans undertake
change voluntarily. Humankind is truly at a crossroads, and the vista
along our future path cannot be accurately predicted but remains to be
seen. What should we/they who apprehend disaster do? What should we/
they try to do? What can be done? Would appointment of a new member
of the U.S. president's Cabinet to explore societal change be feasible? Or

---

[*] Space travel may not be quite as impractical as it first seems. For example, setting aside
innumerable technological problems involving energy and possible collisions, among
others, a book by Hermann Bondi (1980) presents relativistic science applicable to continuing acceleration over ultralong distances. According to relativistic theory involving
time dilation, one could travel 25,000 light-years away in a spacecraft, continuously accel-
erating/decelerating at a value of only the pull of Earth's gravity, in less than 40 years of
travelers' time. However, time at return to the starting point would have lapsed nearly
50,000 years! Note that clocks used in the Global Positioning System are corrected for
relativistic effects, since clocks run slower in the gravitational field that is diminished at
spacecraft altitudes.

would it lead to overwhelming defeat in the next U.S. election? The latter must be of serious concern.

Perhaps a greater effort to enhance public transportation by rail would be a help, copying the example presently set by China, which already has 200+-mph trains and is planning to invest $100 billion in a high-speed rail network over 15,000 miles by 2020. Such an effort in the United States would provide employment to many who presently work in the auto and oil industries, for example. For reduction of carbon dioxide emissions, this immense rebuilding of infrastructure might be accompanied by considerable replacement of coal-fired power plants by natural gas facilities to power the trains. Such initiatives, when completed, might serve for the next hundred years. By then, with depletion of supplies of natural gas, we should hope that electrical power delivered by the sun and/or nuclear fusion would have become sufficiently available. On the other hand, McKibben (2011) presents a somewhat negative picture in his account of serious impediments to China's sincere moves toward a sustainable society.

Various urgent problems—ongoing depletion of several important resources, including phosphates needed for agriculture; loss of biodiversity; the human population problem; religious and social chaos around the world; climate change; and others—still are not receiving the alarmed attention from a majority that they require. A reason for this probably lies in the abilities of science and technology to meet demands even when the demanded products that remain are in short supply. In other words, a store can be depleted at a rapid rate even when the remainder is small. The global situation is immensely complex, opinions differ widely, and as already indicated, our future path is uncertain.

Acknowledgments

The text has been improved by Mary Francis, Marjorie Bedell Greer, Jeffrey Kiehl, David Pimentel, David Sheegog, Kevin Trenberth, and Charles Wright. Richard Hilbert and some other cited authors also contributed to the text. Assistance with computer manipulations by Randy Stafford, referencing by Fannie Bates, and assistance from staff at the History of Science Collection library at the University of Oklahoma is also greatly appreciated. Much of the information in references can be accessed through the Internet. Of course, this text, the points of view, and any errors are mine and my responsibility.

References

ABO (Algal Biomass Organization). 2010. Solazyme delivers 100% algal-derived renewable jet fuel to U.S. Navy: Biotechnology company showcases SolajetTMHRJ-5 jet fuel at the world-famous Farnborough International Air

ShowinUK.Pressrelease,July19.http://www.algalbiomass.org/news/1499/
solazyme-delivers-100-algal-derived-renewable-jet-fuel-to-u-s-navy/.
Abplanap, P. L. 1999. Letter to the editor. *Technology Review*, 102(5): 26.
Anderson, Scott. 2010. The polygamists. *National Geographic*, 217(2): 36–61.
Associated Press. 2010. Texan promotes wood waste for generating power. March
8. http://abclocal.go.com/ktrk/story?section=news/state&id=7316217.
Baran, Jan Witold. 2010. Stampede toward democracy. *New York Times*, January 25.
http://www.nytimes.com/2010/01/26/opinion/26baran.html.
Blunden, J., Arndt, D. S., and Baringer, M. O., eds. 2011. State of the climate in
2010. *Bull. Amer. Meteor. Soc.*, 92(6):S1–S266.
Bondi, Hermann. 1980. *Relativity and common sense*. New York: Dover.
Bontz, Scott, ed. 2011. $695,000 for wheatgrass. *Land Report*, Summer 2011, The
Land Institute, Salina, KS, p. 5–6.
Boyce, James K., and Riddle, Matthew. 2007. Cap and dividend: How to curb global
warming while protecting the incomes of American families. Working Paper
Series, No. 150. Amherst: Political Economy Research Institute, University
of Massachusetts. http://www.peri.umass.edu/fileadmin/pdf/working_
papers/working_papers_101-150/WP150.pdf.
Broder, John M., and Rudolf, John Collins. 2010. Negotiators at global climate
talks continue past the deadline for an agreement. *The New York Times*,
Dec. 11:A12.
Brown, Lester R. 2008. *Plan B 3.0: Mobilizing to save civilization*. New York: Norton.
Bullis, Kevin. 2007. Algae-based fuels set to bloom. *Technology Review*, February 5.
http://www.technologyreview.com/energy/18138/.
———. 2010. The geoengineering gambit. *Technology Review*, 113(1): 50–56.
Catton, William R., Jr. 1980. *Overshoot: The ecological basis of evolutionary change*.
Urbana: University of Illinois Press.
Chell, Kaitlin. 2010. AGU adopts position statement on geoengineering, a hot topic
in Congress and elsewhere. *Eos*, 91(16):146.
Chu, Steven. 2009. Carbon capture and sequestration. *Science*, 325:1599.
Davies, Kate. 2010. Lizards in the lurch. *Defenders*, 85(4):7.
Davies, Paul. 2010. *The eerie silence: Renewing our search for alien intelligence*. Boston:
Houghton Mifflin Harcourt.
Dawson, Terence P., Jackson, Stephen T., House, Joanna I., Prentice, Iain Colin, and
Mace, Georgina M. 2011. Beyond predictions: Biodiversity conservation in a
changing climate. *Science*, 332:53.
Diamond, Jared. 2005. *Collapse: How societies choose to fail or succeed*. New York:
Viking.
Dillard, Carter. 2007. Rethinking the procreative right. *Yale Human Rights and
Development Law Journal*, 10:1–63.
Dodgson, Kathryn. (2010). Searching for balance. *Harvard Divinity Bulletin*, 38(1–
2): 1–2. http://www.hds.harvard.edu/news/bulletin_mag/articles/38-12/
dodgson.html.
Doney, Scott C., Balch, William M., Fabry, Victoria J., and Feely, Richard J. 2009.
Ocean acidification: A critical emerging problem for the ocean sciences.
Oceanography, 22(4): 16–25.
Donner, Simon D. 2011. Making the climate a part of the human world. *Bulletin of
the American Meteorological Society*, 92:1297–1302.
Dowd, Maureen. 2010. A nope for pope. *New York Times*, March 28. http://www
.nytimes.com/2010/03/28/opinion/28dowd.html?ref=maureendowd.

Dukes, Jeffrey S. 2003. Burning buried sunshine: Human consumption of ancient solar energy. *Climate Change*, 61(1–2): 31–44.

Durkheim, Emile. 1951. *Suicide: A study in sociology*, trans. and ed. Spaulding, John A., and Simpson, George. New York: Free Press.

———. 1964. *The division of labor in society*, trans. and ed. Simpson, George. New York: Free Press.

Dworkin, Ronald. 2010. The "devastating" decision. *New York Review of Books*, February 25. http://www.nybooks.com/articles/archives/2010/feb/25/the-devastating-decision/.

Ehrlich, Paul R. 1968. *The population bomb*. New York: Ballantine Books.

Ehrlich, Paul R., and Ehrlich, Anne H. 2008. *The dominant animal: Human evolution and the environment*. Washington, DC: Island Press/Shearwater Books.

Encyclopedia Britannica. 1911. Malthus, Thomas Robert. In *Encyclopedia Britannica*, XVII: 515–516. New York: The Encyclopedia Britannica Company.

EPA (Environmental Protection Agency). 2009. Greenhouse gases threaten public health and the environment. Press release, December 7. http://yosemite .epa.gov/opa/admpress.nsf/0/08D11A451131BCA585257685005BF252.

FAO (Food and Agriculture Organization of the United Nations). 2010. Undernourishment around the world in 2010. In *The State of Food Insecurity in the World, 2010*. 8–11. Rome: Food and Agriculture Organization of the United Nations. http://www.fao.org/docrep/013/i1683e/i1683e.pdf.

Feder, Toni. 2010. ITER collaboration defuses standoff. *Physics Today*, 63(4): 20–21. http://scitation.aip.org/getpdf/servlet/GetPDFServlet?filetype=pdf&id=P HTOAD000063000004000020000001&idtype=cvips&doi=10.1063/1.3397035 &prog=normal.

Feely, Richard A., Doney, Scott C., and Cooley, Sarah B. 2009. Ocean acidification: Present conditions and future changes in a high-CO_2 world. *Oceanography*, 22(4): 37–47.

Fleming, James Rodger. 2010. Gilbert N. Plass: Climate science in perspective. *American Scientist*, 98(1): 60–61.

Fox, Douglas. 2010. New energy, climate change, and sustainability shape a new era. *Christian Science Monitor*, Nov. 8: 26–31.

Friedman, Thomas L. 2011. The Earth is full. *New York Times*, June 7. http://www .nytimes.com/2011/06/08/opinion/08friedman.html.

Geological Society. 2010. Society issues climate change statement. *Geoscientist*, 20(11): 10.

Ghiselin, Dick. 2010. The end of an error. Editorial. *HartEnergy E&P*, June 2. http:// www.epmag.com/Magazine/2010/6/item60843.php.

Goeller, H. E., and Zucker, A. 1984. Infinite resources: The ultimate strategy. *Science*, 223:456–462.

Goodstein, David. 2004. *Out of gas: The end of the age of oil*. New York: Norton.

Gore, Al. 1992. *Earth in the balance: Ecology and the human spirit*. Boston: Houghton Mifflin.

Hall, Charles A. S., and Day, John W., Jr. 2009. Revisiting the limits to growth after peak oil. *American Scientist*, 97(3): 230–237.

Hansen, James E. 2009. Carbon tax & 100% dividend vs. tax & trade. Testimony before the Committee on Ways and Means, U.S. House of Representatives. February 25. http://www.columbia.edu/~jeh1/2009/WaysAndMeans_20090225.pdf.

Haqq-Misra, Jacob, and Baum, Seth. 2010. Where have all the aliens gone? *Sky and Telescope*, 119(3): 86.

Hardin, Garrett., ed. 1964. *Population, evolution, and birth control: A collage of controversial ideas.* San Francisco: W. H. Freeman.
———. 1968. The tragedy of the commons. *Science*, 162:1243–1248.
Hauser, Rachel, and Gordon, Nicole. 2010. Getting serious about geoengineering. *UCAR* (Fall): 9–13. http://www2.ucar.edu/magazine/features/getting-serious-about-geoengineering.
Heinberg, Richard. 2009. *Searching for a miracle: "Net energy" limits and the fate of industrial society.* San Francisco: International Forum on Globalization; Sebastopol, CA: Post Carbon Institute. http://www.postcarbon.org/newsite-files/Reports/Searching_for_a_Miracle_web10nov09.pdf.
Henson, Bob. 2010. Looking within the min. *UCAR* (Winter): 6–8. http://www2.ucar.edu/magazine/features/looking-within-min.
Hilbert, Richard E. 1976. Anomie as an explanation for deviance: Its development and implication for change. In *Issues and ideas in America*, ed. Taylor, Benjamin J., and White, Thurman J., 129–160. Norman: University of Oklahoma Press.
Holman, Craig. 2010. Testimony of Craig Holman (Public Citizen). May 6. http://www.citizen.org/documents/DISCLOSE%20testimony.pdf.
Hvistendahl, Mara. 2011. China's population growing slowly, changing fast. *Science*, 332:650–651.
Interacademy Panel on International Issues. 2009. IAP Statement on Ocean Acidification. http://www.interacademies.net/File.aspx?id=9075.
Jalonick, Mary Clare. 2011. Clinton: Too much ethanol could lead to food riots. *Associated Press*, February 24. http://www.signonsandiego.com/news/2011/feb/24/clinton-too-much-ethanol-could-lead-to-food-riots/.
Kessler, Edwin. 2009. Two decades of deceit. Editorial. *Norman Transcript*, June 20–21. http://normantranscript.com/opinion/x519048901/Two-decades-of-deceit.
Kiehl, Jeffrey. 2011. Lessons from Earth's past. *Science*, 331:158–159.
Kissling, Frances. 2010. The Pope, condoms, and contraception: Let's get this conversation started. RH Reality Check: Reproductive and Sexual Health and Justice. News, Analysis and Commentary. http://www.rhrealitycheck.org/blog/2010/11/28/pope-condoms-contraception-kickstarting-conversation.
Lastoskie, Christian. 2010. Caging carbon dioxide. *Science*, 330:595–596.
Latham, Garl B. 2008. Rail transit: An Oklahoma economic opportunity. Dallas, Texas: Latham Railway Services. http://oklahoma.sierraclub.org/redearth/TrainStudy.pdf.
Lawson, Nigel. 2009. *An appeal to reason: A cool look at global warming.* New York: Duckworth Overlook.
Lempinen, Edward W., ed. 2009. As climate change intensifies, McCarthy urges adaptation focus (10th AAAS James C. Barnard Environmental Lecture). *Science*, 326:680.
———. ed. 2010. Aided by early career scientists, U.S. plans for climate adaptation. *Science*, 330:1195.
Lindzen, Richard S. 2009. The climate science isn't settled: Confident predictions of catastrophe are unwarranted. Editorial. *Wall Street Journal*, November 30. http://online.wsj.com/article/SB10001424052748703939404574567423917025400.html.
Malthus, T. R. 1803. *An Essay on the Principle of Population, or, A view of its past and present effects on Human Happiness with an inquiry into our prospects reflecting the future removal of the evils which it occasions, A new edition, very much*

enlarged. London: Printed for J. Johnson in St. Paul's by T. Bensley, Bolt Court, Fleet Street.

Mason, John. 2010. *The Meteorological Office (1965–83): Reflections by Sir John Mason CB DSc FRS on his time as director-general of the Meteorological Office.* Reading, England: Royal Meteorological Society. http://www.rmets.org/pdf/john-mason.pdf.

McCarthy, Cormac. 2006. *The road.* New York: Vintage.

McDonald, Joe. 2010. Surging food costs hit poorer nations hard; biofuels compound problem. Associated Press, June 6. http://www.cleveland.com/world/index.ssf/2010/06/surging_food_costs_hit_poor_na.html.

McKibben, Bill. 2010. Climate change's O. J. Simpson moment. February 25. http://www.tomdispatch.com/dialogs/print/?id=175211.

———. 2011. Can China go green? *National Geographic,* 219(6): 116–135.

Meadows, Donella, Randers, Jorgen, and Meadows, Dennis. 2004. *Limits to growth: The 30-year update.* White River Junction, VT: Chelsea Green.

Meehl, Gerald, Tebaldi, Claudia, Walton, Guy, Easterling, David, and McDaniel, Larry. 2009. Relative increase of record high maximum temperatures compared to record low minimum temperatures in the U.S. *Geophysical Research Letters,* 36:L23701–L23706.

Merton, Robert King. 1938. Social structure and anomie. *American Sociological Review,* 3(5): 672–682.

Millennium Ecosystem Assessment Board. 2005. Living beyond our means: Natural assets and human well-being. http://www.maweb.org/en/BoardStatement.aspx.

Monbiot, George. 2005. The most destructive crop on Earth is no solution to the energy crisis: By promoting biodiesel as a substitute, we have missed the fact that it is worse than the fossil-fuel burning it replaces. *Guardian,* December 6. http://www.guardian.co.uk/science/2005/dec/06/transport intheuk.comment.

Murphy, Eugene. 2008. *Plan C: Community survival strategies for peak oil and climate change.* Gabriola Island, BC: New Society.

NAS (National Academy of Sciences), National Academy of Engineering, Institute of Medicine, and National Research Council. 2010. *2010 report to Congress.* Washington, DC: National Academies Press. http://www.national academies.org/annualreport/Report_to_Congress_2010.pdf.

NAS, National Academy of Engineering, and National Research Council of the National Academies. 2009. *Liquid transportation fuels from coal and biomass: Technological status, costs, and environmental impacts.* Panel on Alternative Liquid Transportation Fuels. Washington, DC: National Academies Press.

National Solar Observatory. 2010. Sonograms of the sun explain the mystery of the missing sunspots. Press release, June 16. http://gong.nso.edu/news/solarmystery/.

Oreskes, Naomi, and Conway, Erik M. 2010. *Merchants of doubt: How a handful of scientists obscured the truth on issues from tobacco smoke to global warming.* New York: Bloomsbury Press.

Pimentel, David, ed. 2008. *Biofuels, solar and wind as renewable energy systems: Benefits and risks.* Dordrecht, The Netherlands: Springer.

Pimentel, David, and Patzek, Tad. 2008. Ethanol production using corn, switchgrass and wood; biodiesel production using soybean. In *Biofuels, solar and wind as renewable energy systems: Benefits and risks,* ed. Pimentel, David, 373–394. Dordrecht, The Netherlands: Springer.

Rea, Philip A., Zhang, Vivian, and Baras, Yelena S. 2010. Ivermectin and river blindness. *American Scientist*, 98(4): 294–303.

Rees, Martin. 2003. *Our final hour*. New York: Basic Books.

Regelado, Antonio. 2010. Searching for biofuels' sweet spot. *Technology Review*, 113(2): 47–51.

Reid, W. V., Chen, D., Goldfarb, L., Hackman, H., Lee, Y. T., Mokhele, K., Ostrom, E., et al. 2010. Earth system science for global sustainability: Grand challenges. *Science*, 330:916–917.

Revkin, Andrew, and Broder, John M. 2009. Grudging accord on climate along with plenty of discord. *The New York Times*, Dec. 20:1, 4.

Shaw, Jonathan. 2010. After *our* bubble. *Harvard Magazine*, 112(6): 38–43. http:// harvardmagazine.com/2010/07/after-our-bubble.

Shea, Neil. 2010. Africa's last frontier. *National Geographic*, 217(3): 96–123.

Sherwood, Steven. 2011. Science controversies past and present. *Physics Today*, 64(10):39–44.

Simmons, Matthew R. 2005. *Twilight in the desert*. New York: Wiley.

Sinclair, Thomas R. 2009. Taking measure of biofuel limits. *American Scientist*, 97(5): 400–407.

Smith, Jessie, Wigginton, Nick, Ash, Caroline, Fahrenkamp-Uppenbrink, Julia, and Pennisi, Elizabeth. 2010. Introduction: Changing oceans. *Science*, 328:1497.

Somerville, Richard C. J., and Hassol, Susan Joy. 2011. Communicating the science of climate change. *Physics Today*, 64(10):48–53.

Stang, Charles. 2011. Ash and breath: Christ on "The Road." *Harvard Divinity Bulletin*, 39(1, 2):71–76.

Summerhayes, Colin 2011. Geological Society of London issues statement on climate change. *Eos*, 92(5): 39.

Sutley, Nancy H., Lubchenko, Jane, and Abbott, Shere. 2010. Progress report of the Interagency Climate Change Adaptation Task Force: Recommended actions in support of a national climate change adaptation strategy. White House Council on Environmental Quality, October 5. http://www.whitehouse .gov/sites/default/files/microsites/ceq/Interagency-Climate-Change-Adaptation-Progress-Report.pdf.

United Nations. 2008. *UN Collaborative Programme on Reducing Emissions from Deforestation and Forest Degradation in Developing Countries (UNREDD): Framework document*. June 20. New York: UN.

Vaidhyanathan, Ramanathan, Iremonger, Simon S., Shimizu, George K. H., Boyd, Peter G., Alavi, Saman, and Woo, Tom K. 2010. Direct observation and quantification of CO_2 binding within an amine-functionalized nanoporous solid. *Science*, 330:650–653.

White, Lynn, Jr. 1967. The historical roots of our ecological crisis. *Science*, 155:1203–1207.

Whitty, Julia 2010. The last taboo. *Mother Jones*, 35(3): 24–43.

WHO. 2000. *Nutrition for health and development: A global agenda for combating malnutrition*. Department of Nutrition for Health and Development. Geneva: WHO. http://whqlibdoc.who.int/hq/2000/WHO_NHD_00.6.pdf.

chapter six

Uncertain prospects for sustainable energy in the United Kingdom

Andrew R. B. Ferguson

Contents

6.1   The limitations of biofuels due to their very low power density

Three decades ago, a report by the U.S. Energy Research Advisory Board, chaired by David Pimentel, concluded that producing ethanol from corn (maize) was a bad idea (Office of Technology Assessment, 1979). Unfortunately, ever since then, some scientists have tended to obfuscate what is really a fairly obvious conclusion. They have not presented incorrect data, but have presented a false picture by focusing on unimportant matters and ignoring the important ones.

For instance, one extensive report by the U.S. Department of Agriculture (Shapouri et al., 2002) showed that, by including the energy value of the by-products, a substantial excess of energy outputs over the inputs could be demonstrated. What the report omitted was that this conclusion was of little importance because of the very low power density* achieved, even in terms of the gross ethanol output, let alone in terms of excess of output over the input. Neither did the report dwell on the fact that growing corn causes serious soil erosion under present methods of production and, as with other crops that require substantial use of fertilizers, produces serious water pollution problems. The same paper put forward the proposition that, as there was plenty of natural gas in the United States, another way of looking at ethanol production was not in terms of *net* energy, but rather in terms of converting energy from natural gas into a useful *liquid* form. The paper did not explain why the United States imports natural gas from Mexico and Canada if it has plenty of its own to spare!

As with climate change, there will always be dissenting voices, but the consensus now is that liquid biofuels are not going to play a significant part in solving the problem of providing energy from renewable sources. In *Sustainable Energy—without the Hot Air* by David MacKay (2008, p. 44), an overview of such biofuels is set out thus:

> I think one conclusion is clear: *biofuels can't add up—* at least, not in countries like Britain, and not as a replacement for all transport fuels. Even leaving aside biofuels' main defects—that their production competes with food, and that the additional inputs required for farming and processing often cancel out most of the delivered energy—biofuels made from plants, in a European country like Britain, can deliver so little power, I think that they are scarcely worth talking about.

In the following pages, I will often refer to MacKay's book. I don't choose him as the sole authority on renewable energy, but he neatly assembles the data from a large array of scientific papers. MacKay does for the United Kingdom what Howard Hayden (2004) had already done for the United States in *The Solar Fraud: Why Solar Energy Won't Run the World*. The difference between them is chiefly that Hayden appears to think that nuclear power must play a major part, while MacKay puts a lot of hope in solar thermal electricity, of which more later.

Let us look at the reasons MacKay gives for his conclusions about biofuels.

---

* The term *power density* refers to the amount of energy that can be obtained over a period of time, often a year or averaged over a year (the appropriate period is usually obvious), from a specified area of land. Thus if 31.54 gigajoules (about 1.75 tonnes of dry wood) can be collected over a year from 1 hectare of land, then by dividing the number of gigajoules by the number of seconds in a year, 31.54 million, the power density is assessed as 1,000 watts per hectare, or 1 kW/ha.

Britain's main biodiesel crop is *oilseed rape*. He gives the power density as 0.13 W/m².

Sugar beets in the United Kingdom give impressive yields of 53 t/ha/yr. After processing to ethanol, the ethanol produced provides 0.4 W/m²—without accounting for the energy inputs, which likely amount to a substantial part of the ethanol output.

Corn yields in Britain probably do not match those in the United States, but even there, ethanol from corn is produced at a power density of only 0.22 W/m², again not including the fact that some of the liquid ethanol produced would be needed for sowing, harvesting, and transporting the corn (on the order of 15%), and a substantial amount of heat energy would be needed for processing (distillation requires considerable energy).

Cellulosic ethanol can be obtained from *switchgrass*. Again, the data come from the United States. A report in 2008 found that the net energy yield of switchgrass grown over five years on marginal cropland on 10 farms in the U.S. Midwest was 0.2 W/m² (Perrin et al., 2008). Note that this is a *net* energy yield, unlike the previous figures given.

Jatropha is an oil-bearing crop that grows best in dry tropical regions with between 300 and 1,000 mm of rain per year. Jatropha likes temperatures of between 20°C and 28°C. The power density that might be obtained on good land is about 0.18 W/m² and on wasteland 0.065 W/m².

Where bioethanol can be produced from *sugarcane*, a significant ethanol yield can be achieved, giving a *gross* energy density, without considering the inputs, of 1.2 W/m². There is the additional problem that sugarcane is one of the crops that is particularly liable to cause soil erosion (Pimentel, 1993).

Neither jatropha nor bioethanol from sugarcane can be produced in the United Kingdom, so the gross power density that is likely to be achieved for liquid biofuels is in the region of 0.25 W/m², again without accounting for the energy inputs required. Ignoring the energy inputs, let us see what contribution 0.25 W/m² could make to British energy requirements.

The total U.K. land area provides 4,000 m² per person, and cropland and pasture amount to about 3,000 m² per person. We should bear in mind that the United Kingdom produces only about 70% of its food, but for the sake of this exercise, suppose that all U.K. citizens become vegetarian and thus somehow manage to not only grow sufficient food but also keep, say, 800 m² per capita of average ecologically productive land for producing liquid biofuels. At 0.25 W/m², this would provide 200 watts, which is 4% of the total per capita current energy use in the United Kingdom—5 kilowatts per person (or 24 × 5 = 120 kWh per day). This 4% might be enough to carry on some vital agricultural, industrial, and distribution activities, but would not support a modern form of society. Thus MacKay's summary paragraph quoted earlier seems well justified.

The loss of energy in converting plant material into liquid fuels is only part of the problem. Let us suppose that the need for liquid fuels could be somehow eliminated, so the mooted 800 m^2 of average ecologically productive land—freed up by a vegetarian diet—could be used to produce biomass, and that the very high yield of 9 dry tons of plant matter per hectare could be achieved (see note b to Table 6.1). As there are about 20 million households in the United Kingdom, each one would have 2,400 m^2 for this purpose, which, based on this optimistic yield, would produce 2.2 tons of dry matter. The winter 2009 issue of *Clean Slate* (Kemp, 2009), the practical journal of the Centre for Alternative Technology, tells

Table 6.1 Power per Unit Land or Water Area

Source	Power density	Notes
Wind	2 W/m^2	Total area over which turbines are spread
Offshore wind	3 W/m^2	Ditto
Tidal pools	3 W/m^2	Partially controllable
Tidal stream	6 W/m^2	
Solar photovoltaic (PV) panels	5–20 W/m^2	Panel area only—18 W/m^2 in UK[a]
Green plants	0.5 W/m^2	But not electricity, as for the other sources[b]
Rainwater (highlands)	0.24 W/m^2	A highland catchment area
Hydroelectric facility	11 W/m^2	Using a high Scottish lake[c]
Solar chimney	0.1 W/m^2	
Concentrating solar power (desert)	15 W/m^2	

Source: Data from MacKay, D. J. C., *Sustainable energy—without the hot air* (Cambridge, England: UIT, 2008).

[a] The power density range for PV panels needs to extend higher than suggested in the table. One extensive U.K. domestic system, of 26 m^2 panel area, situated 50 miles to the west of London, delivered 18 W/m^2, averaged over two years. That is impressive for British insolation (at most 140 W/m^2), but power density is not the main problem with PV, but rather capacity factor. The rated power of those panels is 175 W/m^2, which makes their capacity factor 18/175 or about 10%. That shows the variability of the output: 10% on average but at least 100% at times, and hence the difficulty of integration into the grid.

[b] If the plant matter were turned into electricity, the power density would be at best 30% of the 0.5 W/m^2 shown, which is the heat value of the dry wood. Moreover, 0.5 W/m^2 assumes a generous yield of 9 dry t/ha/yr (calculated on the basis of a calorific value of 17.5 GJ/t). *Natural* forest growth in the United Kingdom is probably closer to the world average for forests of 3 t/ha/yr, but note that 9 t/ha/yr. It is hard to see 9 t/ha/yr being achieved continually without significant inputs.

[c] The figure of 11 W/m^2 is the result of a calculation MacKay does based on a high lake in Scotland. But note that when 50 hydropower reservoirs were selected at random in the United States, ranging in area from 481 to 763,000 ha, the power density was only 0.15 W/m^2 (Pimentel and Pimentel, 2008). High power densities are obtainable only in places that allow a large drop from the top of the dam.

us that a fully automated central heating system would require about 4 tons of wood pellets per year. So, even if we did manage to get 44 million tons of dry matter distributed to our 20 million households, those houses would not be kept very warm although how much heat would be generated would depend on the heating capacity of the dry matter versus wood pellets.

MacKay's proposed solution to the problem is to try to do everything with electricity—heating houses by using heat pumps and using electrical vehicles. The question that arises is whether renewable energy sources could provide the electricity. Table 6.1 shows the data that MacKay provides by way of a summary of power densities available from sources of renewable energy that could provide electricity.

The legend to MacKay's table reads, "Renewable facilities have to be country-sized because all renewables are so diffuse." That sums up the essential difficulty when trying to capture sufficient energy from renewable sources, even when using devices with the power densities shown in Table 6.1, and those power densities are mostly much higher than for liquid biofuels. Having noted this problem, MacKay concludes that the United Kingdom cannot be self-sufficient and proposes that the only answer is to go to North Africa and use the solar energy there with a technology known to concentrate solar power—called *solar thermal electricity*. Apart from a relatively good power density (15 W/m^2), solar thermal electricity has the advantage of being partially controllable, at least in most systems—those that involve storing substantial volumes of hot liquids.

But before moving on to consider controllability problems, which we will do in the next sections, let us consider why the apparently attractive photovoltaic (PV) option is not of much use. Although PV looks attractive on the basis of its power density, it is very difficult to integrate into the grid because of its low capacity factor, and hence the huge difference between average and peak output (see note a to Table 6.1; for further detail, see also Ferguson, 2007).

The major question is whether it will be possible to integrate various uncontrollable and partially controllable renewable energy sources to produce all the electricity that is going to be required not only to drive heat pumps to provide heating but also to charge the batteries that will, according to this plan, power electrical vehicles. That is the question to address next. What evidence can be brought to bear on the question?

6.2   The limits to using uncontrollables are set by storage limits

There are clear indications of the limited potential for integrating uncontrollable inputs from wind turbines into an electrical system. Denmark encounters problems when 8.5% of its electrical consumption comes directly from wind (it exports and perhaps later imports, as hydropower, the rest of its wind output). Similar problems are being encountered in the United States. On October 2, 2009, the *Wall Street Journal* reported:

> The Bonneville Power Administration, a government-owned utility based in Portland, Oregon, taps one of the biggest collections of wind farms in the country. Between January and August, average wind-power production accounted for 12% of average electricity consumption in Bonneville's service area. (Ball, 2009)

The article then went on to show that, despite the fact that most of the electricity in the Bonneville service area comes from hydroelectricity (the ideal plant for dealing with uncontrollable inputs), Bonneville was encountering difficulties in coping with the occasional high output from the wind turbines, so that on occasion no electricity demand was left over for the dam turbines to satisfy, so to prevent the dams from overflowing, water had to be released from them without using it to generate electricity.

Spain is another country in which the installation of wind turbines is sufficiently advanced for its experience to start to confirm theoretical calculations that, without a storage facility, the amount of electricity generated from wind power is in the region of 25% of the total electrical demand (Ferguson, 2008).

During some periods on the night of November 8/9, 2009, wind provided 53% of Spain's electricity demand. While not immediately obvious, the implication of this is that the best that could be done—to make maximum use of wind—would be to double the existing wind turbine capacity. Given that in 2008 wind power supplied 11.5% of Spain's electricity, this would mean that the amount of electricity that wind would contribute to the grid after doubling the wind capacity would be about 23% of the average electricity demand (assuming that not too much has to go to waste).

Because the implications are not immediately obvious, let us dwell on the matter. To be more precise, let us dwell on what the implications would be if there were to be twice as much installed wind capacity, so that sometimes the output from the wind turbines would provide 2 × 53% or 106% of the level of low demand in Spain's electricity system. The main effects would be these:

1. Even supposing that all controllable plants could be switched off—so as to make it possible to make use of the wind output—6% of the electricity being produced by the wind turbines would have to go to waste.
2. One thing that sometimes happens with wind turbines is that the wind exceeds the safe operation limit of the turbines and, for safety

reasons, they have to be switched off. Thus it can happen, and has, that a large part of the output from the wind turbines is suddenly lost. When the wind output peaks at 53% of low demand, controllable plants can be brought on line quickly enough to cope with the loss from the wind turbines, but were wind output to reach 100% of demand, it is unlikely that there would be enough *highly* controllable capacity (e.g., hydropower and open-cycle gas turbines) to replace such a sudden loss of wind output (for evidence of this, see below for a description of just such an incident).

3. From the previous observation, it is manifest that all the controllable plants need to be *highly* controllable, so before considering installing so much wind capacity that peak output from wind approaches 100% of low demand, it is necessary to ensure that if there is to be any input into the grid from nuclear power—which is only weakly controllable and undesirable to have its output wasted—then the installed wind capacity should be diminished to take account of this inflexibility. Note, too, that the highly controllable plant is likely to be less efficient.

4. In the scenario being contemplated, we have used up all the uncontrollable capacity that could be incorporated. For instance, if a tidal flow resource (predictable but uncontrollable) were to be added to the system, the high output from the wind turbines could be accompanied by high output from the tidal flow, compounding the problem of excessive uncontrollable output.

In giving consideration to these four effects, the point might be raised that high output from the wind turbines will not happen at night (i.e., periods of low demand) too often, and so one could allow the excess wind output to go to waste. Against that is the argument that the output from wind turbines has already been curtailed as much as the manufacturers think wise. In practice, once the wind speed reaches about 13 m/s, further increases in wind are not allowed to produce a higher output—despite the fact that at 17 m/s they could produce twice as much electricity and at 19 m/s three times as much, and they don't have to be shut down until about 25 m/s (Hayden, 2004) The reason for this "waste" is to lessen the extent of the problem that we are considering here, namely, that high peaks of output create difficulties for the rest of the grid system. We are now considering making further curtailments during periods of low demands; admittedly, it is hard to pin down exactly what proportion of electricity would be lost were one to accept this additional curtailment of the potential of wind turbines to extract energy from the wind, but we hardly need to pin it down because of the problems of dealing with very large rapid changes, something that we will come back to.

Point 2 above indicated the problem associated with unanticipated high winds. This can be a problem even when wind is producing only 11.5% of Spain's electricity, as the following incident makes clear.

On January 23, 2009, the Spanish grid was expecting very high winds from an Atlantic storm, and projections were for the country's turbines to produce about 11 gigawatts of electricity for most of that day and that night. This 11 GW would be something like 75% of the wind turbines' full capacity, which is about what one would expect from widely spread wind turbines. The wind arrived late, but when it did, it was stronger than forecast. By 4 p.m. on the 23rd, the output from the wind turbines peaked, and some turbines started to close down because of excessive winds. The shutdown problem continued, and at the time of minimum output from the wind turbines, early next morning, the gap between the forecast and actual wind turbine production exceeded 7 GW.

The challenge was met by using hydroelectric power, of which Spain has a lot, and pumped storage hydropower and by temporarily importing power through international connections. Hydro and pumped storage are the easiest type of plant to ramp up, but clearly there is a question as to whether there would still be sufficient rapidly controllable plants available in Spain if the gap were to have been 14 GW instead of 7, as would happen with a doubling of the existing wind turbines (so that they could supply 23% of Spain's electricity). Moreover, many other countries do not have as high a proportion of their electricity supplied by hydroelectricity.

What might allow a significant expansion in the use of uncontrollables is the availability of pumped storage. An experiment funded by the German Economics Ministry involving three companies and Kassel University throws some light on the extent of pumped storage required (Ferguson, 2009). The experiment, called the Combined Renewable Energy Power Plant, aimed to assess the potential contribution of a combined pumped storage, wind, solar, and biogas plant, scaling it to represent a 10,000th part of the electricity demand in Germany.

A report on the experiment stated that the power plant "was capable of generating 41 gigawatt hours of electricity a year" (Elliott, 2009, p. 23). As the experiment did not run for a year, presumably the meaning is that the network would manage to generate 41 GWh were it to be run for a whole year, that is, an *average* 4.7 MW, that is, 4.7 MWy during the course of year. However, note that 4.7 MWy/y is actually closer to a 14,000th part, as electrical production in Germany is of the order of an average 65 GW (i.e., 65 GWy/y) (Elliott, 2009).

The report explains that 36 decentralised plants were linked and controlled by a central control unit and that pumped storage was available (Elliott, 2009). Let us now look at the details given regarding the plant that was used in this trial.

"Over the period of the experiment 61% of the electricity came from eleven wind turbines (total 12.6 MW) [installed capacity]" (Elliott, 2009, p. 23). The use of wind power on that scale immediately makes the need for pumped storage evident. The peak output from the wind turbines, even when as widely spread as those over the 800 km of the wind turbine network of the large-scale utility operator E.ON Netz, is likely to be at least 80% of full capacity of all the wind turbines, that is, $0.8 \times 12.6 = 10$ MW, which is more than double the average demand of 4.7 MW.

The next figure referred to biogas, and it was stated that 25% of the electricity came from "four biogas CHP (combined heat and power) plants (total 4 MW capacity)." Of course, there is no problem with excess electrical output from biogas provided that the CHP plant is fully controllable (which it rarely is). CHP can probably be regarded as a substantially controllable plant, and perhaps the main point to make is that to produce 25% of electricity from biogas does not appear at all likely to be possible on a nationwide basis (the power density of producing biogas is too low) (Elliott, 2009, p. 23).

There is also a considerable amount of PV generation within the mix of power plants: 14% of the average demand of 4.7 MW (i.e., an average of 0.66 MW) came from 20 PV installations having an installed capacity of 5.5 MW (indicating a capacity factor of 12%). Once again, the need for pumped storage is evident from that figure. The capacity factor of PV in Germany is unlikely to be higher than 12%, so 5.5 MW of capacity is required to achieve a 14% contribution. Countrywide sunshine is not too unusual, so peak output from the system can be full capacity (individual plants can go above the nominal capacity), and as 5.5 MW is above even the *average* demand of 4.7 MW, there is an evident need for more pumped storage (Elliott, 2009).

What was not reported was the amount of pumped storage that was available. I wrote to Erneuerbare Energien e.V. and was told that the pumped storage available to the project was a simulated 84.8 MWh. If that is scaled up by a factor of 14,000, it becomes 1,170 GWh. The largest pumped storage facility in the United Kingdom is Dinorwig, offering 9 GWh of storage, and all three U.K. pumped storage facilities combined offer 30 GWh. Thus the requirement for scaling nationwide in Germany would be 130 Dinorwigs, or 39 times the total pumped storage currently available in the United Kingdom.

It is possible to generalize from the experiment as follows: Biogas is controllable, so in that experiment, only 75% of the input was uncontrollable. Thus, we can make an analysis relevant to trying to achieve 75% of the input to the grid from uncontrollables. How much pumped storage does that amount of electricity from uncontrollables require?

The miniature (aiming to be a 10,000th part of the real supply) experimental system in Germany produced an average of 4.7 MW. Of this, 75% amounts to an average 3.525 MW. The pumped storage capacity needed to support this was 84.8 MWh. On that basis, 1 GWh of pumped storage capacity allows an uncontrollable input into the grid of 3.525 × 1,000/84.8 = 41.6 MW/GWh on average.

The United Kingdom has a pumped storage capacity of about 30 GWh. If all of that was to be used to deal with uncontrollables, it would be able to support 30 × 41.6 = 1,250 MW, or 1.25 GW, on average. U.K. electricity consumption averages about 45 GW, so that represents 1.25/45 = 2.8% of the whole. There are possibilities for increasing pumped storage, but not without difficulty and, of course, expense. Other countries may be better placed. Anyhow, the ratio of 41.6 average MW per GWh of pumped storage capacity provides a quick estimate of the ability of pumped storage to raise the proportion of two uncontrollables (wind turbines and PV) that can be introduced into an electrical grid.

The above analysis is just one useful way of looking at the results of the German trial. The 41.6 average MW/GWh can be inverted to give 0.0240 GWh/MW or 24.0 GWh/GW. That ratio can be used to calculate the pumped storage needed for the United Kingdom to produce 75% of its electricity from uncontrollables, namely, 0.75 × 45 [average GW] × 24.0 = 810 GWh pumped storage capacity. That is about 27 times the amount presently available.

Pumped storage is clearly a limited solution to the problem of storage of electricity. Solar thermal electricity has some capacity for storage, so that source of renewable energy has two attractions: a relatively high power density and some storage capacity. We need to take a look at that.

6.3 Solar thermal electricity: Many unanswered questions about its viability

Solar thermal electricity is also called *concentrating solar power*. Useful information on this has been published by Howard Hayden (2004) and Ted Trainer (2007). We have already noted that MacKay (2008) comes to the conclusion that solar thermal electricity from North Africa is an essential part of the mix. But does he adequately cover the problems associated with introducing yet another uncontrollable into the system?

Many more questions than answers surround the subject, and probably the best that can be done is to look at a 2009 article on solar thermal in *New Scientist* by Fred Pearce, *New Scientist*'s chief environment correspondent. Pearce (2009) gives some useful information on the subject, but is not nearly sufficient to decide whether this project has much chance of success. Describing a project that goes by the name Desertec, Pearce explains:

> The current plan, outlined by the German Aerospace
> Centre (DLR) in a report to the federal government,
> envisages that the project will meet 15 per cent of

Europe's electricity by 2050, with a peak output of
100 gigawatts—roughly equivalent to 100 coal-fired
power stations. (Pearce, 2009, p. 38)

Pearce is right that the peak output of an average coal-fired power
station is around 1 GW, and therefore there is an equivalence between the
peak output of 100 GW and the peak output of 100 coal-fired power stations. But to mention that without further explanation tends to mislead,
since the capacity factor (*average* output compared to *maximum* output) of
coal-fired power stations is around 60%, far higher than solar thermal.
Moreover the main reason that it is as low as 60% is that power output is
varied according to requirements.

Hayden (2004, p. 190) estimates the capacity factor of the large Solar
Electric Generating System (SEGS), located between Los Angeles and Las
Vegas, at 22%.[*] Moreover, SEGS, which incidentally is a "trough mirror"
design, does not attempt to store the hot liquid so that electricity can be
generated during the evening and night. That raises the first question relevant to evaluating the Desertec proposal:

*Q1. What will be the capacity factor after sufficient hot liquid has been stored to
allow power to be delivered on demand (as with a coal-fired power station)?*

Not making any allowance for storage or for the losses in
transmission—but as a rough indication of what would be a more useful
comparison with coal-fired power plants, since the power output of the
system would be 22% of the 100 GW, that is, an average 22 GW—the average number of coal-fired power stations needed, operating at a "normal"
60% capacity factor, would be 22/0.6 = 37 (not 100). Pearce goes on to say:

Preliminary designs in the German report show
electricity reaching Europe via 20 high-voltage direct
current power lines, which will keep transmission
losses below 10 per cent. Trans-Mediterranean links
will cross from Morocco to Spain across the strait of
Gibraltar; from Algeria to France via the Balearic
Islands; from Tunisia to Italy; from Libya to Greece; and
from Egypt to Turkey via Cyprus. (Pearce, 2009, p. 38)

That poses a further question:

*Q2. What is the energy embodied in making all this hardware and in maintaining
it, and how does that relate to the output?*

According to Pearce, "In the Mojave desert in California, an inter-
linked system of nine solar thermal plants which use trough mirrors have

---

[*] Hayden, 2004, p. 190, shows the capacity factor calculation. He correctly subtracts the
output due to the natural gas turbines before making the calculation.

been generating up to 300 megawatts of electrical power for more than two decades" (Pearce, 2009, p. 38). Pearce is referring to SEGS. But once again the "up to 300 megawatts" tends to mislead. The fact that power output peaks as high as 300 MW is an embarrassment when it is considered that the *average* power output captured from the sun is 300 × 0.22 = 66 MW. It is not easy to accommodate into the electrical system a device that, at times of its own choosing, produces 300 MW, but on average produces only 66 MW. Pearce here makes use of the trap for the unwary that is almost invariably used by renewable energy industries, namely, to constantly quote *peak* outputs while avoiding any mention of average outputs.

In the case of SEGS, the variable output from the sun is smoothed by allowing natural gas turbines to produce 25% of the electrical output of the plant. As we have seen, the plan for North Africa, Desertec, is for the variable output to be smoothed by storing hot liquids. Pearce does note this advantage of solar thermal over photovoltaics, observing that solar thermal can "feed electricity into the grid at night as well as by day. This is done by storing the heated fluid in an insulated container and releasing it hours later when the energy is required" (Pearce, 2009, p. 38). We have already noted that this gives rise to Question 1. There are now two further questions:
Q3. Are there sometimes several days in a row when it is sufficiently cloudy that it is necessary to rely on hot fluid stored several days previously?
Q4. What is the energy embodied in this further requirement for hardware to cope with storage and the energy needed for installation?

There is a far bigger problem associated with solar thermal: namely, the difference between the high output in summer and the much lower output in winter. For SEGS, Trainer (2007, p. 168) tells us that "On a typical winter day output from SEGS VI reaches only one-quarter that of a typical summer day," and that the U.S. National Renewable Energy Laboratory estimates SEGS VI's performance in winter to be only 20% of summer output.[*] This gives rise to the most important question, one which Pearce omits altogether:
Q5. What plans are there to deal with the seasonal difference in output?

The power density of solar thermal is possibly[†] as high as 15 W/m². From this, we can estimate the area needed to supply this average 22 GW

---

[*] Note that the whole of SEGS comprises nine units. SEGS VI is presumably one for which some precise summer and winter data has been collected.

[†] Based on National Renewable Energy Laboratory data, Hayden (2004, p. 190), after making due allowance for the gas used to maintain optimum temperatures, calculates that the power density of SEGS is about 11 W/m², but MacKay (2008, p. 177) gives the power density of "concentrating solar thermal (desert)" as 15 W/m². It is that 15 W/m² figure that I use, although it is likely to be optimistic. Those operating Stirling dishes claim on the basis of experience, rather than projections, that they obtain the highest power density at 14 W/m² (MacKay, 2008, p. 184), but Stirling dishes should be highly efficient because they can track the sun the year-round. The disadvantage is that their design makes it difficult to store the hot liquid in suitable tanks.

as being about 1,500 km². Hayden (2004, p. 189) throws light on the problems associated with servicing this area:

> The *optical efficiency* varies from 71% (units I and II) to 80% (units VIII and IX). That is, between 71% and 80% of the sunlight that strikes the mirrors is actually reflected to the pipes containing the therminol. They achieve this high efficiency by washing the mirrors every five or so days, and with a high-pressure wash every ten-to-twenty days.*

This brings up two further questions:
Q6. How much water is going to be required for washing the mirrors?
Q7. Using currently available equipment, covering an area of 1,500 km² with suitable washing equipment would require a substantial amount of oil, but oil is likely to be prohibitively expensive before 2050, so will it be possible to use electrically powered vehicles for this purpose, and if so, how much electricity will be used for this purpose, and what will be the embodied energy in the batteries?

Pearce fails to mention the mirror-washing aspect of water use, but he does quote some figures related to SEGS with regard to the water consumed by the turbines: 3 liters per kWh. This is a significant problem in dry areas. Pearce (2009, p. 40) notes that the U.S. National Park Service warned in April 2009 that "solar thermal projects proposed for the Mohave desert could destroy its limited underground water reserves." The problem of farmers selling their water based on irrigation rights to energy companies is already apparent, and a matter of debate.

One solution to the turbine water-demand problem is to use air cooling, for that only uses a 10th of the water, but at a cost of about 10% loss of efficiency. A further problem suggests itself with regard to air cooling:
Q8. Would air cooling result in an intolerable increase in local air temperature?
There is a political problem in this project insofar as the countries in which the solar thermal units are placed are likely to require some benefit for themselves. Pearce tells us that the Desertec planners suggest that heat from the plants could be used to produce distilled water, but that gives rise to a further question not addressed by Pearce:
Q9. What would be the capacity factor of the plant when some of the heat is used to produce the amount of desalinated water needed to wash the mirrors, drive the turbines, and supply sufficient water to satisfy the host country?

Pearce points out that the plant cannot be situated near the sea since there are usually clouds in the vicinity of the coast. Thus, there would be a

---

* Hayden, too, is amazed by the work involved in washing mirrors and says, "Let's repeat that: they wash several million square meters of mirror—much more than the 2.3 million m² of aperture—about 25 times a year!" (p. 189). He omits looking at the problem of finding the water for this in the dry areas in which such plants are normally situated.

need to pump the water a considerable distance to the solar thermal plant as well as to pump the desalinated water back to where it is needed. That gives rise to these additional questions:

Q10. *What is the energy needed to produce and install the water pipes, and what is the ongoing energy requirement of pumping?*

Many questions need to be answered before we can consider that solar thermal in North Africa will make a useful contribution to electrical energy requirements in Europe. The largest problem is the seasonal difference in output, but there are many other possible pitfalls.

6.4 Almost insurmountable difficulties with EROEI and net power densities

For many decades, scientists have made efforts to include in their energy calculations an estimate of the amount of energy required as input to produce some desired output, such as a specified amount of food. To assist in this task, Boustead and Hancock (1979) published their *Handbook of Industrial Energy Analysis*, which gives the results of calculations of the energy required to produce various materials used in construction, for example, the stainless steel and reinforced concrete. When the desired output is energy itself—as when producing ethanol from corn, or electricity from photovoltaics—quantifying the inputs to achieve the output becomes vitally important. There has been some recent awakening to this fact with revival of interest in what is known as energy return on energy invested (EROEI). However, few people appreciate that the methods that are appropriate while fossil fuels are the normal energy input will not be applicable when dealing with a renewable energy future. The difficulties surrounding EROEI analyses introduce huge uncertainties about the viability of any renewable energy project. As we will see, even a rough analysis suggests dire problems for the United Kingdom.

The first issue to look at is electrical input and output. Electricity is generated from fossil fuel with an efficiency of around 33%, so in making an energy analysis, the actual energy input required is three times the amount of electrical energy used. Does it logically follow that the electricity produced by wind turbines should be assessed at three times the electrical output? This was done in the only analysis that I have seen making a fairly thorough attempt to assess the EROEI of wind turbines. It is arguably valid in the present world, while fossil fuels are used for almost everything, but not in a world where nearly all the energy produced is electricity, and where electricity is used whenever possible, even to just produce heat.

To make the machinery needed to construct, install, and maintain wind turbines, heat would be required. Presently that would be supplied mainly by gas or coal, but in the future it would have to be supplied by electricity. Thus, when the requirement is for heat, then it is only to the extent that heat can be produced from electricity more efficiently than from gas or coal that electricity has any advantage. In other words, there may be an argument for slightly augmenting the energy value of electricity, but not by three times.

An even more severe problem arises from the requirements for liquid fuel inputs. In order to revise a *gross* power density to produce a *net* power density, we have to know the area of land needed to produce the input. In conventional analyses, *net* energy has been assessed simplistically. For instance, if a third of the output of PV panels—over their lifetime—is required as input (e.g. for construction, transport, installation, and maintenance), then without regard to the type of energy, the *net* power density would be assessed as two-thirds of the *gross* power density. But that is simplistic, because it takes no account of the very low power density of liquid fuels, some of which would inevitably be required. The almost unfathomable complexity of assessing EROEI and net power density—when being more realistic by taking the *type* of energy needed into account—is perhaps best explained by following through an attempt at such an analysis, looking at the interaction between the high power density of wind turbines and the low power density of liquid fuels.

Earlier, we arrived at a gross power density for liquid fuels produced in the United Kingdom of 0.25 W/m^2. For present purposes, it will be convenient to use larger units, so let us use an equivalent 2.5 kW/ha. The first step in reducing that to a net power density is to consider how much of that liquid fuel output will be needed as input. In the previously cited report by the U.S. Department of Agriculture (Shapouri et al., 2002), liquid input was estimated at 15% of the liquid fuel output, which could not represent the actual total liquid input, as the paper did not account for amortization of the energy embodied in machinery. Of course, mining and transporting of ores to produce the metals needed to make the machinery requires liquid inputs; thus, liquid inputs properly assessed would be more than 15%. Boustead and Hancock's (1979) book needs updating to apportion the amount of liquid fuel required. Nevertheless, as a rough simplification, let us accept 15% as our figure for liquids, and accordingly (assuming this required input is satisfied from the output) reduce the 2.5 kW/ha to a "useful" liquid output of 2.13 kW/ha.

A lot of the energy needed to produce ethanol is heat for distillation. Everyone agrees that, not counting by-products, the energy inputs when producing ethanol from corn are in close balance with the ethanol output. Thus, nonliquid inputs, which are essentially heat inputs, amount to 85% of the gross output of 2.5 kW/ha, namely, 2.13 kW/ha; let us round that to 2 kW/ha. To recapitulate, we have noted that the heat input required amounts to the same as the liquid output minus the liquid input, or roughly 2 kW/ha.

If that heat could be produced at the optimistic power density given by MacKay (see Table 6.1) of 0.5 W/m², or 5 kW/ha, the land required for heat would be 2/5 = 0.4 ha. Thus the fully net power density for liquid fuels would be the net power from the liquid fuel (2.13 kW/ha) divided by the total area needed, that is, 2.13/(1 + 0.4) = 1.52 kW/ha.

It might be that we could improve on that by getting our heat from a renewable energy source that would provide a better net power density than the 5 kW/ha from green plants, but as we will see later, it is not clear that there is such a source.

With an approximate figure for liquid fuels, we can start to look at wind turbines. MacKay (Table 6.1) gives the power density of onshore wind as 2 W/m² (20 kW/ha). The land that is actually used for installation and roads giving access is usually estimated at between 2% and 5% of the protected area around the turbine needed to prevent interference from turbulence. Using the lower figure of 2%, the gross power density of the wind turbine becomes 50 times greater, that is, 1,000 kW/ha of *land actually used* by the wind turbine.

But to manufacture, transport, install, and maintain the wind turbine is going to take inputs. Taking inputs as 5% of the output might be in the ballpark,* making the required input 50 kW/ha. When the problems of raw materials, transport, installation, and maintenance are included, it would be reasonable to assume that half of that (i.e., 25 kW/ha) needs to be in the form of liquid fuel, while for the other 25 kW/ha the requirement is simply for heat (or electricity). Subtracting for the heat (i.e., assuming it to be derived from the electrical output) reduces the net output to 925 kW/ha. For the 25 kW/ha in the form of liquids, using the liquid net power density of 1.52 kW/ha as calculated above, the area needed is 25/1.52 = 16.4 ha. The area needed for each kW is therefore 16.4/925 = 0.018 hectares per kilowatt (a power density of 56 kW/ha).

The eagle-eyed reader will have noted that in the last calculation we did not include land used by the wind turbines themselves, but onshore the wind turbines *actually use* a mere 0.001 hectares per kilowatt, and that amount would make no significant difference to the calculation. Moreover, MacKay estimates the *absolute limits* of wind output, per person, at 20 kWh/day from onshore turbines, with 16 kWh/day from shallow

---

* Heinberg (2009) says that operational assessments of the EROEI for wind turbines is 18:1— that is, the input is about 5%— but I am doubtful about that figure for three main reasons: (1) In operating wind turbines extensively, it is necessary to have controllable backup plant equal to the power output (*not* the capacity, as is sometimes implied) of the wind turbines. (2) To try to even out the input from wind turbines, it is necessary to have a very extensive grid that can cope with the occasional very high input from the wind turbines. (3) In making these EROEI assessments, the output is usually uprated by a factor of about 3, to the fossil fuel equivalent. This is a dubious proposition in a renewable energy world as explained in Section 6.4.

seas and 32 kWh/day from deep sea. With most of the wind coming from offshore, we can certainly ignore the insignificant amount of land used by the wind turbines and concentrate on the land needed as inputs to produce liquid fuels for construction, transport, installation, and maintenance.

The net power density of 56 kW/ha, in the form of electricity, is a huge improvement for providing heat compared to the 5 kW/ha available from green plants, but *producing sufficient controllable electricity to balance the uncontrollable output of the wind turbines is the biggest problem, and it has a very large effect on net power density.* That is the subject that we will turn to next.

6.4.1 The need to incorporate controllables ruins the power density

As discussed in earlier sections, the uncontrollable nature of wind turbine output means that wind can supply only a fraction of the total power requirement to the grid. That fraction is likely to be less than a third, but one-third will do well enough as an approximation for our next task of estimating the power density that results from the need to have controllable electrical input.

As just implied, for each kilowatt of *uncontrollable* input to the grid, another 2 kW of *controllable* input is required. We have already established that pumped storage is only a very partial solution and hydroelectricity is confined to a subsidiary role, so let us assume that the 2 kW is provided by burning green plants of one kind or another. It is very hard to get wood and such plant matter to burn nearly as efficiently as pulverized coal, but for the sake of this analysis, let us assume that a 33.3% efficiency can be achieved when burning dry wood. That means that the input required to produce 2 kW of electricity is 6 kW of plant matter. MacKay (2008) gives the power density of green plants as 0.5 W/m² (5 kW/ha), so the area needed is 1.2 hectares. Thus, total production of 3 kW by wind turbines backed by the necessary amount of controllable inputs is 1.2 + 0.018 = 1.218 ha, or 1.218/3 = 0.406 ha/kW, which is a power density of 1/0.406 = 2.5 kW/ha. That is a large change from 56 kW/ha.

Note, too, that 2.5 kW/ha is lower than the 5 kW/ha available from green plants, which we used to provide heat when establishing the net power density of liquid fuel from renewable sources. In fact, the two figures are closer than it appears, because the 5 kW/ha was not a *net* power density, and also because electricity can be converted into heat with almost perfect efficiency, whereas that is far from the case with plant matter. Importantly, it remains the case that we were being optimistic in using 5 kW/ha to provide heat.

MacKay's (2008) estimates for all possible wind turbine contributions add up to 68 kWh per day per person, or 68/24 = 2.83 kW/person.

That looks like a respectable part of the approximate 5 kW/person of total energy that is used in the United Kingdom. But the area needed, per person, would be 2.83/2.5 = 1.13 ha, or 11,300 m². By dividing the area of the United Kingdom by the population, it works out that each person has 4,000 m² of land, of which 3,000 m² is cropland and pasture (Duguid and Parsons, 2004); including all agricultural land and forest, each person has 3,500 m². The 3,500 m² might be called "ecologically productive" land. Bearing in mind that the United Kingdom imports 30% of its food and 85% of its timber, it is clear that only a fraction of the present U.K. population could be supported while being provided with 2.83 kW per person, using renewable energy, and relying on such sources as appear to be definitely viable at the present time.

Let's try to improve the situation. Making use of data presented by Smil (1993), it seems that a civilized life could be preserved based on 2 kW/person, at least in temperate climates like Great Britain's, even though that is only about 40% of present energy use. This is not too surprising, knowing that we currently use energy extravagantly—albeit at half the rate of energy use in the United States. Prior to 1890, energy use in the United States is estimated at 3.7 kW/person (Hayden, 2004), and the United Kingdom has a more temperate climate.

The reduction to 2 kW/person would reduce the demand for land to satisfy energy to 8,000 m² per person. If we reduced population to a third, 20 million, each person would have 10,500 m² of ecologically productive land. But as food, timber, and some additional liquid fuel are required, even those reductions in population look to be insufficient.

Despite the unfathomable complexity, even a simple analysis reveals enough of the truth to make it clear that, without fossil fuels, the United Kingdom could—based on technology that is presently known to work—support only a fraction of its present population.

6.4.2 David MacKay's five plans—which won't work!

Although I have drawn the wind turbine figure of 68 kWh/d/person from MacKay's (2008) marvelous book *Sustainable Energy—without the Hot Air*, I need to make it clear that he never suggests trying to make such extensive use of wind turbines. The figure of 68 kWh/d/person was just his estimate of a *theoretical* limit to wind power.

While wind has a high power density, at least in terms of the land monopolized by the wind turbines, its high power density is fatally undermined partly by the need for liquid fuels and partly by the need for a controllable input. Instead of trying to use as much wind as possible, MacKay (2008) draws up five possible plans for allowing Britain to live off renewable energy, each using different mixtures of renewable sources, with some of them using the nonrenewable nuclear option. The plan that most relies on wind calls for taking only 32 kWh/d/person (1.3kW/person) from wind turbines. What he says about all of the plans is illuminating:

> All these plans are absurd.
>
> If you don't like these plans, I'm not surprised. I agree that there is something unpalatable about every one of them. Feel free to make another plan that is more to your liking. But make sure it adds up!
>
> Perhaps you'll conclude that a viable plan has to involve less power consumption per capita. I might agree with that, but it is a difficult policy to sell—recall Tony Blair's response when someone suggested that he should fly overseas for holidays less frequently!
>
> Alternatively, you may conclude that we have a too high a population density, and that a viable plan requires fewer people. Again, a difficult policy to sell. (MacKay, 2008, p. 212–213)

What we have shown above is that, on present evidence, it seems highly unlikely that *any* plan will add up, even with a third of our present population. The essence of all MacKay's plans is to try to do almost everything using electricity. But that would be possible only if there is a way to produce the electricity. One weakness in MacKay's book is that there is a failure to give adequate consideration to the problems of integrating uncontrollable power sources. However, we can surely agree with him when he says:

> This heated debate is fundamentally about numbers. How much energy could each source deliver, at what economic and social cost, and with what risks? But actual numbers are rarely mentioned. In public debates, people just say "Nuclear is a money pit" or "We have a *huge* amount of wave and wind." The trouble with this sort of language is that it's not sufficient to know that something is huge: we need to know how the one "huge" compares with another "huge," namely *our huge energy consumption*. To make this comparison, we need numbers, not adjectives. (MacKay, 2008, p. 3)

We do indeed have a *huge* energy consumption that arises because we have a *huge* population. It is to be hoped that the numbers given above will assist in getting some reality into this heated debate.

References

Ball, J. 2009. Unbridled energy: Predicting volatile wind and sun. *Wall Street Journal*, October 5. http://online.wsj.com/article/SB125443333547957485.html.

Boustead, I., and Hancock, G. F. 1979. *Handbook of industrial energy analysis.* Chichester, England: Ellis Horwood.

Duguid, J. P., and Parsons, J. 2004. *Population, resources and the quality of life.* Llantrisant, Wales: Population Policy Press.

Elliott, D. 2009. Integrated renewables—German style. *Renew, NATTA Newsletter*, (March–April 2009) No. 178:23.

Ferguson, A. R. B. 2007. Choosing the best uncontrollable between wind and PV. *Optimum Population Trust Journal*, 7(2): 31–32. http://populationmatters.org/journal/j72.pdf.

———. 2008. Wind power: Benefits and limitations. In *Biofuels, solar and wind as renewable energy systems: Benefits and risks*, ed. Pimentel, D., 133–151. Dordrecht, The Netherlands: Springer.

———. 2009. Scalability of a Trial Renewable Energy Electricity System. *Optimum Population Trust Journal*, 9(2): 28.

Hayden, H. C. 2004. *The solar fraud: Why solar energy won't run the world.* 2nd ed. Pueblo, CO: Vales Lake.

Heinberg, R. 2009. *Searching for a miracle: "Net energy" limits and the fate of industrial society.* San Francisco: International Forum on Globalization; Sebastopol, CA: Post Carbon Institute. Accessed July 14, 2010. http://www.postcarbon.org/new-site-files/Reports/Searching_for_a_Miracle_web10nov09.pdf.

Kemp, M. 2009. Powering up. *Clean Slate* 74(Winter). http://www.cat.org.uk/membership/cs_back.tmpl.

MacKay, D. J. C. 2008. *Sustainable energy—without the hot air.* Cambridge, England: UIT. http://www.withouthotair.com.

Office of Technology Assessment. 1979. *Gasohol.* Washington, DC: Congress of the United States. http://www.fas.org/ota/reports/7908.pdf.

Perrin, R. K., Vogel, K. P., Schmer, M. R., and Mitchell, R. B. 2008. Farm-scale production cost of switchgrass for biomass. *BioEnergy Research*, 1: 91–97.

Pearce, F. 2009. Sunshine superpower. *New Scientist*, 204(2731): 38–41.

Pimentel, D., ed. 1993. *World soil erosion and conservation.* Cambridge: Cambridge University Press.

Pimentel, D., and Pimentel, M. 2008. *Food, energy, and society.* 3rd ed. Boca Raton, FL: CRC Press.

Shapouri, H., Duffield, J. A., and Wang, M. 2002. The energy balance of corn ethanol: An update. U.S. Department of Agriculture. Agricultural Economic Report, No. 813. http://www.usda.gov/oce/reports/energy/aer-814.pdf.

Smil, V. 1993. *Global ecology: Environmental change and social flexibility.* London: Routledge.

Trainer, F. E. 2007. *Renewable energy cannot sustain a consumer society.* Dordrecht, The Netherlands: Springer.

chapter seven

Net energy balance and carbon footprint of biofuel from corn and sugarcane

Claudinei Andreoli, David Pimentel, and Simone Pereira de Souza

Contents

7.1 Introduction

On December 17, 2007, President George W. Bush signed into law the Energy Independence and Security Act of 2007 (EISA), which requires an expanded renewable fuel standard (RFS2) increasing biofuels production to 36 billion gallons by 2022. Of this amount, 15 billion gallons is to come from conventional (cornstarch-based) ethanol. EISA established several categories of biofuels, determined by their reductions in life-cycle green-house gases (GHGs) versus baseline fuel (gasoline or diesel fuel).

In April 2009, the California Air Resources Board (CARB) adopted the Low Carbon Fuel Standard (LCFS) (Farrel and Sperling, 2007a, 2007b; CARB, 2009c), and in May 2009, as part of proposed revisions to the National Renewable Fuel Standard (RFS) program, the U.S. Environmental Protection Agency (EPA) analyzed life-cycle GHG emissions for corn ethanol and sugarcane ethanol (EPA, 2009). Later, CARB and the EPA revised the estimations of GHG emissions for corn and sugarcane ethanol.

Section 201 of EISA defines *life-cycle greenhouse gas emissions* as

> the aggregate quantity of GHG emissions (includ-ing direct and indirect emissions such as significant emissions from land use changes), as determined by the Administrator, related to the full fuel life-cycle, including all stages of fuel and feedstock production and distribution, from feedstock gen-eration or extraction through the distribution and delivery and use of the finished fuel to the ultimate consumer, where the mass values for all greenhouse gases are adjusted to account for their relative global warming potential.[*]

The life-cycle GHG emissions of renewable fuels are compared to the life cycle assessment (LCA) emissions for gasoline or diesel.

The energy balance and GHG emissions of biofuels remain a con-troversial topic in the popular press, with government policy makers, and within the academic community. The energy balance and life-cycle GHG emissions of corn and sugarcane plants have been studied for a long time, resulting in a wide range of data (Andreoli, 2006; Andreoli and Souza, 2006–2007; Boddey et al., 2008; CARB, 2009a, 2009b, 2009c; de Oliveira et al., 2005; de Oliveira, 2008; EPA, 2010; Farrel et al., 2006; Hill et al., 2006; Liska et al., 2008; Macedo et al., 2004, 2008; Macedo and Seabra, 2008; Pimentel and Patzek, 2005, 2008a, 2008b; Shapouri et al., 2003; Wang et al., 2007, 2008).

CARB and the EPA released the final rules for the LCFS in July 2009 and the Regulations for the National Renewable Fuel Standard Program for 2010 and Beyond (RFS2 Final Rule) on February 3, 2010. Thereafter, they included the international land-use change (ILUC) in the total GHG emission calculation for corn and sugarcane ethanol. Table 7.1 summa-rizes the corn and sugarcane life-cycle GHG emissions from CARB and the EPA.

---

[*] Clean Air Act § 211(o)(1)(H).

Table 7.1 Comparison of Corn and Sugarcane Ethanol Rules on GHG Emissions from CARB and the EPA

	Corn		Sugarcane	
	EPA[a]	CARB[b]	EPA[a]	CARB[b]
Lifecycle phase	(g CO_2e/MJ over 30 years)			
Net domestic agriculture	3.8	35.85	—	18.6
Net international agriculture	11.4	—	40.8	—
Domestic land-use change	−1.9	−11.51	—	—
International land-use change (ILUC)	30.3	30.0	3.3	46.0
Fuel production	26.5	38.3	−11.4	1.9
Fuel and feedstock transport	3.8	4.92	4.1	6.1
Tailpipe emissions	0.9	0.0	0.9	0.0
Net total emission	74.8	97.6	36.2	72.6
Gasoline	92.9	96.0	92.9	96.0
% GHG reduction	21	−1.1	61	24.4

Sources:

[a] Environmental Protection Agency, *Renewable Fuel Standard Program: Final regulatory impact analysis*, EPA-420-R-10-006 (Washington, DC: EPA, 2010).
[b] California Air Resources Board, *Detailed California-modified GREET pathway for Brazilian sugar cane ethanol*, Version 2.0, released January 12, 2009.

The EPA's analysis showed GHG reductions for dry mill corn ethanol using natural gas for ethanol processing ranging from 7% to 32%, with the midpoint of the range being a 21% reduction. The results for the sugarcane ethanol scenario are that the midpoint of the range of results is a 61% reduction in GHG emissions compared to the gasoline baseline (using 30-year, 0% discount).

We call attention to the fact that, for the same product, there are huge variations over the GHG emission life cycle, such as GHG emissions for U.S. average dry mill and wet mill corn and Brazilian sugarcane ethanol in different phases. For instance, for corn ethanol, CARB estimated 35.85 g CO_2e/MJ for net domestic Agriculture, while the EPA calculated only 3.8. The same is seen for the controversial ILUC: for sugarcane, CARB devoted 46.0 g CO_2e/MJ and EPA just 3.3. Moreover, the total carbon footprint was estimated by CARB at 97.6 g CO_2e/MJ for corn ethanol and 72.6 g CO2e/MJ for sugarcane ethanol. Corn ethanol offset 21% of CO_2 under the United States Environmental Protection Agency's rules, but adds 1.1% GHGs under LCSR in California; sugarcane ethanol reduces GHG emissions 61% and 24% , respectively, as displaced by fossil fuel (see Table 7.1). Therefore, we can conclude that corn ethanol cannot be considered an advanced biofuel according to LCSR rules; however, sugarcane ethanol, even including ILUC, is considered an advanced biofuel under EPA's rules.

Table 7.2 Comparison of Baseline Pathway with Two Additional Scenarios
Analyzed for Ethanol Sugarcane

	Pathway Baseline	Scenario 1	Scenario 2
Mechanized harvest with coproduct electricity	No	Yes	No
Electricity credit	No	Yes	Yes
Total GHG emissions (g CO_2e/MJ)	27.40 (73.4)[a]	12.20 (58.2)[a]	20.40 (66.4)[a]
% GHG reduction	23.5	40.0	31.0

Sources: California Air Resources Board, *Detailed California-modified GREET pathway for Brazilian sugar cane ethanol*, Version 2.0, released January 12, 2009; California Air Resources Board, *Detailed California-modified GREET pathway for Brazilian sugar cane ethanol: Average Brazilian ethanol, with mechanized harvesting and electricity co-product credit, with electricity co-product credit*, Version 2.3, released September 23, 2009.

[a] Land-use change of 46 g CO_2e/MJ was included.

On July 2009, CARB revised the life cycle of sugarcane ethanol and included two scenarios on the baseline pathway, as described in Table 7.2:

- Scenario 1: with mechanized harvest and export coproduct electricity
- Scenario 2: with electricity credit

Table 7.2 shows the result of CARB's calculations for the two scenarios. Even with credit carbon from mechanical harvesting and electricity cogeneration from burning bagasse, sugarcane ethanol cannot be considered an advanced biofuel. Indeed, even including Scenario 1 (mechanical harvest and electricity credit from bagasse) in the ethanol sugarcane LCA, the total GHG direct emissions (12.2 g CO_2e/MJ) were below CARB's criteria for advanced biofuels when ILUC was included (CARB, 2009b).

The objective of this chapter is to evaluate, through life-cycle accounting, the energetic and environmental benefits of ethanol from corn in the United States and from sugarcane in Brazil. We will attempt to examine the implication of expanding areas for the production of these renewable fuels.

7.2   Corn and ethanol production: Net energy ratio and greenhouse gas emissions

The United States leads the world in ethanol production, with 39.3 BL (billion liters) in 2009. Table 7.3 shows ethanol production in the United States from 2000 to 2009 (RFA, 2010). Ethanol production is expanding rapidly with the adoption of improved technologies to increase energy efficiency, ethanol conversion, and coproduct uses in response to supportive federal subsidies. Corn ethanol production started in 2000 at 6.2 BL, rising to 39.3 BL in 2009, and the government and industry goals are to reach 56.8 BL in 2015.

Table 7.3 Evolution of Ethanol Production
in the United States, 2000–2009

Year	Ethanol production (million L)
2000	6,169.5
2001	6,699.5
2002	8,062.0
2003	10,598.0
2004	12,869.0
2005	14,776.6
2006	18,376.2
2007	24,602.5
2008	34,065.0
2009	39,269.4
2015[a]	56,775.0

Source: Renewable Fuels Association, Industry statistics: Ethanol industry overview (Washington, DC: RFA, 2010).

[a] RFS2 estimation (EISA, 2007).

LCA can evaluate the impact of these technological and production changes on energetic and environmental issues. To better understand the net energy and environmental impacts of corn ethanol, we surveyed the academic literature and information produced by public institutions such as CARB and the EPA. Estimates of net energy and GHG emissions are highly sensitive to assumptions of system boundaries and key parameters (Farrel et al., 2006).

Recent published results indicate that corn ethanol energy content is from −39% to +59% renewable (CARB, 2009c; de Oliveira et al., 2005; EPA, 2010; Farrel et al., 2006; Hill et al., 2006; Liska et al., 2008; Shapouri et al., 2003; Wang et al., 2007). Pimentel and Patzek (2008b) report that, based on a net energy loss of 2,244 kcal of ethanol produced, 46% more fossil energy is expended than is produced by ethanol. Direct-effect GHG emission values range from a 20% increase to a decrease of 59% compared with gasoline, depending on the energy source used for ethanol processing (Farrel et al., 2006; Liska et al., 2008; Wang et al., 2007). It is worthwhile to note in Table 7.1 that CARB and the EPA calculated a reduction of 52% and 30% for corn ethanol, respectively, as land-use change was not included (CARB, 2009c; EPA, 2010).

More recently, Hertel, Golub, et al. (2010) reported that direct CO_2 releases for corn ethanol produced in the United Sates amounted to 65 CO_2 g/MJ and, in April, Tyner et al. (2010), modifying and improving the analysis, estimated on average 63.6 CO_2 g/MJ. These values mean a 32% reduction in GHG emissions compared with gasoline.

Many authors in the United States have suggested that corn ethanol systems have great potential to mitigate GHG emissions and reduce dependence on imported petroleum for transport fuels. However, the energetic, economic, and environmental benefits of corn ethanol are very low compared with sugarcane ethanol production, mainly due to the lower fossil energy inputs for ethanol processing and the additional energy supply and export from burning the sugarcane bagasse (Andreoli, 2006; Macedo et al., 2008; Souza, 2010). Moreover, at present oil prices, corn farming and the ethanol industry in the United Sates are very dependent upon federal and state subsidies. On the other hand, the United States imposes two duties on ethanol imports: a 2.5% *ad valorem* tariff plus an additional "other duty or charge" of $0.54 per gallon. According to data from the U.S. International Trade Commission, the combined duties have amounted to about a 30% tariff on ethanol imports. Without the subsidies or import tariff, the corn ethanol industry would fail.

7.3 Sugarcane and ethanol production

Brazil is the leading sugarcane producer in the world, with 7.1 million ha in production in 2008 (about 0.3% of total arable area). Figure 7.1 displays the growth of sugarcane production in Brazil from 1990 to 2008. According to the National Supply Company (CONAB, 2010), the sugarcane acreage of 7.1

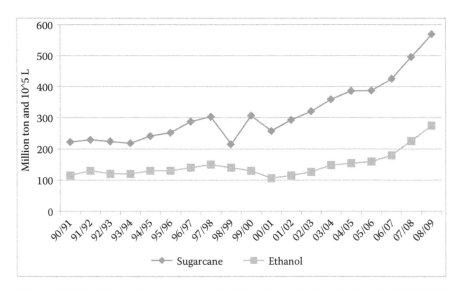

Figure 7.1 Evolution of sugarcane and ethanol production in Brazil, 1990–2008. (From UNICA. 2011. Dados e Cotações – Estatisticas. União da Indústria de Cana-de-açúcar. Accessed November 16, 2011. http://www.unica.com.br/dadosCotacao/estatistica.)

Table 7.4 Acreage, Yield, and Ethanol Production of Sugarcane by Different Regions and the State of São Paulo, Brazil, 2009–2010

Region	Area (1,000 ha)	Yield (t/ha)	Production (1,000 t)	Ethanol (million L)
North	17.2	57.6	991.6	53.36
Northeast	1,082.6	56.0	60,677.2	1,972.2
Center-West	940.3	82.3	77,435.9	4,287.2
Southeast	4,832.6	86.9	419,857.7	17,565.9
South	537.0	84.8	45,551.3	1,883.8
Northeast	1,099.8	56.1	61,668.8	2025.6
Center-South	6,309.8	86.0	542,844.8	23,737.0
São Paulo	4,129.9	87.8	362,664.7	16,214.1
Brazil	*7,409.6*	*81.6*	*604,513.6*	*25,762.6*

Source: Companhia Brasileira de Abastecimento, 2011. Cana-de-Acucar, Safra 2011/2012, Segundo Levantamento. http://www.conab.gov.br/OlalaCMS/uploads/arquivos/11_08_30_13_41_19_boletim_cana_portugues_-_agosto_2011_2o_lev.pdf.

million ha produced 572 million tons of sugarcane and 26.7 BL of ethanol in the crop year 2008–2009. The Center-South region, which includes the states of Southeast, South, and Center West, account for 90% of domestic production (see Table 7.4).

The state of São Paulo ranks first in production (342.9 million tons harvested from 3.29 million ha) and provides 60% of the total sugarcane and ethanol production in Brazil. Recent satellite images confirm the area committed to sugarcane for the state of São Paulo (see Figure 7.2). The present sugarcane yield for the Center-South region is, on average, 85.5 t/ha, and this value was used in this study (Table 7.5) (CONAB, 2010).

The harvested sugarcane used for ethanol production was 325.9 million tons in 2009, an average mix of 45% for sugar and of 55% for ethanol production (CONAB, 2010). In 2008, ethanol production reached 26.7 BL—19.4 BL of hydrated ethanol, which is used in the flex fleet, and 7.3 BL of anhydrous ethanol, which it is mixed with gasoline at 20–25% (Walter, 2009). The present conversion of ethanol per Mg of fresh cane in the Center-South region is on average 85.0 L (CONAB, 2010), and this ethanol conversion yield was used in this chapter.

7.3.1 Net energy ratio

It is important to point out that the database used in this chapter comes from mills in the Center-South region, which mowed 296.3 million tons of sugarcane for ethanol production in 2009 (CONAB, 2010). The data for sugarcane yield and for fertilizer, herbicide, insecticide, lime, and diesel consumption for all crop practices, harvesting, cane transport, and delivering

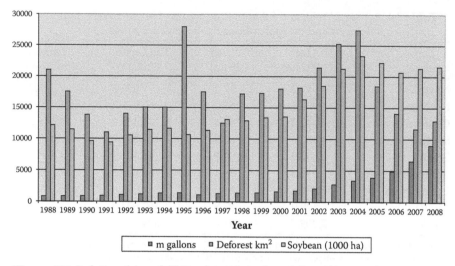

Figure 7.2 Relationship of U.S. corn ethanol and Brazilian soybean expansion with Amazon deforestation rate, 1988–2008. (Data of ethanol from Renewable Fuels Association, 2010, Industry statistics: Ethanol industry overview, Washington, DC: RFA, 2010; of soybean from Companhia Brasileira de Abastecimento, 9° levantamento de cana-de açúcar, Brasilia, D.F., 2010; and of Amazon deforestation from Instituto Nacional de Pesquisas Espaciais, *Estimativa das emissões de CO$_2$ por desmatamento na Amazônia Brasileira*, Brasilia, D.F., 2009.)

of coproducts to the field are found in Tables 7.5 and 7.6. The energy needed for each of these agricultural inputs was derived from the studies by Souza (2010) and the IPCC (2006a). Embodied energy used in operating agricultural equipment and for manufacturing this equipment were estimates based on Souza's data. For planting stock, we estimated 2.28% of total energy for sugarcane production, as calculated by Boddey et al. (2008) and Macedo et al. (2008). All these estimates of per hectare energy use were transformed into MJ per ton based on a crop yield of 85.5 tons per hectare. The total farming energy inputs are shown in Table 7.7.

The fossil energy cost of transport for alcohol production from sugarcane is high and includes cane loading and transport from the field to the mill, coproduct deliveries, and ethanol transport from mills to the pump stations. For cane transport, we have assumed heavy trucks (average 49 tons of cane per load) will require 2.0 loads per hectare. For a round trip of 58.3 km with a mean of 1.85 km/L, the diesel consumption to fetch 1 ha of cane would be 64.2 liters. Considering that five cuts in six cycles (83.3%) of the transport occur, the cane transport consumes 53.5 L/ha. At an energy value of 47.6 MJ/L, the mean fossil energy consumption to transport cane to the mill becomes 29.82 MJ/ton (see Table 7.7). The same

Table 7.5 Yield of Sugarcane and Coproducts

Sugarcane	
Cane yield	85.5 t/ha/year
Seed stock	12 t/ha
Cycle	6 years
Number of cuts	5
Ethanol	
Ethanol	7,267.5 L/ha/year
Ethanol	85 L/t cane
Coproducts	
Bagasse	0.14 dry t/t cane
Filter cake	23.9 kg/t cane
Stillage	14.0 L/L ethanol
Stillage	101.7 m³/ha

Source: Adapted from Souza, S. P., Produção integrada de biocombustíveis: Uma proposta para reduzir o uso de combustível fóssil no ciclo de vida do etanol de cana-de-açúcar [Integrated production of biofuels: A proposal to reduce fossil fuels in the life cycle of sugarcane ethanol], thesis, University of São Paulo, São Carlos, 2010.

estimations were used for coproducts (stillage and filter cake delivering to the field), providing an energy cost of 94.62 MJ/ton (Table 7.7; see Souza, 2010, for details).

As seen in Table 7.7, the largest energy inputs in sugarcane production are fertilizer and diesel fuel for cane harvesting, transport, and coproduct disposal deliveries. The total farming energy input was 30.3 GJ/ha or 354.3 MJ/ton, and 89% of the energy input is fossil fuel. In contrast, 160–210 MJ/ton of agricultural energy inputs (see Table 7.10, below) have been estimated for sugarcane in Brazil (Boddey et al., 2008; Macedo and Seabra, 2008; Macedo et al., 2008; Pereira and Ortega, 2010; Wang et al., 2008).

The main contributing pathway is the farming stage, accounting for 88.9% of all energy flow to produce ethanol. The industrial steps account for just 2.0%, and ethanol transport to the pumps makes up the remaining 9.1%. Another important operation in the sugarcane ethanol industry is coproduct deliveries to the field, which has been neglected in the life cycle and contribute 23.8% of the energy total input (see Table 7.8).

The total energy use for ethanol production (the processing phase to convert sugar into ethanol) is 6.32 MJ/ton. This energy input is low because all the mills in the sugar-ethanol industry in Brazil burn bagasse to supply the energy for steam and electricity used during processing. Based on 140 kg of bagasse produced per ton of sugarcane (Table 7.6), it cogenerates 87.5 kWh.

Table 7.6 Database for Sugarcane Production, Harvesting and Transport

Factor	Unit	Amount used	Energy intensity (MJ/kg or L)
Sugarcane yield	t/ha	85.5	
Ethanol yield	L/t	85.0	22.21
Straw yield (bagasse)	Dry t/t sugarcane	0.14	13.8
Diesel	L/ha	277.6	47.6
Nitrogen			67.6
Planting cane	kg/ha	60	
Ratoon with stillage	kg/ha	30	
Ratoon w/o stillage	kg/ha	100	
P_2O_5			34.7
Planting cane	kg/ha	90	
Ratoon with stillage	kg/ha	60	
Ratoon w/o stillage	kg/ha	60	
K_2O			9.64
Planting cane	kg/ha	120	
Ratoon with stillage	kg/ha	0.0	
Ratoon w/o stillage	kg/ha	120	
Lime	kg/ha	2,000	0.38
Herbicide	kg/ha	5.0	407
Insecticide	kg/ha	2.0	405
Filter cake application	kg/ha	2,200	
Stillage application	m³/ha	140	
Mechanical harvesting	% area	40	
Burning field	% area	60	

Source: Adapted from Souza, S. P., Produção integrada de biocombustíveis: Uma proposta para reduzir o uso de combustível fóssil no ciclo de vida do etanol de cana-de-açúcar [Integrated production of biofuels: A proposal to reduce fossil fuels in the life cycle of sugarcane ethanol], thesis, University of São Paulo, São Carlos, 2010.

For ethanol distribution (from mill to gas station), the assumptions were: 100% road heavy trucks, a mean distance of 350 km, truckloads of 30,000 liters for each trip, and diesel consumption of 0.54 L/km (average load and empty). Therefore, the energy cost for alcohol transport was 36.14 MJ/t (see Table 7.9).

The energy needed to produce a ton of sugarcane is 361.5 MJ. Adding the energy for ethanol distribution of 36.1MJ/ton, the total energy to produce and distribute sugarcane ethanol is 397.7 MJ/ton (see Table 7.8). The total energy output of the sugarcane, including the bagasse supply, is 2,032.1 MJ/ton (see Table 7.9). It is important to note that there is an energy

Table 7.7 Total Energy Input for Sugarcane Farming Production in Brazil

	GJ/ha/yr	MJ/t
Agricultural phase		
Seeds	0.69	8.07
Fertilizer		
Nitrogen (N)	4.73	46.67
Phosphate (P_2O_5)	0.90	8.54
Potassium (K_2O)	0.40	3.97
Lime	0.76	8.89
Total	*6.79*	*68.07*
Pesticides		
Herbicide	2.04	23.86
Insecticide	0.81	9.47
Total	*2.85*	*33.33*
Fuel consumption		
Cultural treatments	7.09	82.92
Harvesting	2.69	31.46
Cane transportation	2.55	29.82
Coproducts disposal	8.09	94.62
Total	*20.42*	*238.82*
Machinery and equipment		
Machinery	0.14	1.64
Implements	0.13	1.52
Trucks, harvesting, loading	0.24	2.81
Total	*0.51*	*5.96*
Total farming input	30.29	354.26

gain of 230 MJ/ton of sugarcane as electricity surplus from burning the bagasse (Table 7.9).

The summary of energy flow and the net energy ratio is presented in Table 7.9. The total energy cost of the annual mean per hectare of sugarcane (7,267 liters of ethanol) is 173.74 GJ/ha, or 397.66 MJ/ton of sugarcane. As noted above, farming and cane transport energy costs represent the largest part, 88.9% of the entire energy inputs for sugarcane ethanol production in Brazil. The total sugarcane energy output, including electricity from burning bagasse, is 2,032 MJ/ton. Based on the net energy gain of 1,634.4 MJ (output – input) ethanol produced, the net energy ratio is 1:5.11; in other words, for each MJ of fossil energy input in the sugarcane ethanol production, there is a gain of 5.11 MJ of ethanol produced.

Table 7.8 Total Energy Input for Sugarcane Ethanol Production

	GJ/ha/yr	MJ/t
Industrial phase		
Chemicals and lubricants		
CaO	0.03	0.35
Phosphoric acid	0.00	0.00
Sulfuric acid	0.11	1.29
Lubricants	0.00	0.00
Total	*0.14*	*1.64*
Equipment and construction		
Cement	0.01	0.117
Mild steel	0.24	2.807
Mild steel in light equipment	0.21	2.456
Stainless steel	0.08	0.936
Total	*0.54*	*6.32*
Electricity consumption	4.97	58.13
Total industry	*0.68*	*8.06*
Total input (farming + industry)	30.97	362.32
Total input allocated	30.91	361.52
Ethanol transport	3.09	36.14
Total input allocated	34.0	397.66

Table 7.9 Energy Flow and Net Energy Ratio for Sugarcane Ethanol in Brazil

Sugarcane ethanol components	GJ/ha/yr	MJ/t
Fossil input		
Farming and cane transport	30.22	353.45
Ethanol production	0.69	8.07
Ethanol transport	3.09	36.14
Total	*34.0*	*397.66*
Sugarcane output		
Ethanol[a]	154.07	1,802.00
Electricity produced	24.02	280.94
Electricity surplus[b]	19.67	230.06
Bagasse surplus	0.00	0.00
Total output gross	*173.74*	*2,032.06*
Net energy ratio	1:5.11	1:5.11

[a] Based on LHV (lower heating value) = 21.20ML/L.

[b] Considering the replacement of biomass electricity for natural gas electricity.

Our net energy balance (NEB) estimate is less than those reported previously for sugarcane ethanol in Brazil (Boddey et al., 2008; Macedo et al., 2008, Macedo and Seabra, 2008; Pereira and Ortega, 2010; Wang et al., 2008). Comparison of our NEB results with those of other authors is shown in Table 7.10. The main reasons for the difference are the underestimated farming energy costs and the other authors' failure to account for the energy cost of delivering coproducts into the field (see Table 7.7). Yet, these authors have published similar results, because they have used the same database from Macedo et al. (2004, 2008). They have reported between 160 and 210 MJ/ton to produce and transport sugarcane, whereas we have estimated 354 MJ/ton. Production, harvesting, and transport pathways of sugarcane consume a large amount of fossil fuel (oil), and this makes a large difference in the net energy balance. Boddey et al. (2008), Macedo et al. (2004, 2008), Macedo and Seabra (2008), and Pereira and Ortega (2010) estimated much less diesel oil to be needed during all farming operations and transport than we have used in this study, 306 L/ha. Nonetheless, the output energy for ethanol of all papers is almost identical (roughly 2.0 GJ/ton of sugarcane).

In summary, we can conclude that sugarcane is the most profitable and the best alternative to produce renewable fuel for transportation as has been reported previously (Andreoli and Souza, 2006–2007; Andreoli, 2006; Boddey et al., 2008; EPA, 2010; Goldemberg, 2007; Goldemberg et al., 2008; Goldemberg and Guardabassi, 2009; Macedo et al., 2008; Macedo and Seabra, 2008; Pereira and Ortega, 2010; Souza, 2010; Wang et al., 2007). However, Brazilian ethanol has been criticized by many authors who do not have the complete data and the Brazilian approach to sugarcane and ethanol production (Fargione et al., 2008; Lapola et al., 2010; Laurance, 2007;

Table 7.10 Comparison of Estimates of Energy Flow and Balance of Sugarcane Ethanol Production in Brazil

Lifecycle phase	Macedo et al. (2008)	Boddey et al. (2008)	Wang et al. (2008)	This study
	(MJ/ton)			
Farming and transport	210.2	159.70	194.2	389.59
Fuel production	23.6	34.04	37.2	8.07
Fossil input	233.8	193.74	231.4	397.66
Ethanol	1,926.0	1,756.8	1,952.0	1,802.0
Bagasse surplus	176.0	0.0	0.0	0.0
Electricity surplus	82.8	0.0	279.0	230.06
Renewable output	2,185.0	1,756.8	2,231.0	2,032.06
Energy balance	9.3	9.07	9.6	5.11

Melillo et al. 2009; Pimentel and Patzek, 2008a; Sawyer, 2008; Scharlemann and Laurance, 2008; Searchinger et al., 2008).

7.3.2   Greenhouse gas emissions

The use of fossil fuels and soil management in sugarcane and ethanol production results in the emissions of greenhouse gases. The amount of energy in each component of the ethanol life cycle was used to estimate GHG emissions based on emission factors of each component (IPCC, 2006b; see Souza, 2010 for details). We estimated carbon dioxide (CO_2), methane (CH_4), and nitrous oxide (N_2O) emissions for sugarcane farming (including cane transport, straw burning in the field before harvest, bagasse burning for steam and energy use in the mill, and soil management—N_2O emissions from N application and CO_2 from lime), as well as for ethanol production and transport. The GHG emissions from coproduct deliveries (stillage and filter cake) to the field were also estimated. A summary of the results is displayed in Tables 7.11 and 7.12.

The GHG emissions from farming sugarcane total 13.30 g CO_2e/MJ, of which fertilizer use and diesel fuel consumption account for 91% (see Table 7.11). The GHG emissions from ethanol production (processing, fermentation, and distillation) in the mills are only 0.32 g CO_2e/MJ (see Table 7.13). The emissions from fertilizer and soil management, which include emissions from fertilizer, herbicide, pesticide, soil N-N_2O, and lime application, amount to 6.35 g CO_2e/MJ (see Table 7.14).

7.3.2.1   GHG emissions from straw burning in the field

The sugarcane field is burned prior to manual harvest and CO_2, CH_4, and N_2O gases are emitted into the atmosphere. For the 2008–2009 harvest, we assume that 60% of the sugarcane field was burned and the straw yield was 0.19 dry kg of straw per kg of cane. The emission factors used for CH_4 and N_2O, derived from IPCC (2006a), were 2.7 g and 0.07 g per kg of straw burned, respectively. The IPCC global warming potential factors to calculate CO_2-equivalent (CO_2e) values for CH_4 and N_2O were 25 and 298, respectively. The CO_2 emissions were not considered, because they are recycled and renewable during photosynthesis.

7.3.2.2   Soil N_2O emissions resulting from nitrogen
            fertilizer use in sugarcane production

Nitrogen is an essential nutrient for plant growth. Soil organic nitrogen decomposes into ammonia; soil microorganisms convert ammonia to nitrate (nitrification) and may further utilize the nitrate for respiration and convert it into N_2 (denitrification). In the sugarcane production process, some N_2O is released and acts as a potential greenhouse gas.

Table 7.11 Total GHG Emissions from Sugarcane Farming in Brazil

	GJ/ha/yr	MJ/t
Agricultural phase		
Seeds	61.63	0.40
Fertilizer		
Nitrogen (N)[a]	727.20	4.720
Phosphate (P_2O_5)	52.08	0.338
Potassium (K_2O)	21.12	0.137
Lime[a]	171.80	1.115
Total	*972.18*	*6.31*
Pesticides		
Herbicide	77.50	0.503
Insecticide	27.40	0.178
Total	*104.98*	*0.681*
Fuel consumption		
Cultural treatments	114.98	0.746
Harvesting	193.70	1.257
Cane transportation	183.63	1.192
Coproducts disposal	408.38	2.651
Total	*900.69*	*5.85*
Machinery and equipment		
Machinery	2.13	0.014
Implements	2.53	0.016
Trucks, harvesting, loading	4.16	0.027
Total	*8.82*	*0.057*
Total farming input	2,048.3	13.30

[a] GHGs emissions from Soil N_2O-N and $CaCO_3$ are included

Table 7.12 Total GHG Emissions from Sugarcane Fields

Field emissions	kg CO_2e/ha/yr	g CO_2e/MJ
Stillage	278.94	1.81
Field burning	610.20	3.96
Filter cake	52.24	0.34
Bagasse burning	287.78	1.87
Total	*1,229.16*	*7.98*

Source: From Souza, S. P., Produção integrada de biocombustíveis: Uma proposta para reduzir o uso de combustível fóssil no ciclo de vida do etanol de cana-de-açúcar [Integrated production of biofuels: A proposal to reduce fossil fuels in the life cycle of sugarcane ethanol], thesis, University of São Paulo, São Carlos, 2010.

Table 7.13 Total GHG Emissions from Ethanol Production

	GJ/ha/yr	MJ/t
Industrial phase		
Chemicals and lubricants		
CaO	1.155	0.0075
Phosphoric acid	2.280	0.0148
Sulfuric acid	15.58	0.1011
Lubricants	0.832	0.0054
Subtotal	*19.847*	*0.129*
Equipment and construction		
Cement	2.74	0.0178
Mild steel	13.02	0.0845
Mild steel in light equipment	11.69	0.0759
Stainless steel	1.86	0.0121
Subtotal	*29.31*	*0.190*
Total industry	*49.16*	*0.319*

All agricultural production using a N fertilizer input will inevitably contribute to N_2O emissions. N_2O is now an element of the LCA, but a clear understanding of the N_2O factor is lacking. The IPCC (2006b) proposes a release rate based on the extent of N fertilizer added to soil, recommending that 1.3% of each N addition will be converted to N_2O.

Crutzen et al. (2008), assuming biofuels are produced in the same manner as conventional crops, have determined the need for adding fertilizer N from plant content, then related the associated N_2O emissions to the fossil fuel replaced and resulting CO_2 emissions. They concluded that GHG emissions of commonly used biofuels can contribute as much or more to global warming from N_2O emissions than the emissions from the consumption of fossil fuels. Even assuming 3–5% of N fertilizer is lost as N_2O, their data show that ethanol sugarcane has the lowest relative climate warming among different energy crops (see Crutzen et al.,

Table 7.14 Total GHG Emissions from Agricultural Chemical Use for Sugarcane Ethanol in Brazil

Farming components	Fertilizer	Herbicide	Insecticide	Soil N_2O and NO	CO_2 from lime	Total
GHGs ($g CO_2e/MJ$)	1.96	0.50	0.18	2.76	0.95	6.35

2008, Table 1). By using the assumptions of Crutzen et al., our relative warming was much lower by adding the energy surplus from bagasse for steam and electricity cogeneration. Crutzen et al. did not account for the carbon present in the bagasse. Therefore, the N_2O release from sugarcane ethanol production does not negate global warming reduction by replacing fossil fuels as postulated by Crutzen et al. (2008).

In our study, assuming the IPCC estimate of 1.3% of nitrogen fertilizer converted into N_2O, we calculated the direct field N_2O emissions resulting from nitrogen fertilizer application in sugarcane for ethanol production in Brazil. Table 7.15 shows total GHG impacts from N_2O emissions from two inputs: N fertilizer (kg/ha) and percentage conversion of N input to N_2O. The N_2O emissions for sugarcane production were estimated to be 2.76 g CO_2e/MJ. N_2O-N emission values up to 6 g CO_2e/MJ have been reported for ethanol production (Hill et al., 2006; EPA, 2010). Drastically, Melillo et al. (2009) report that, over the next century (100 years), the N_2O emissions become larger in CO_2e than carbon emissions from land use.

7.3.2.3 Effect of lime added to soil on GHG emissions

We assumed that all of the carbon in added lime is converted into CO_2. Thus, soil CO_2 emissions = 3,898.6 g $CaCO_3$/ton × (44 g CO_2/100g $CaCO_3$) = 1,715.4 g CO_2/ton. This converts to 1,715.4/85 × 21.2 = 0.95 g CO_2/MJ. Carbon intensity emissions from lime have been estimated between 1.2 and 2.5 g CO_2e/MJ (CARB, 2009a; Hill et al., 2006; Macedo et al., 2008).

7.3.2.4 Total GHGs emissions for sugarcane ethanol

All direct emissions in terms of CO_2-equivalents total 22.76 g CO_2/MJ (see Table 7.16), which represents a reduction of 76% compared with fossil fuel (see Table 7.17). Values between 10 and 26.6 g CO_2e/MJ have been reported elsewhere for ethanol sugarcane (Boddey et al., 2008; EPA, 2010; Goldemberg, 2007; Macedo et al., 2008; Macedo and Seabra, 2008; Pereira and Ortega, 2010; Souza, 2010; Wang et al., 2007).

Compared with other crop feedstocks, sugarcane ethanol is certainly the most prominent and efficient renewable liquid fuel for fleet transport, and it is considered an advanced biofuel by the EPA's calculation (EPA, 2010), comparable only with cellulose ethanol. However, the second-generation technology for commercial-scale production of ethanol is not available yet.

7.3.2.5 GHG emission savings from electricity generation credit

The GHG emissions mitigation from surplus bagasse and surplus electricity from burning bagasse was estimated for a boiler of 65 kgf/cm² and a turbogenerator that generates 312.5 kWh per ton of bagasse (0.28 kg bagasse/kg cane, 50% moisture content) or 87.5 kWh per ton of fresh cane.

Table 7.15 Inputs and Soil NO and N_2O Emissions for Sugarcane Farming in Brazil

	Fertilizer N input (g/ton)	Percent conversion to N_2O-N	N_2O formed/ N_2O-N (g/g)	N converted (g/ton)	N_2O or NO emissions (g/ton)	GHG emissions (g CO_2e/ MJ)
N_2O	818.7	1.3%	44/28	10.64	16.72	2.76

For its final rule, the EPA (2010) chose to model both low (40 kWh/MT cane) and high (135 kWh/MT cane) surplus electricity scenarios.

In our scenario, 19% of all energy produced is utilized for ethanol production as steam and electricity, and 81% is exported as electricity surplus. The exported electricity is assumed to displace power from energetic mix generation, which in Brazil is calculated at 73 g CO_2e/kWh (IEA, 2009). The saving CO_2 credit through electrical generation displaces 7.96 g CO_2e/MJ of the baseline (see Table 7.17). Therefore, subtracting this credit from the baseline, the total GHG emission is 14.7 g CO_2e/MJ for sugarcane ethanol production in Brazil. Relative to the fossil fuel displaced, the carbon intensity is reduced 84.3% by farming production, transport, and combustion of sugarcane ethanol (see Table 7.18). This estimate is very close to CARB's result (12.2 g CO_2e/MJ) and that of Macedo et al. (2008) (10.2 g CO_2/MJ), not including ILUC data (see Tables 7.1 and 7.18).

7.4 Land-use change and biofuel production

The need for LCA to include indirect effects has been widely recognized. However, the international land-use change is very controversial, uncertain, and inconclusive as described by many authors (Darlington, 2009, Hertel, Golub, et al., 2010; Hertel, Tyner, et al., 2010; Kim et al., 2009; Pfuderer et al., 2009; Sylvester-Bradley, 2008; Tyner et al., 2010; Wang and Haq, 2008).

Table 7.16 GHG Emissions Summary for Sugarcane Ethanol in Brazil

Sugarcane ethanol components	GHGs (kg CO_2e/ ha/yr)	GHGs (g CO2e/MJ)
Sugarcane farming	2,048.3	13.30
Ethanol production	49.16	0.32
Ethanol transport	178.72	1.16
Field emissions	1,229.16	7.98
Total output emissions	3,505.34	22.76
Gasoline (g CO_2e/MJ)[a]		94.0
% Emission reduction		76%

[a] GHG emissions for gasoline (EPA, 2010).

Table 7.17 GHG Savings Using Electricity Credit from
Burning of Bagasse

Pathway	g CO_2e/MJ	% saving as compared with gasoline[a]
Baseline	22.76	76%
With electricity surplus	−7.96	
Total GHG emissions	14.70	84.3%

[a] 94 g CO_2e/MJ for gasoline.

Two articles published in *Science* in early 2008 received a great deal of attention from the public because they imply that biofuels will emit more GHGs than petroleum-based gasoline due to indirect land-use changes (Fargione et al., 2008; Searchinger et al., 2008). Accordingly, they claimed that energy-crop ethanol should not be considered renewable fuels.

Based on their study, Searchinger et al. (2008, p. 1239) wrote: "Using a world agricultural model to estimate land-use change, we found that corn-based ethanol, instead of producing a 20% savings, nearly doubles GHG emissions over 30 years and increases greenhouse gases for 167 years." The study estimated the land-use change due to expansion of corn-based ethanol at 104 g CO_2e/MJ. CARB and the EPA, by comparison, have estimated 30 and 30.3 g CO_2e/MJ. Recently, Hertel, Tyner, et al. (2010, p. 223), analyzing these releases for corn ethanol produced in the United States, concluded:

> Factoring market-mediated responses and by-product use into our analysis reduces cropland conversion by 72% from the land used for the ethanol feedstock. Consequently, the associated GHG release estimated in our framework is 800 grams of carbon dioxide per megajoule (MJ); *27 grams per MJ per year*, over 30 years of ethanol production, or *roughly a quarter of the only other published* estimate of releases attributable to changes in indirect land use. Nonetheless, 800 grams are enough to cancel out the benefits that corn ethanol has on global warming, thereby limiting its potential contribution in the context of California's Low Carbon Fuel Standard.

Searchinger et al. (2008) also reported that the projected increase in biofuel from 15 billion to 30 billion gallons a year by 2015, would lead to the clearing of 10.8 million hectares of forest and grass to be converted to crops: 2.8 million in Brazil, 2.3 million in China and India, 2.2 million in the United States, and the rest in other countries. The type of land cleared

Table 7.18 Comparison of Estimates of GHG Emissions of Sugarcane-Based Ethanol

Lifecycle phase[a]	Macedo et al. (2008)	Wang et al. (2008)	CARB (2009)	This study
	(g CO_2e/MJ)			
Farming	15.8	9.00	10.4	13.30
Fuel production	−9.4	3.47	1.9	−7.64
Fuel transport	0.0	3.07	6.1	1.16
Field burning	3.8	4.90	8.2	7.98
Total	10.2	20.44	26.6	14.70
Gasoline	94.0	92.9	96.0	94.0
GHG reduction	89%	78%	72%	84%

[a] Land-use change is not included.

can have a major influence on any analysis of GHG emissions, though. Searchinger et al. did not explain what land-use changes would be needed to meet 10.8 million ha of new cropland. On average, the greenhouse emissions ascribed to the loss of forest is four times the emissions from the loss of grasslands in a similar climate (Fargione et al., 2008). The assumptions of Searchinger et al. are flawed by predicting deforestation in the Amazon and savanna in Brazil and conversion of grassland into cropland in China, India, and the United States.

In addition, indirect land-use changes, especially those pushing the rangeland frontier into the Amazonian forests, could offset the carbon savings from biofuels (Lapola et al., 2010; Laurance, 2007; Melillo et al., 2009; Righelato and Spracklen, 2007; Scharlemann and Laurance, 2008). Specifically for sugarcane ethanol and soybean biodiesel, Lapola et al. (2010) reported that each contributes nearly half of the projected indirect deforestation of 121,970 km² by 2020, creating a carbon debt that would take about 250 years to be repaid using these biofuels instead of fossil fuels.

On other hand, these two articles on indirect land-use change have generated a flurry of comments and rebuttals. While scientific assessment of land-use change issues is urgently needed for biofuel production, conclusions regarding the GHG-emission effects of biofuels based on speculative, limited land-use change modeling may misguide biofuel policy development (Darlington, 2009; Hertel, Golub, et al., 2010; Hertel, Tyner, et al., 2010; Sylvester-Bradley, 2008; Tyner et al., 2010; Wang and Haq, 2008).

In response to Searchinger et al.'s paper, Wang and Haq (2008) concluded that there has been no indication that U.S. corn ethanol expansion has so far caused indirect land-use change worldwide. After

Searchinger et al.'s and Fargione et al.'s publications, Pfuderer et al. (2009) postulated:

> There has been a significant and sustained increase in biofuel production in the last ten years. Evidence suggests that this rise in biofuel production was driven primarily by government policies rather than by increasing energy prices. In terms of maize [corn], even though this increase in demand was largely anticipated and has led to a proportionate increase in the production of maize, it seems clear that maize prices are higher than they would otherwise have been. As biofuel production accelerated in 2006, *it also seems fair to conclude that part of the increase in maize prices in 2008 was due to biofuels.*
>
> The above discussion also suggests that, in the run up to the spike in prices in 2007/08, the increase in the price of maize did have some knock-on effects to soybean planting decisions and therefore the price of soybeans. On the more important question of whether biofuel demand led to increases in maize prices which then impacted indirectly on the wheat price, we find no evidence that maize was substituted for wheat and indeed, find that the steep price increase was led by wheat and the relative prices suggest demand substitution was into maize rather than away from it. The increased demand for bio-fuels has had some effect on soybean prices but probably a very small effect on wheat prices. Finally, the fact that production of bio-fuels has remained steady since mid-2008, whilst prices of the food commodities used directly for bio-fuels have fallen dramatically (between June 2008 and December 2008, maize prices fell by 45%) suggests that *bio-fuels were not the key driver, even in those feedstocks used directly in biofuel production.* (emphasis in original)

A recently completed state-of-the-art analysis from Purdue University using the Global Trade Analysis Project (GTAP) model concludes that land-use change emissions potentially associated with corn ethanol expansion are likely less than half of the level estimated by previous papers (Tyner et al., 2010). It found that the total ethanol CO_2 emissions ranged from 77.5 to 84.4 g CO_2e/MJ and the emissions due to indirect land-use change contributed 14 to 21 g CO_2e /MJ—as compared with gasoline emissions of 93.3 g CO_2/MJ. Finally, the study's authors said they were confident that corn ethanol would meet a 10% savings standard. In February 2010, however, the EPA had estimated a 21% reduction relative to petroleum gasoline. Indeed, the uncertainties remain.

The fact is that deforestation of Amazon has been declining in the last five years and has not been associated with biofuel expansion from corn and sugarcane ethanol (see Figure 7.3). A working paper from the World Bank entitled "Causes of deforestation of the Brazilian Amazon" (Margulis, 2004) and another *Science* article (Phillips et al., 2009) clearly demonstrate that Amazon clearing is not directly related to biofuel expansion in Brazil or the United States. Nassar et al. (2008) showed that 53% of sugarcane expansion in the Center-South region of Brazil occurred by displacing other crops (cotton, beans, corn, rice) and 45% displaced pastureland, not Amazon forest or Cerrado biomes.

A recent paper using sophisticated satellite images and complete data demonstrates that for the entire state of São Paulo (see Figure 2), from 2003 to 2008, an additional area of 1.88 million ha of sugarcane was grown. Of this, 56.5% occurred over pastureland and 40.2% on existing agricultural land. Moreover, Brazilian policies have recently prohibited sugarcane

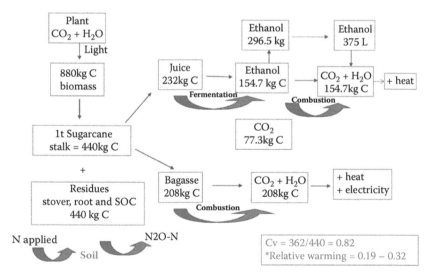

*Relative warming = 1, there is neither climate warming nor cooling by biofuel production.

Figure 7.3 Net climate cooling by fossil fuel CO_2 savings from sugarcane ethanol production due to N_2O emissions. The carbon of the bagasse was added. (Adapted from Crutzen, P. J., Mosier, A. R., Smith, K. A., and Winiwarter, W., 2008, *Atmospheric Chemistry and Physics*, 8:389–395.)

production in the Legal Amazon and Pantanal Wetlands biomes. The map and image land surveys have clearly showed that 40 million hectares of land is available in the Center-South region for sugarcane and ethanol production in the future (INPE, 2009).

In addition, Macedo and Seabra (2008), referring to Nassar et al. (2008) and Amaral (2008) data, have described the international land-use change impacts on sugarcane and ethanol expansion impact in Brazil. They conclude that the expansion of sugarcane until 2020 is not expected to contribute to ILUC greenhouse gas emissions. However, an estimated increase of the ILUC estimation for Brazilian sugarcane ethanol of 46.0 g CO_2e/MJ has been calculated by CARB (2009a). Tyner et al. (2010) have reported that CARB overestimated the ILUC impact of grain-based ethanol by a factor of two in developing its Low Carbon Fuels Standard in 2009. We can deduce that the impact of ILUC emissions of sugarcane ethanol as calculated by CARB may indeed be overestimated.

Converting Cerradão grassland and Cerradão wooded biomes to produce sugarcane ethanol in Brazil incurs a carbon debt of 1.92 and 3.75 t $CO_2e/ha/yr$, respectively, or requires 5.66 and 11.06 years to repay the biofuel carbon debt (Souza, 2010). These figures show three to four times less CO_2 released than the annual GHG reductions that this biofuel would provide by displacing fossil fuels as claimed by Fargione et al. (2008) and 40 to 50 times less than reported by Lapola et al. (2010), Mellilo et al. (2009), and Searchinger et al. (2008). Nevertheless, no report has yet demonstrated the carbon payback time of CO_2 debt for converting degraded pastureland into sugarcane production in Brazil. Actually, with the high level of farming and industrial technology of sugarcane production in Brazil, there may be a huge *reduction* in GHG emissions instead of a carbon debt by producing sugarcane ethanol in degraded pastureland (Amaral, 2008).

For the future, a total of 60 BL of ethanol sugarcane is estimated to be produced in Brazil in 2020 (44 BL for domestic use and 16 BL for export). In this scenario, there is a need of 34 BL more than was produced in 2008, which will require an additional of 4.6 million ha of land, based on the ethanol yield calculated earlier. However, if we use the assumptions of Tyner et al. (2010) for Brazilian sugarcane ethanol, the land requirement would be reduced to 600,000 ha. This land requirement for biofuel expansion is much less than those previously reported (Fargione et al., 2008; Lapola et al., 2010; Melillo et al., 2009; Searchinger et al., 2008, Tyner et al., 2010), which represents a small fraction of the total available land (40 million ha). The newest area of sugarcane will occur over pastureland. As a result, we envision that the expansion of sugarcane ethanol in Brazil will not compete with food/fiber and will definitely not result in clearing the Amazon Forest and Pantanal Wetlands in the next decade. It was the intense 2005 drought in the Amazon forest that had a total biomass carbon impact of 1.2 to 1.6×10^9 tons (Phillips et al., 2009), not clearing the forest for ethanol production.

7.5 Conclusions

Renewable fuels will account for a small fraction of the total fossil energy use in the next decades. However, a global biofuel program will lead to expanded allocation of land for fuel stocks and may increase GHG emissions from land-use changes. Previous authors have devoted special attention to these policies on a country-by-country basis. In this chapter, we have examined the direct and indirect impacts of ethanol production from corn and sugarcane in the United States and Brazil, respectively.

For corn ethanol, the recent literature shows that the net energy balance is positive, around 20%, and the total ethanol emissions savings would meet a 10% goal as compared with gasoline. However, because the uncertainties of the ILUC analysis and potential increase in direct emissions associated with corn farming and ethanol production and use, we cannot conclude that corn ethanol would meet a 20% standard. In addition to environmental concerns, corn ethanol faces economic problems and without subsidies would fail.

After the article by Searchinger et al. (2008), who estimated the ILUC of 104 g CO_2/MJ for ethanol expanded production in the United States, the ILUC emissions have dropped down to 14 g CO_2/MJ, a figure half of what CARB and the EPA have estimated. In addition, from 2004 to 2009, the improvements in corn farming production and energy cost savings in ethanol conversion have increased substantially. Therefore, it is premature to say that biofuels do not contribute to a reduction in global climate change by replacing fossil fuels.

In the case of sugarcane ethanol, previous authors have estimated the net energy balance to be between 1:9 and 1:12. In this chapter, with more precise data, we have calculated 1:5.1. Based on all LCA studies, we can conclude that Brazilian alcohol is by far the most important and efficient renewable liquid fuel for fleet transport among all the feedstocks. In addition, it has been recently designated an advanced biofuel by the EPA.

Many authors have criticized the Brazilian ethanol program by claiming that Amazon forests will be cleared due to land-use change impacts. However, during the great expansion of ethanol production in Brazil from 2002 to 2009, deforestation declined by 60%, total crop production increased by 40% and total coarse grain acreage remained the same at about 46 million ha. Furthermore, for the sustainability of ethanol production, in a great effort, the Brazilian government and industry decided to prohibit sugarcane plantations in the Legal Amazon and Pantanal Wetlands biomes in 2008.

On the other hand, the GHG emissions due to ILUC-impact studies vary from 104 to as little as 14 g CO_2/MJ, and the payback time has dropped since 2008 (Searchinger et al., 2008; Hertel, Golub, et al., 2010; Hertel, Tyner, and Birur, 2010; Souza, 2010). Clearly, uncertainties remain.

Finally, as Tyner et al. (2010, p. 47) concluded, "Land use change impact emissions are not zero, but measuring them with high precision is not yet possible." With all this in mind, it is premature to conclude that global biofuel program will devastate the Amazon and savanna biomes.

Acknowledgments

We wish to express our sincere gratitude to the Cornell Association of Professors Emeriti for the partial support of our research through the Albert Podell Grant Program.

References

Amaral, W. A. N. 2008. Environmental sustainability of sugarcane ethanol in Brazil. In *Sugarcane ethanol: Contributions to climate change mitigation and the environment*, ed. Zuurbier, P., and Vooren, J., 113–138. Wageningen, The Netherlands: Academic.

Andreoli, C. 2006. Etanol pode contribuir com as metas energéticas e ambientais. *Biomassa & Energia*, 3(2): 193–196.

Andreoli, C., and Souza, P. S. 2006–2007. Sugarcane: The best alternative to convert solar and fossil energy into ethanol. *Economia & Energia*, No. 59: 27–32.

Boddey, R. M., Soares, L. H. de B., Alves, B. J. R., and Urquiaga, S. 2008. Bio-ethanol production in Brazil. In *Biofuels, solar and wind as renewable energy systems: Benefits and risks*, ed. Pimentel, D., 321–356. New York: Springer.

CARB (California Air Resources Board). 2009a. *Detailed California-modified GREET pathway for Brazilian sugar cane ethanol.* Version 2.0, released January 12. http://www.arb.ca.gov/fuels/lcfs/lcfs.htm.

———. 2009b. *Detailed California-modified GREET pathway for Brazilian sugar cane ethanol: Average Brazilian ethanol, with mechanized harvesting and electricity co-product credit, with electricity co-product credit.* Version 2.3, released September 23. http://www.arb.ca.gov/fuels/lcfs/lcfs.htm.

———. 2009c. *Detailed California-modified GREET pathway for corn ethanol.* Version 2.1, released February 27. http://www.arb.ca.gov/fuels/lcfs/022709lcfs_cornetoh.pdf.

CONAB (Companhia Brasileira de Abastecimento). 2011. Cana-de-Acucar, Safra 2011/2012, Segundo Levantamento. http://www.conab.gov.br/OlalaCMS/uploads/arquivos/11_08_30_13_41_19_boletim_cana_portugues_-_agosto_2011_2o_lev..pdf.

Crutzen, P. J., Mosier, A. R., Smith, K. A., and Winiwarter, W. 2008. N_2O release from agro-biofuel production negates global warming reduction by replacing fossil fuels. *Atmospheric Chemistry and Physics*, 8:389–395.

Darlington, L. T. 2009. *Land use effects of U.S. corn-based ethanol.* Novi, MI: Air Improvement Resource. http://www.airimprovement.com/reports/land_use_effects_of_us_corn.pdf.

de Oliveira, M. E. D. 2008. Sugarcane and ethanol production and carbon dioxide balances. In *Biofuels, solar and wind as renewable energy systems: Benefits and risks*, ed. Pimentel, D., 215–230. New York: Springer.

de Oliveira, M. E. D., Vaughan, B. E., and Rykiel, E. J. 2005. Ethanol as fuel: Energy, carbon dioxide balances, and ecological footprint. *Bioscience*, 55:593–602.

EPA (U.S. Environmental Protection Agency). 2009. *EPA lifecycle analysis of greenhouse emissions from renewable fuels.* EPA-420-F-09-24. Washington, DC: U.S. EPA. http://www.epa.gov/otaq/renewablefuels/420f09024.pdf.

———. 2010. *Renewable Fuel Standard Program: Final regulatory impact analysis.* EPA-420-R-10-006. Washington, DC: U.S. EPA. http://www.epa.gov/oms/fuels/renewablefuels/regulations.htm.

Fargione, J., Hill, J., Tillman, D., Polasky, S., and Hawthorne, P. 2008. Land clearing and the biofuel cabon debt. *Science*, 319:1235–1238.

Farrel, A. E., Plevin, R. J., Turner, B. T., Jones, A. D., O'Hare, M., and Kammen, D. M. 2006. Ethanol can contribute to energy and environmental goals. *Science*, 311:506–508.

Farrel, A. E., and Sperling, D. 2007a. *A low-carbon fuel standard for California.* Part 1, *Technical analysis.* Berkeley: University of California. http://www.energy.ca.gov/low_carbon_fuel_standard/UC_LCFS_study_Part_1-FINAL.pdf.

———. 2007b. *A low-carbon fuel standard for California.* Part 2, *Policy analysis.* Berkeley: University of California. http://www.energy.ca.gov/low_carbon_fuel_standard/UC_LCFS_study_Part_2-FINAL.pdf.

Goldemberg, J. 2007. Ethanol as a sustainable energy future. *Science*, 315:808–810.

Goldemberg, J., Coelho, S. T., and Guardabassi, P. 2008. The sustainability of ethanol production from sugarcane. *Energy Policy*, 36:2086–2097.

Goldemberg, J., and Guardabassi, P. 2009. Are biofuels a feasible option? *Energy Policy*, 37:10–14.

Hertel, T., Golub, A., Jone, A., O'Hare, M., Plevin, R., and Kammen, D. 2010. Effects of US maize ethanol on global land use change and greenhouse gas emissions: Estimating market-mediated responses. *BioScience*, 60(3): 223–231.

Hertel, T., Tyner, W., and Birur, D. 2010. The global impacts of biofuels mandates. *Energy Journal*, 31(1): 75–100.

Hill, J., Nelson, E., Tilman, D., Polasky, S., and Tiffany, D. 2006. Environmental, economic, and energetic costs and benefits of biodiesel and ethanol biofuels. *Proceedings of the National Academy of Sciences*, 103:11,206–11,210.

IEA (International Energy Agency). 2009. CO_2 emissions from fuel combustion. Paris: IEA. http://www.iea.org/about/copyright.asp.

INPE (Instituto Nacional de Pesquisas Espaciais). 2009. *Estimativa das emissões de CO_2 por desmatamento na Amazônia Brasileira.* Brasilia, D.F. http://www.inpe.br/noticias/arquivos/pdf/Emissoes_CO2_2009.pdf.

IPCC (Intergovernmental Panel on Climate Change). 2006a. National Greenhouse Gas Inventories Programme. *2006 IPCC guidelines for national greenhouse gas inventories.* Ed. Eggleston, H. S., Buendia, L., Miwa, K., Ngara, T., and Tanabe, K. Hayama, Japan: IGES. http://www.ipcc-nggip.iges.or.jp/public/2006gl/.

———. 2006b. N_2O emissions from managed soils, and CO_2 emissions from lime and urea application. In IPCC (2006a), Vol. 4, Chap. 11. http://www.ipccnggip.iges.or.jp/public/2006gl/pdf/4_Volume4/V4_11_Ch11_N2O&CO2.pdf.

Kim, H., Kim, S., and Dale, B. E. 2009. Biofuels, land use change, and greenhouse gas emissions: Some unexplored variables. *Environmental Science & Technology*, 43(3): 961–967.

Lapola, D. M., Schaldach, R., Alcamo, J., Bondeau, A., Koch, J., Christina Koelking, C., Joerg, A., and Priess, J. A. 2010. Indirect land-use changes can overcome carbon savings from biofuels in Brazil. *Proceedings of the National Academy of Sciences*, 107(8): 3388–3393. http://www.pnas.org/content/early/2010/02/02/0907318107.full.pdf+html.

Laurance, W. F. 2007. Switch to corn promotes Amazon deforestation. *Science*, 318:1721.

Liska, A. J., Yang, H. S., Bremer, V. R., Klopfenstein, T. J., Walters, D. T., Erickson, G. E., and Cassman, K. G. 2008. Improvements in the life cycle energy efficiency and greenhouse gas emissions of corn-ethanol. *Journal of Industrial Ecology*, 13(1): 58–74.

Macedo, I. D. C., Leal, M. R. L. V., and Seabra, J. E. A. 2004. Assessment of greenhouse gas emissions in the production and use of fuel ethanol in Brazil. [In Portuguese.] April. São Paulo: Secretaria do Meio Ambiente, Governo de São Paulo.

Macedo, I. D. C., and Seabra, J. E. A. 2008. Mitigation of GHG emissions using sugarcane ethanol. In *Sugarcane ethanol: Contributions to climate change mitigation and the environment*, ed. Zuurbier, P., and Vooren, J., 95–111. Wageningen, The Netherlands: Academic.

Macedo, I. D. C., Seabra, J. E. A., and Silva, J. E. A. R. 2008. Greenhouse gases emissions in the production and use of ethanol from sugarcane in Brazil: The 2005/2006 averages and a prediction for 2020. *Biomass and Bioenergy*, 32(7): 582–595.

Margulis, S. 2004. *Causes of deforestation of the Brazilian Amazon*. Working Paper, No. 22. Washington, DC: World Bank. http://www-wds.worldbank.org/external/default/WDSContentServer/WDSP/IB/2004/02/02/000090341_20040202130625/Rendered/PDF/277150PAPER0wbwp0no1022.pdf.

Melillo, J. M., et al. 2009. Indirect emissions from biofuels: How important? *Science*, 326:1397–1399.

Nassar, A. M., et al. 2008. Sugarcane ethanol: Contributions to climate change mitigation and the environment. In *Sugarcane ethanol: Contributions to climate change mitigation and the environment*, ed. Zuurbier, P., and Vooren, J., 63–94. Wageningen, The Netherlands: Academic.

Pereira, C. L. F., and Ortega, E. 2010. Sustainability assessment of large-scale ethanol production from sugarcane. *Journal of Cleaner Production*, 18:77–82.

Pfuderer, S., Davies, G., and Mitchell, I. 2009. *The role of demand for biofuel in the agricultural commodity price spikes of 2007/08*. Annex 5 of *The 2007/08 agricultural price spikes: Causes and policy implications*. London: Department for Environment, Food and Rural Affairs (DEFRA). http://www.defra.gov.uk/foodfarm/food/pdf/ag-price-annex%205.pdf.

Phillips, O. L., et al. 2009. Drought sensitivity of the Amazon rainforest. *Science*, 323:1344–1347.

Pimentel, D., and Patzek, T. W. 2005. Ethanol production using corn, switchgrass, and wood; biodiesel production using soybean and sunflower. *Natural Resources Research*, 14(1): 65–76.

———. 2008a. Ethanol production: Energy and economic issues related to U.S. and Brazilian sugarcane. In *Biofuels, solar and wind as renewable energy systems: Benefits and risks*, ed. Pimentel, D., 357–371. New York: Springer.

————. 2008b. Ethanol production using corn, switchgrass, and wood; biodiesel production using soybean, In *Biofuels, solar and wind as renewable energy systems: Benefits and risks*, ed. Pimentel, D., 373–394. New York: Springer.

RFA (Renewable Fuels Association). 2010. Industry statistics: Ethanol industry overview. Washington, DC: RFA. http://www.ethanolrfa.org/industry/statistics.

Righelato, R., and Spracklen, D. V. 2007. Carbon mitigation by biofuels or by saving and restoring forests? *Science*, 317:902.

Rudorff, B. F. T., Aguiar, D. A., da Silva, W. F, Sugawara, L. M, Adami, M., and Moreira, M. A. 2010. Studies on the rapid expansion of sugarcane for ethanol production in São Paulo State (Brazil) using Landsat data. *Remote Sensing of Environment*, 2:1057–1076.

Sawyer, D 2008. Climate change, biofuels and eco-social impacts in the Brazilian Amazon and Cerrado. *Philos. Trans. R. Soc. London Biol. Sci.*, 363:1747–1752.

Scharlemann, J. P. W., and Laurance, W. F. 2008. How green are biofuels? *Science*, 319:43–44.

Searchinger, T., Heimlich, R., Houghton, R. A., Dong, F., Elobeid, A., Fabiosa, J., Tokgoz, S., Hayes, D., and Yu, T.-H. 2008. Use of US croplands for biofuels increases greenhouse gases through emissions from land-use change. *Science*, 319:1238–1240.

Shapouri, H. J., Duffield, J., and Wang, M. 2003. The energy balance of corn ethanol revisited. *Transactions of the ASAE*, 46:959–968.

Souza, S. P. 2010. Produção integrada de biocombustíveis: Uma proposta para reduzir o uso de combustível fóssil no ciclo de vida do etanol de cana-de-açúcar [Integrated production of biofuels: A proposal to reduce fossil fuels in the life cycle of sugarcane ethanol]. Thesis, University of São Paulo, São Carlos.

Sylvester-Bradley, R. 2008. Critique of Searchinger (2008) and related papers assessing indirect effects of biofuels on land-use change. Version 3.2, June 12. In *Gallagher Biofuels Review for Renewable Fuels Agency*, Department of Transport. Boxworth, Cambs., England: ADAS UK. http://www.globalbioenergy.org/uploads/media/0806_ADAS_-_Seachinger_critique.pdf.

Tyner, W. E., Taherpour, F., Zhuang, Q., Birur, D., and Baldos, U. 2010. *Land use changes and consequent CO$_2$ emissions due to US corn ethanol production: A comprehensive analysis.* Presentation at the University of Chicago, April. http://www.transportation.anl.gov/pdfs/MC/625.PDF.

UNICA. 2011. Dados e Cotações – Estatisticas. União da Indústria de Cana-de-açúcar. Accessed November 16, 2011. http://www.unica.com.br/dadosCotacao/estatistica.

Walter, A. 2009. Bioethanol development(s) in Brazil. In *Biofuels*, ed. Soetaert, W., and Vandamme, E. J., 55–75. Chippenham, England: John Wiley and Sons.

Wang, M., and Haq, Z. 2008. Letter to *Science*. Rev. March 14. http://www.transportation.anl.gov/pdfs/letter_to_science_anldoe_03_14_08.pdf.

Wang, M., Wu, M., and Huo, H. 2007. Life-cycle energy and greenhouse gas impacts of different corn ethanol plant types. *Environmental Research Letters*, 2(2): 024001.

Wang, M., Wu, M., Huo H., and Liu J. 2008. Life-cycle energy use and greenhouse gas emissions implications of Brazilian sugarcane ethanol simulated with the GREET model. *International Sugar Journal*, 110:527–545.

chapter eight

Water, food, and biofuels

Claudinei Andreoli and David Pimentel

Contents

8.1 Introduction

Water and food are vital to humans and all other animals, plants, and microbes on Earth. Per capita food supplies (cereal grains) for humans have been declining for about 20 years, in part because of shortages of freshwater and cropland and the concurrent increase in human numbers. Shortages in food supplies have contributed to the increase in the number of malnourished people in the world to more than 4.4 billion, two-thirds of the population (WHO, 2000; Pimentel and Satkiewicz, forthcoming).

Two of the most serious malnutrition problems include iron deficiency, affecting 2 billion people, and protein/calorie deficiency, affecting 925 million (WHO, 2000; FAO, 2009). The iron and protein/calorie deficiencies each result in about a million deaths a year.

Humans obtain almost all their nutrients from crops and livestock, and these nutrient sources require water, land, and energy for production (Pimentel and Pimentel, 2008). Food and biofuels are dependent on these same resources for production. Diverse conflicts exist in the use of land, water, energy, and other environmental resources for food and biofuel production. In the United States, about 19% of all fossil energy is utilized in the food system, nearly 40% of all land is used for food production, and 80% of all water is used for agriculture (Pimentel et al., 2004).

The world population currently numbers 7 billion, with more than a quarter of a million people being added each day (PRB, 2010). The world population is projected to reach about 13 billion people by 2070 based on a rate of population increase of 1.2% per year. In addition, freshwater demand worldwide has been increasing rapidly as population and economies grow (Shiklomanov and Rodda, 2003; UNEP, 2003a, 2003b; Gleick, 2004). Population growth, accompanied by increased water use, will not only severely reduce water availability per person but also stress all aspects of the entire global ecosystem (Vörösmarty et al., 2000).

Major factors influence water availability, including rainfall, temperature, evaporation rates, soil quality, vegetation type, and water runoff. In addition, serious difficulties already exist in fairly allocating the world's freshwater resources between and within countries. Overall, water shortages severely reduce food production; negatively impact both terrestrial and aquatic environments; reduce plant, animal, and microbe biodiversity; facilitate the spread of serious human diseases; and diminish water quality (Vörösmarty et al., 2000, 2004; Pimentel and Pimentel, 2008).

In this chapter, water utilization—especially for food production— and the impacts of corn and sugarcane ethanol production on water use and food supply are analyzed.

8.2   Water resources

8.2.1   Hydrologic cycle

Of the estimated 1.4×10^{18} m^3 of water on the Earth, more than 97% is in the oceans (Shiklomanov and Rodda, 2003). Approximately 35×10^{15} m^3, or only 3% of Earth's water, is freshwater, of which about 0.3% is held in rivers, lakes, and reservoirs (Shiklomanov and Rodda, 2003). The remainder of the freshwater is stored in glaciers, permanent snow, and groundwater aquifers. The Earth's atmosphere contains about 13×10^{12} m^3 of water and is the source of all the rain that falls on Earth (Shiklomanov and Rodda, 2003).

Yearly, about 151,000 quads (1 quad = 10^{15} BTU) of solar energy cause evaporation and move about 577×10^{12} m^3 (0.04%) of the water from the Earth's surface into the atmosphere. Of this evaporation, 86% is from oceans (Shiklomanov, 1993). Although only 14% of the water evaporation is from land, about 20% (115×10^{12} m^3 per year) of the world's precipitation falls on land with the surplus water returning to the oceans via rivers (Shiklomanov, 1993). Thus, each year, solar energy transfers 6% of water from the oceans to land areas. This aspect of the hydrologic cycle is vital not only to agriculture but also to human life and natural ecosystems (Jackson et al., 2001).

8.2.2   Availability of water

Although water is considered a renewable resource because it is replenished by rainfall, its availability is finite in terms of the amount available per unit time in any one region. The average precipitation for most continents is about 700 mm/yr (7 million liters/ha/yr), but varies among and within them (Shiklomanov and Rodda, 2003). In general, a nation is considered water scarce when the availability of water drops below 1,000 m^3/yr per capita (Engelman and LeRoy, 1993). Africa, despite having an average of 640 mm/yr of rainfall, is relatively arid since its high temperatures and winds foster rapid evaporation (Vörösmarty et al., 2000; Ashton, 2002)

The Corn Belt and the Northeast region of the United States receive about 1,000 mm of rainfall per year. The average rainfall for the United States is 750 mm per year and, among urban areas, ranges from 1,650 mm for Mobile, Alabama, to only 175 mm for Reno, Nevada (USCB, 2007). World regions that receive low rainfall (less than 500 mm/yr) experience serious water shortages and inadequate crop yields. For example, 9 of the 14 Middle Eastern countries (including Egypt, Jordan, Israel, Syria, Iraq, Iran, and Saudi Arabia) have insufficient freshwater (Myers and Kent, 2001; UNEP, 2003a, 2003b).

Substantial withdrawals from lakes, rivers, groundwater, and reservoirs are used to meet the needs of individuals, cities, farms, and industries, and these withdrawals already stress the availability of water in some parts of the United States (Alley et al., 1999). When managing water resources, the total agricultural, societal, and environmental system must be considered. Legislation is sometimes required to ensure a fair allocation of water. For example, laws determine the amount of water that must be left in the Pecos River in New Mexico to ensure that sufficient water flows into Texas (Washington State Department of Ecology and West Water Research, 2004).

8.2.3   Groundwater resources

Approximately 30% (11×10^{15} m³) of all freshwater on Earth is stored as groundwater. The amount of water held as groundwater is more than 100 times the total amount collected in rivers and lakes (Shiklomanov and Rodda, 2003). Most groundwater has accumulated over millions of years in vast aquifers located below the surface of the Earth. Aquifers are replenished slowly by rainfall, with an average recharge rate that ranges from 0.1% to 3% per year (Covich, 1993; La Salle et al., 2001). Assuming an average 1% recharge rate, only 110×10^{12} m³ of water per year are available for sustainable use worldwide. At present, world groundwater aquifers provide approximately 23% of all water used throughout the world (USGS, 2003a). Irrigation for U.S. agriculture relies heavily upon groundwater, with 65% of irrigation water being pumped from aquifers (McCray, 2001).

Population growth, increased irrigated agriculture, and other water uses are mining groundwater resources. Specifically, the uncontrolled rate of water withdrawal from aquifers is significantly faster than the natural rate of recharge, causing water tables to fall by more than 30 m in some U.S. regions between 1950 and 1990 (Brown, 2002b). The withdrawal of global groundwater has increased from 35 cubic miles (126 cubic kilometers) in 1960 to 68 cubic miles (283 cubic kilometers) in 2000, a rate that researchers at Utrecht University who made the calculations felt was clearly unsustainable. The greatest rates of depletion occurred in the world's largest agricultural regions including northwest India, northeastern China, and California's Central Valley (Wada et al., 2010).

For example, the capacity of the U.S. Ogallala aquifer, which underlies parts of Nebraska, South Dakota, Colorado, Kansas, Oklahoma, New Mexico, and Texas, has decreased 33% since about 1950 (Opie, 2000). Withdrawal of water from the Ogallala is three times faster than the recharge rate (Gleick et al., 2002). Water is being withdrawn more than 10 times faster than the recharge rate for aquifers in parts of Arizona (Gleick et al., 2002).

Similar problems exist throughout the world. For instance, in the agriculturally productive Chenaran Plain in northeastern Iran, the water table has been declining by 2.8 m per year since the late 1990s (Brown, 2002b). Water withdrawal in Guanajuato, Mexico, has caused the water table to fall by as much as 3.3 m per year (Brown, 2002b). The rapid depletion of groundwater poses a serious threat to water supplies in world agricultural regions, especially for irrigation. Furthermore, when some aquifers are mined for their water, the land surface over aquifers is prone to sink, resulting in an inability of the aquifer to be recharged (Youngquist, 1997; Glennon, 2002).

8.2.4   Stored water resources

In the United States, many dams were built during the early 20th century in arid regions in an effort to increase the available supply of water. Although the era of constructing large dams and associated conveyance systems to meet water demand has slowed down in the United States (Coles, 2000), dam construction continues in many developing countries worldwide.

The expected life of a dam is about 50 years, and 85% of U.S. dams will be more than 50 years old by 2020 (Prial, 2009). Prospects for the construction of new dams in the United States do not appear encouraging. Furthermore, over time, the capacity of all dams is reduced as silt accumulates behind them. Estimates are that 1% of the storage capacity of the world's dams is lost due to silt each year (Kirby, 2001).

8.3 Water use

Water from different sources is withdrawn both for *use* and *consumption* in diverse human activities. The term *use* refers to all human activities for which some of the withdrawn water is returned for reuse, such as cooking water, wash water, and wastewater. In contrast, *consumption* means that the withdrawn water is nonrecoverable. For example, evapotranspiration of water by plants is released into the atmosphere and is considered nonrecoverable.

The water content of living organisms ranges from 60% to 95%; humans are about 60% water (American Museum of Natural History, 2011). To sustain health, humans should drink from 1.5 to 2.5 liters of water per person per day (NAS, 1968). In addition to drinking water, Americans use about 1,410 liters/person/day of water for cooking, washing, disposing of wastes, and other personal uses (USCB, 2007). In contrast, 83 other countries report an average below 100 liters/person/day of water for personal use (Gleick et al., 2002). The availability and consumption of water (m³ per capita per year) for different locations are shown in the Figure 8.1.

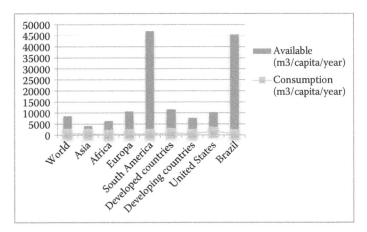

Figure 8.1 Mean availability and consumption of water in different locations. (From Gleick, P. H., *The world's water, Volume 7: The biennial report on freshwater resources*. Washington, DC: Island Press, 2011.)

Currently the U.S. freshwater withdrawals, including that for irrigation, total about 1,650 billion liters/day or about 5,400 liters/person/day. Of this amount, about 80% comes from surface water and 20% is from groundwater resources (USCB, 2007). Worldwide, the average withdrawal is 1,970 liters/person/day for all purposes (Gleick et al., 2002). Approximately 70% of the water withdrawn worldwide is consumed and nonrecoverable.

8.4　Food production and water

8.4.1　Water in crop production

Plants require water for photosynthesis, growth, and reproduction. Water used by plants is nonrecoverable because some water becomes a part of the plant chemically and the remainder is released into the atmosphere. The processes of carbon dioxide fixation and temperature control require plants to transpire enormous amounts of water. Various crops transpire water at rates between 600 and 2,000 liters of water per kilogram of dry matter of crops produced. The average global transfer of water into the atmosphere from the terrestrial ecosystems by vegetation transpiration is estimated to be about 64% of all precipitation that falls to Earth (Schlesinger, 1997).

The minimum soil moisture essential for crop growth varies. For instance, U.S. potatoes require 25–50%, alfalfa 30–50%, and corn 50–70% (Broner, 2005), while rice in China is reported to require at least 80% soil moisture (Zhi, 2000). Rainfall patterns, temperature, vegetative cover, high levels of soil organic matter, active soil biota, and water runoff all effect the percolation of water into the soil where it will be used by plants.

A hectare of U.S. corn, with a yield of approximately 9,000 kg/ha, transpires about 9 million liters of water during the growing season (Smil, 2008), while an additional 1 to 2.5 million liters/ha of soil moisture evaporates into the atmosphere (Desborough et al., 1996). This means that about 1,000 mm (10 million liters/ha) of rainfall are required for corn production during the growing season. Even with about 1,000 mm of annual rainfall in the U.S. Corn Belt, corn frequently suffers from insufficient water during the critical summer growing period (Troeh and Thompson, 2004).

A hectare of high-yielding rice requires approximately 11 million liters of water annually to yield on average 7 metric tons per hectare (Snyder, 2000). On average, soybeans require about 6 million L/ha of water for a yield of 3 t/ha (CSIRO, 2000). In contrast, wheat, which produces less plant biomass than either corn or rice, requires only about 730,000 L/ha of water for a yield of 1 t/ha (USDA, 1997). Note that under semiarid conditions, yields of nonirrigated crops such as corn are low (1 to 2.5 t/ha) even when ample amounts of fertilizers are applied. In the tropics, a hectare requires

Table 8.1 Comparison of Water Requirement of Corn and Sugarcane Ethanol (Liters of Water Required to Produce 1 Liter of Ethanol)

Source	Crop	Process water	Irrigated water	Total water
Pimentel, 2003	Corn	15[a]	248	263
Pimentel and Patzek, 2005	Corn	40	248	288
National Research Council, 2008	Corn	3.3–4[b]	780	783.3–784
De Fraiture et al., 2008	Corn	NA	400	400[e]
Berndes, 2008	Corn	NA	NA	1,628–7,716[c]
Berndes, 2008	Sugarcane	NA	NA	825–3,456[c]
Gerbens-Leenes et al., 2009	Corn	NA	1,013	2,570[c]
Gerbens-Leenes et al., 2009	Sugarcane	NA	1,364	2,516[c]
Chiu et al., 2009	Corn	51	91	142[d]
King and Weber, 2008	Corn	NA	497	497[e]

[a] Total water required for the process of fermentation and distillation.
[b] Net consumption.
[c] Water losses to evapotranspiration in biomass production.
[d] EWe (embodied water in ethanol), defined as the sum of irrigated and process water, averaging 19 ethanol-producing states in 2007.
[e] Irrigation water.

12 million to 16 million liters of water to produce 85 tonnes of sugarcane, or a consumption of 1,500 liters of water to produce 1 kilogram of sugar (Farias et al., 2008). In this chapter, we estimated that, for the Central-Southeast region of Brazil, 150 m^3 of water is used to produce a ton of sugarcane.

8.4.2   Irrigated crops and land use

World agriculture consumes approximately 70% of freshwater withdrawn per year (UNESCO, 2001a, 2001b). Approximately 17% of the world's cropland is irrigated, but that land produces 40% of the world's food (FAO, 2002). Worldwide, the amount of irrigated land is slowly expanding, even though salinization, waterlogging, and siltation continue to decrease its productivity (Gleick, 2002). Despite a small annual increase in total irrigated areas, the per capita irrigated area has been declining since 1990, due to rapid population growth (Gleick, 2002). Specifically, global irrigation per capita has declined nearly 10% from 2000 to 2010 (Gleick, 2011), while in the United States irrigated land per capita has remained constant at about 0.08 ha (USDA, 2007).

Irrigated U.S. agricultural production accounts for about 40% of freshwater withdrawn (USGS, 2003b) and more than 80% of the water consumed (EPA, 2003b). California agriculture accounts for 3% of the state's economic production, but consumes 85% of the water withdrawn (Myers and Kent, 2001).

8.4.3   Energy use in irrigation

Irrigation requires a significant expenditure of fossil energy for pumping and delivering water to crops. Annually in the United States, we estimate 15% of the total energy expended for all crop production is used to pump irrigation water (Hodges et al., 1994). Overall, the amount of energy consumed in irrigated crop production is substantially greater than that expended for rain-fed crops. For example, irrigated wheat requires the expenditure of more than three times as much energy as rain-fed wheat. Specifically, about 4.2 million kcal/ha/yr are the required energy input for rain-fed wheat, while irrigated wheat requires 14.3 million kcal/ha/yr to apply an average of 5.5 million liters of water (USDA, 1997; Pimentel et al., 2002).

Delivering the 10 million liters of irrigation water needed by a hectare of irrigated corn from surface water sources requires the expenditure of about 880 kWh of fossil fuel (Batty and Keller, 1980). In contrast, when irrigation water must be pumped from a depth of 100 m, the energy cost increases up to 28,500 kWh/ha, or more than 32 times the cost of surface water (Gleick, 1993).

The costs of irrigation for energy and capital are significant. The average cost to develop irrigated land ranges from $3,800 to $7,700 per hectare (Postel, 1999). Thus, farmers must not only evaluate the dollar cost of developing irrigated land but also consider the annual costs of irrigation pumping. For example, delivering 7–10 million liters/ha of water costs from $750 to $1,000 (Larson et al., 2002; Pitts et al., 2002). About 150,000 ha of agricultural land have already been abandoned in the United States due to high pumping costs (Youngquist, 1997).

The large quantities of energy required to pump irrigation water are significant considerations from the standpoint of both energy and water resource management. For example, approximately 8 million kcal of fossil energy is expended for machinery, fuel, fertilizers, pesticides, and partial (15%) irrigation to produce one hectare of rain-fed U.S. corn (Pimentel et al., 2002, Pimentel and Patzek, 2008). In contrast, if the corn crop were fully irrigated, the total energy inputs would rise to nearly 25 million kcal/ha (2,500 liters of oil equivalents) (Gleick, 1993). In the future, this energy dependency will influence not only the overall economics of irrigated crops but also the selection of specific crops worth irrigating. While a low-value crop like alfalfa may be uneconomical, other crops might use less water and have a higher market value.

The efficiency varies with irrigation technologies (Postel, 1992, 1993). The most common irrigation methods—flood irrigation and sprinkler irrigation—frequently waste water. In contrast, the use of more focused application methods, such as drip or "micro-irrigation," has found favor because of their increased water efficiency. Drip irrigation delivers water to individual plants by plastic tubes and uses from 30% to 50% less water than surface irrigation. In addition to conserving water, drip irrigation reduces the problems of salinization and waterlogging. Although drip systems achieve up to 95% water efficiency, they are expensive, may be energy intensive, and require clean water to prevent the clogging of the fine delivery tubes (Shock, 2006).

8.4.4   Soil salinization and waterlogging in irrigation

With rain-fed crops, salinization is not a problem because the salts are naturally flushed away. But when irrigation water applied to crops returns to the atmosphere via plant transpiration and evaporation, dissolved salts concentrate in the soil where they inhibit plant growth. The practice of applying about 10 million liters of irrigation water per hectare each year, results in approximately 5 t/ha of salts being added to the soil (Bouwer, 2002). The deposited salt can be flushed away with additional irrigation water, but at a significant cost (Bouwer, 2002). Worldwide, approximately half of all existing irrigated soils are adversely affected by salinization (UNESCO, 2006). Each year the amount of world agricultural land abandoned due to salinized and waterlogged soil is estimated to be 5% of irrigated land worldwide (FAO, 2009).

In addition, drainage water from irrigated cropland collects large quantities of salt. For instance, as the Colorado River flows through Grand Valley, Colorado, it picks up 580,000 tons of salts per year (USDI, 2001). Based on the drainage area of 20,000 ha, the water returned to the Colorado River contains an estimated 30 t/ha of salts per year (Pugh, 2001). The Colorado River delivers a total of 1.6 million tons of salt into south-central Arizona each year (USGS, 2009).

Waterlogging is another problem associated with irrigation. Over time, seepage from irrigation canals and irrigated fields causes water to accumulate in the upper soil levels. Due to water losses during pumping and transport, approximately 60% of the water intended for crop irrigation never reaches the crop (Wallace, 2000). In the absence of adequate drainage, water tables rise in the upper soil levels, including the plant root zone, and crop growth is impaired. Such irrigated fields are rendered unproductive (Postel, 1993) and are sometimes referred to as "wet deserts." For example, in India, waterlogging adversely affects 8.5 million hectares of cropland and results in the loss of as much as 2 million tons of grain every year (Nagdeve, 2006). To prevent both salinization and waterlogging, sufficient water along with adequate soil drainage must be available to ensure salts and excess water are drained from the soil.

8.4.5   Water runoff and soil erosion

Because more than 99% of the world food supply comes from the land, an adequate world food supply depends on the continued availability of productive soils (FAO, 2011). Soil erosion adversely affects crop productivity by reducing the availability of water; diminishing soil nutrients, soil biota, and soil organic matter; and decreasing soil depth (Pimentel, 2006). The reduction in the amount of water available to the growing plants is considered the most harmful effect of erosion, because eroded soil absorbs 87% less water by infiltration than uneroded soils (Nolan and Goddard, 2002). Soybean and oat plantings intercept approximately 10% of the rainfall, whereas tree canopies intercept 15–35% (Pimentel et al., 2004). Thus, the removal of trees increases water runoff and reduces water availability.

A water runoff rate of about 30% of total rainfall of 800 mm/yr causes significant water shortages for growing crops like corn and ultimately lowers crop yields (Troeh and Thompson, 2004). In addition, water runoff, which carries sediments, nutrients, and pesticides from agricultural fields into surface and ground waters, is the leading cause of nonpoint-source water pollution in the United States (EPA, 2003a). Thus, soil erosion is a self-degrading cycle on agricultural land. As erosion removes topsoil and organic matter, water runoff is intensified and crop yields decrease. The cycle is repeated again with even greater intensity during subsequent rains.

Increasing soil organic matter by applying manure or similar materials can improve the water infiltration rate by as much as 150% (Nolan and Goddard, 2002). In addition, using vegetative cover, such as intercropping and grass strips, helps slow both water runoff and erosion (Troeh and Thompson, 2004). For example, when silage corn is interplanted with red clover, water runoff can be reduced by as much as 87% and soil loss can be reduced by 78% (Wall et al., 1991). Reducing water runoff in these and other ways is an important step in increasing water availability to crops, conserving water resources, decreasing nonpoint-source pollution, and ultimately decreasing water shortages.

Planting fast-growing trees to serve as shelter belts in north China dryland regions reduced the wind velocity by 20–40% and evaporation by 9–25% in the agrosilvicultural systems (Qi and Tishun, 2004). In the farmland shelter belt, the water content of the soil increased to 8.7% from the original level of 5.9%. The crop yield for wheat in the experimental treatments increased from a range of 9.7% to 30.0% (Qi and Tishun, 2004).

8.4.6   Water use in livestock production

The production of animal protein requires significantly more water than the production of plant protein (Pimentel, 2004). Although U.S. livestock directly uses only 2% of the total agricultural water (Solley et al., 1998), the water inputs for livestock production are substantial because water is also required for the forage and grain crops.

Each year, 15 million tons of grain is fed to U.S. livestock, which requires a total of about 250×10^{12} liters of water (World Resources Institute, 2007). Worldwide grain production specifically for livestock requires nearly three times the amount of grain that is fed U.S. livestock and three times the amount of water used in the United States to produce the grain feed (Brown, 2002a).

Animal products vary in the amounts of water required for their production. For example, producing 1 kg of chicken requires 3,500 liters of water, while 1 kg of sheep requires approximately 51,000 liters of water to produce the required 21 kg of grain and 30 kg of forage (USDA, 2001; Buchanan-Smith, 2001). For open rangeland (instead of confined feedlot production), from 120 kg to 200 kg of forage is required to produce 1 kg of beef, and this requires 120,000–200,000 liters of water per kilogram of beef (Rangeland, 1994; Dorsett, 2003). Beef cattle can be produced on rangeland, but a minimum of 200 mm per year of rainfall is needed (Hays and White, 1998).

U.S. agricultural production is projected to expand in order to meet the increased food needs of a U.S. population that is projected to double in the next 70 years (USCB, 2007). The food situation is expected to be more serious in developing countries, such as Egypt and Kenya, because of more rapidly growing populations (Rosengrant et al., 2002). Increasing

Table 8.2 Water Use Efficiency of Six Energy Crops (per Liter of Ethanol/Biodiesel Produced)

Crop	Water use efficiency	
	m^3 per GJ ethanol	L of water per L ethanol
Ethanol[a]		
Sugarcane	75	1.765
Corn	113	2.655
Cassava	90	2.116
Biodiesel[a]		
Canola	339	12.778
Palm oil	90	3.409
Soybean	324	12.222

Source: From Andreoli, C., *The ethanol industry perspective for 2015–2020* (Londrina, PR, Brazil: Embrapa Soja, 2009).

[a] High heat value of ethanol, 29.7 MJ/kg and biodiesel, 37.7 MJ/kg.

crop yields necessitates a parallel increase in agricultural freshwater utilization. Therefore, increased crop and livestock production during the next five to seven decades will significantly increase the demand on all water resources, especially in the western, southern, and central United States (USDA, 2001), as well as in many regions of the world with low rainfall.

8.5   Water and biofuels

Global shortages of fossil energy, especially oil and natural gas, have increased interest in biomass and biofuels worldwide (Santa Barbara, 2007). This emphasis on biofuels as renewable energy sources has developed globally, especially those biofuels made from crops such as corn, sugarcane, canola, rape, and soybean (Pimentel, 2008). Although it may seem beneficial to use renewable plant materials for biofuel, the use of crops and other biomass materials for biofuels raises many environmental and ethical concerns (Pimentel, 2006, 2008; Pimentel et al., 2008). Diverse conflicts exist in the use of land, water, energy, and other environmental resources for food and biofuel production. Food and biofuels are dependent on the same resources for production: land, water, and energy. Food production in the United States uses about 40% of all the land, 80% of the water consumed, and 19% of the fossil energy (Pimentel et al., 2008).

8.5.1   Corn ethanol

In the United States, corn ethanol constitutes 99% of all biofuels. For capital expenditures, new plant construction costs from $1.05 to $3.00 per gallon of ethanol (Shapouri and Gallagher, 2005). Fermenting and distilling corn ethanol requires large amounts of water. The corn is finely ground and approximately 1 liter of water is added per 0.18 kg of ground corn. After fermentation, to obtain a liter of 95% pure ethanol from the 10% ethanol and 90% water mixture, 1 liter of ethanol must be extracted from approximately 10 liters of the ethanol/water mixture. To be mixed with gasoline, the 95% ethanol must be further processed and more water must be removed, requiring additional fossil energy inputs to achieve 99.5% pure ethanol (Pimentel et al, 2010). Thus, a total of about 12 liters of wastewater must be removed per liter of ethanol produced, and this relatively large amount of sewage effluent has to be disposed of at an energetic, economic, and environmental cost. The total water required to produce 1 gallon of ethanol (including irrigation) is 2,655 gallons (1,720 gallons of water without irrigation) (Pimentel et al., 2010). In contrast, other studies have reported 248–780 gallons of irrigation water per gallon of corn-based ethanol (see Table 8.1).

8.5.2 Water use in sugarcane and ethanol production

The expansion in ethanol production in the world will lead to a dramatic increase in water required both for agricultural production of corn and sugarcane and for ethanol conversion from the biomass. Ethanol production increased from 20 BL (billion liters) in 2000 to 80 BL in 2008, with 90% of this increase due to sugarcane in Brazil and corn in the United States (Pfuderer et al., 2009). The challenges in the future to biofuel development that have not received appropriate attention are water use and water quality. The central questions are how water use and water quality are expected to change as the world agricultural portfolio shifts to include more bioenergy, fiber, and food.

To illuminate these issues, we report here some recent data on water use for sugarcane ethanol in Brazil and the implications of a large-scale expansion of biomass for energy from a water perspective. We will attempt to calculate the amount of blue and green water required to grow biomass and convert it into biofuel in the sugar-ethanol industry in Brazil.

8.5.2.1 Water use for sugarcane production

Brazilian ethanol production increased from 10 BL in 2000 to 26.7 BL in 2008 (CONAB, 2010). Production is growing rapidly and is expected to reach 50 BL in 2017, but it still provides only a small portion of the total U.S. liquid transportation fuels (Andreoli, 2009).

Although ethanol's climate-change benefits, energy efficiency, and impacts on environmental quality have been the main focus of recent studies, the impact of ethanol production on water use and quality has raised little concern among the important players: the research community, national policy makers, and industry.

Highlighting ethanol dependence on water, prior studies for corn estimated the total water use to produce 1 liter of ethanol to be between 263 and 784 liters (see Table 8.1). However, these estimates have not been described for sugarcane biomass and for the process water consumed within the biorefineries.

Sugarcane is a C_4 plant and requires temperatures between 20°C and 35°C and large amounts of water to convert solar energy into biomass. The maximum conversion efficiency of solar energy to biomass is 6% for C_4 plants, or the minimum energy required to convert 1 mol CO_2 into carbohydrates represents 2,052 kJ (Zhu et al., 2008).

Information is lacking on the total water (rain-fed + irrigation) use for sugarcane biomass production and on what the impact of biofuel production is on water use and water quality in Brazil. Farias et al. (2008), using supplemental irrigation in the Northeast region of Brazil, have reported that sugarcane consumes 1,500 L of water to produce 1 kg of sugar. In 2002, Berndes (2008), using de Oliveira's (de Oliveira et al., 2005) data, estimated 825–3,456 L of water per L of biofuel output (see Table 8.1). Gerbens-Leenes et al. (2009), using the water footprint concept based on crop water requirements by adding up daily crop evapotranspiration in different countries and biomass chemical composition, calculated that the sugarcane crop used 1,152 L of green water and 1,364 L of blue water to produce a liter of ethanol, 2,516 L of water overall (see Table 8.1). The *green water* and *blue water* refer to rain-fed and surface water for irrigation, respectively, which evaporate during crop production.

More recently, Mulder et al. (2010), using a comparative analysis for estimating the net energy return on water invested (EROWI), have showed that the best net EROWI for renewables is sugarcane ethanol at 0.903, more than two orders of magnitude lower than the most water-efficient fossil-energy sources. The EROWI calculation for corn was 0.024, using data from Shapouri et al. (2004) and Pimentel and Patzek (2005).

Based upon average water requirements for sugarcane production of 150–200 m^3/t, sugarcane yield of 85.5 t/ha, and industrial conversion of 85 L/t ethanol in the Central-Southeast region of Brazil, we have estimated that sugarcane for biomass production consumes 75 m^3/GJ or 1,765 liters of water per liter of ethanol produced (see Table 8.2). Compared with other energy crops—canola, cassava, corn, oil palm, and soybean (see Table 8.2)—sugarcane, so far, is the most water-use-efficient crop to produce biofuel and energy in the world, as it has been confirmed by the research reported above.

8.5.2.2 *Water use for ethanol production*

There are no publicly available records on water use by ethanol plants in Brazil. In a review of ethanol-producing states, only the Sugarcane Technological Center (CTC) has published data on water use by mills. Recently, Elia-Neto (2009) published a guideline for water use and conservation in the sugar and ethanol industry in Brazil. Table 8.3 summarizes the average water use in the mills for ethanol and sugar production.

The mean water consumption in the mills for a production mix (50% sugar and 50% ethanol) is 22 m^3 of water per ton of fresh sugarcane. If we consider only the water utilized for ethanol production, water consumption is reduced to 13.7 m^3/ton. Although the total water cycle system for sugar and ethanol processing consumes 22 m^3/ton of sugarcane, only 1.85 m^3 of water per ton is captured as freshwater from streams or rivers (Elia-Neto, 2009). Elia-Neto also points out that CTC's goal is to capture 1.0 m^3 of water per ton of cane. This goal will be possible to achieve because of government policies, environmental agreements between the ethanol industry and the Secretary of Environmental Issues, and a large investment in research and technology on water use.

A decade before, water capture from rivers was 10 m^3 per ton of cane (Elia-Neto, 2005). This value varies among mills, depending upon mill type, production mix, and the amount of electricity cogeneration. Figure 8.2

Table 8.3 Average Water Use in Mills for Sugar and Ethanol
Production in Brazil

Sector	Purpose	Specific use	Average use m³/t cane	%
Feeding,	Sugarcane washing	2.200 m³/t cane	2.200	9.9
washing, and	Imbibitions	0.250 m³/t cane	0.250	1.1
extraction	Cooling	0.035 m³/t cane	0.035	0.2
(milling and diffuser)	Oil cooling	0.130 m³/t cane	0.130	0.6
Subtotal			*2.615*	*11.8*
Juice	Cooling column sulfites[a]	0.100 m³/t cane	0.050	0.2
treatment	Lime preparation	0.030 m³/t cane	0.030	0.1
	Juice heating For sugar[a]	160 kg steam/t cane	0.080	0.4
	For ethanol[b]	50 kg steam//t cane	0.025	0.1
	Cake washing	0.030 m³/t cane	0.030	0.1
	Condensers of filters	0.30–0.35 m³/tcl	0.350	1.6
Subtotal			*0.573*	*2.6*
Sugar factory[a]	Steam for evaporation	0.414 t/t cane	0.207	0.9
	Condensers/multijects evaporation	4–5 m³/t cane	2.250	10.2
	Steam for cooking	0.170 t/t cane	0.085	0.4
	Condensers/multiject cooking	8–15 m³/t cane	5.750	26.0
	Molasses dilution	0.050 m³/t cane	0.030	0.1
	Cooking delay	0.020 m³/t cane	0.010	0.0
	Sugar washing (1/3 water and 2/3 steam)	0.030 m³/t cane	0.015	0.1
	Retention of sugar fine	0.040 m³/t cane	0.020	0.1
Subtotal			*8.367*	*37.8*
Fermentation[b]	Mush preparation	0–10 m³/m³ ethanol	0.100	0.5
	Juice cooling	30 m³/m³ ethanol	1.250	5.6
	Yeast preparation	0.010 m³/m³ ethanol	0.001	0.0
	CO_2 gas washing	1.5–3.6 m³/m³ ethanol	0.015	0.1
	Tanks cooling	60–80 m³/m³ ethanol	3.000	13.6
Subtotal			*4.366*	*19.7*
Distillation[b]	Heating (steam)	3.5–5 kg/m³ ethanol	0.360	1.6
	Cooling tower	80–120 m³/m³ ethanol	3.500	15.8
Subtotal			*3.860*	*17.4*

continued

Table 8.3 Average Water Use in Mills for Sugar and Ethanol
Production in Brazil (Continued)

Sector	Purpose	Specific use	m³/t cane	%
			Average use	
Cogeneration of energy	Production of direct steam	400–600 kg/t cane	0.500	2.3
	Supercooling	0.030 l/kg steam	0.015	0.1
	Gas washing from boiler	2.0 m³/t steam	1.000	4.5
	Ash cleaning	0.500 m³/t steam	0.250	1.1
	Cooling oil and air of turbines	15 L/kW	0.500	2.3
	Tower water of condensation[c]	38 m³/t steam	6.00[c]	27.1
Subtotal			*2.265*	*10.2*
Others	Floor and equipment cleaning	0.050 m³/t cane	0.050	0.2
	Drinking water	70 L/day	0.030	0.1
Subtotal			*0.080*	*0.4*
TOTAL			22.126	100.0

[a] Not included in the ethanol process.
[b] Not included in the sugar process.
[c] Water used only for surplus electricity generation.

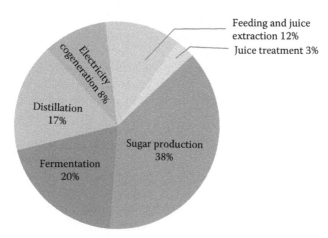

Figure 8.2 Mean distribution of water use by sectors in the sugarcane mills. The total water use is 22.1 m³/ton of sugarcane. (Data from Table 8.3.)

shows that the production of sugar, on average, uses 38% of the total water, and the fermentation and distillation processes for ethanol production both consume 37%. The remaining process utilizes 25%. It is important to emphasize that water use for electricity production from boilers and turbogenerators can reach 6 m^3/ton cane or more (see Table 8.3), which is a substantial amount (27.1%) of the total water consumed for sugar and ethanol production.

Many of these ethanol plants have little or no wastewater discharge. They recycle a significant portion of their process water through a combination of centrifuges, evaporation, and anaerobic digestion. Therefore, water demand is primarily related to energy production, specifically the cooling tower and boiler systems.

8.6   Conclusions

Water use has been drastically reduced in ethanol production for both corn and sugarcane mills. Even though the use of energy-crop irrigation is crucial, the influence of rain-fed energy-crop cultivation can also be significant in water-scarce regions, such as China, India, Africa, and the Northeast and Central regions of Brazil. An important topic for future research is the global potential of biofuel production given the basin-scale competition for water (corn production in the United States).

Acknowledgments

We wish to express our sincere gratitude to the Cornell Association of Professors Emeriti for the partial support of our research through the Albert Podell Grant Program.

References

Alley, W. M., Reilly, T. E., and Franke, O. L. 1999. Sustainability of ground-water resources. Circular 1186. Denver: USGS. http://pubs.usgs.gov/circ/circ1186/.

American Museum of Natural History. 2011. Life in water: Humans in water. New York: American Museum of Natural History. http://www.amnh.org/exhibitions/water/?section=lifeinwater&page=lifeinwater_h.

Andreoli, C. 2009. *The ethanol industry perspective for 2015–2020.* Londrina, PR, Brazil: Embrapa Soja.

Ashton, P. J. 2002. Avoiding conflicts over Africa's water resources. *Ambio*, 31:236–242.

Batty, J. C., and Keller, J. 1980. Energy requirements for irrigation. In *Handbook of energy utilization in agriculture*, ed. Pimentel, D., 35–44. Boca Raton, FL: CRC Press.

Berndes, G. 2008. Future biomass energy supply: The consumptive water use perspective. *Water Resources Development*, 24(2): 235–245.

Bouwer, H. 2002. Integrated water management for the 21st century: Problems and solutions. *Journal of Irrigation and Drainage Engineering*, 128(4): 193–202.

Broner, I. 2005. *Irrigation scheduling: The water balance approach*. Colorado State University Extension Report, No. 4,707. http://www.ext.colostate.edu/pubs/crops/04707.html.

Brown, L. R. 2002a. Plan B updates: Rising temperatures and falling water tables raising food prices. Washington, DC: Earth Policy Institute. http://www.earth-policy.org/plan_b_updates/2002/update16.

———. 2002b. Water deficits growing in many countries: Water shortages may cause food shortages. Washington, DC: Earth Policy Institute. http://www.populationpress.org/publication/2002-9-brown.html.

Buchanan-Smith, J. 2001. Budgeting feed requirements for beef cattle on pasture. Guelph, ON: Department of Animal and Poultry Science, University of Guelph. http://www1.foragebeef.ca/$foragebeef/frgebeef.nsf/all/frg39/$FILE/pasturestockingrates.pdf.

Chiu, Y. W., Walseth, B., and Suh, Sangwon. 2009. Water embodied in bioethanol in the United States. *Environmental Science & Technology*, 43:2688–2692.

Coles, P. 2000. Large dams: End of an era? *UNESCO Courier*, 53(4): 10–11. http://www.petercoles.net/LargeDams.pdf.

CONAB (Companhia Brasileira de Abastecimento). 2011. Cana-de-Acucar, Safra 2011/2012, Segundo Levantamento. http://www.conab.gov.br/OlalaCMS/uploads/arquivos/11_08_30_13_41_19_boletim_cana_portugues_-_agosto_2011_2o_lev.pdf.

Covich, A. P. 1993. Water and ecosystems. In *Water in crisis: A guide to the world's fresh water resources*, ed. Gleick, P. H., 40–55. New York: Oxford University Press.

CSIRO. 2000. Water for food. http://www.clw.csiro.au/issues/water/water_for_food.html.

De Fraiture, C., Giordano, M., and Liao, Y. 2008. Biofuels and implications for agriculture water use: Blue impacts of green energy. *Water Policy*, 10(S1): 67–81.

de Oliveira, M. E. D., Vaughan, B. E., and Rykiel, E. J. 2005. Ethanol as fuel: Energy, carbon dioxide balances, and ecological footprint. *Bioscience*, 55:593–602.

Desborough, C. E., Pitman, A. J., and Irannejad, P. 1996. Analysis of the relationship between bare soil evaporation and soil moisture simulated by 13 land surface schemes for a simple non-vegetated site. *Global & Planetary Change*, 13(1–4): 47–56.

Dorsett, D. J. 2003. Pasture management for beef cattle production. College Station: Texas Agricultural Extension Service, Texas A&M University. http://agfacts.tamu.edu/D11/NLO/Pasture%20Management/PASTURE%20MANAGEMENTFORBEEFCATTLEPRODUCTION.pdf.

Elia-Neto, A. 2005. Captação e uso de água no processamento da cana-de-açúcar. In *A energia da cana-de-açúcar: Doze estudos sobre a agroindústria da cana-de-açúcar no Brasil e sua sustentabilidade*, ed. Macedo, I. C., et al., 104–110. São Paulo: UNICA.

———. 2009. Manual de conservação e reuso na agroindústria sucroenergética. Brasília, DF, Brazil. http://www.ctcanavieira.com.br.

Engelman, R., and LeRoy, P. 1993. *Conserving land: Population and sustainable food production*. Washington, DC: Population Action International.

EPA (Environmental Protection Agency). 2003a. *National management measures to control nonpoint source pollution from agriculture*. EPA-841-B-03-004, Office of Water (4503T). Washington, DC: EPA. http://water.epa.gov/polwaste/nps/agriculture/agmm_index.cfm.

———. 2003b. Water. Washington, DC: EPA. http://www.epa.gov/water/you/chap1.html.

FAO (Food and Agriculture Organization). 2002. *Crops and drops: Making the best use of water for agriculture.* Rome: FAO. http://www.fao.org/docrep/005/y3918e/y3918e00.htm.

———. 2009. *The state of food insecurity in the world.* Rome: FAO. http://www.fao.org/docrep/012/i0876e/i0876e00.htm.

———. 2011. Food balance sheets. [Online database accessed May 26, 2011.] http://faostat.fao.org/site/368/default.aspx.

Farias, C. H. A., Fernandes, P. D., Neto, J. D., and Gheyi, H. R. 2008. Eficiência no uso da água na cana-de-açúcar sob diferentes laminas de irrigação e níveis de zinco no litoral paraibano. *Eng. Agric.,* 28(3): 1–11.

Gerbens-Leenes, W., Hoekstra, A. Y., and van der Meer, T. H. 2009. The water footprint of bioenergy. *Proceedings of the National Academy of Sciences,* 106(25): 10,219–10,223.

Gleick, P. H. 1993. *Water in crisis: A guide to the world's fresh water resources.* New York: Oxford University Press.

———. 2002. Soft water paths. *Nature,* 418:373.

———. 2004. Global freshwater resources: Soft-path solutions for the 21st century. *Science,* 302:1524–1528.

———. 2011. *The world's water, Volume 7: The biennial report on freshwater resources.* Pacific Institute for Studies in Development, Environment and Security. Washington, DC: Island Press.

Gleick, P. H., Wolff, E. L., and Chalecki, R. R. 2002. *The new economy of water: The risks and benefits of globalization and privatization of freshwater.* Oakland, CA: Pacific Institute for Studies in Development, Environment, and Security.

Glennon, R. 2002. The perils of groundwater pumping. *Issues in Science and Technology,* 19(1): 73–79.

Hays, K. B., and White, L. D. 1998. Healthy range watersheds critical to the future of Texas. College Station: Texas Agricultural Extension Service, Texas A&M University. http://rangeweb.tamu.edu/trm/docs/waterseries2.pdf.

Hodges, A. W., Lynne, G. D., Rahmani, M., and Casey, C. F. 1994. Adoption of energy and water-conserving irrigation technologies in Florida. Fact Sheet EES 103. Gainesville: Florida Cooperative Extension Service, Institute of Food and Agricultural Sciences, University of Florida. http://www.p2pays.org/ref/12/11357.pdf.

Jackson, R. B., Carpenter, S. R., Dahm, C. N., McKnight, D. M., Naiman, R. J., Postel, S. L. and Running, S. W. 2001. Water in a changing world. *Ecological Applications,* 11(4): 1027–1045.

King, C. W., and Webber, M. E. 2008. Water intensity of transportation. *Environmental Science & Technology,* 42(21): 7866–7872.

Kirby, A. 2001. Silt behind dams "worsens water shortage." *BBC News,* December 4. http://news.bbc.co.uk/2/hi/science/nature/1691732.stm.

Larson, K., Thompson, D., and Harn, D. 2002. Limited and full irrigation comparison for corn and grain sorghum. http://www.Colostate.edu/depts/prc/pubs/.

La Salle, C., Marlin, C., Leduc, C., Taupin, J. D., Massault, M., and Favreau, G. 2001. Renewal rate estimation of groundwater based on radioactive tracers (^{3}H, ^{14}C) in an unconfirmed aquifer in a semi-arid area, Iullemeden Basin, Niger. *Journal of Hydrology,* 254(1–4): 145–156.

McCray, K. 2001. American agriculture and ground water: Bringing issues to the surface. *Irrigation Journal*, 51(5): 27–29.

Mulder, K., Hagens, N., and Fisher, B. 2010. Burning water: A comparative analysis of the energy return on water invested. *Ambio*, 39:30–39.

Myers, N., and Kent, J. 2001. *Perverse subsidies: How tax dollars can undercut the environment and the economy*. Washington, DC: Island Press.

Nagdeve, D. A. 2006. Population, poverty and environment in India. Mumbai: ENVIS Centre on Population and Environment, International Institute for Population Services. (IIPS), , India. Accessed May 26, 2011. http://www.iipsenvis.nic.in/Newsletters/vol3no3/DANagdave.htm.

NAS (National Academy of Sciences). 1968. *Recommended dietary allowances*. 7th ed. Food and Nutrition Board Publ. 1694. Washington, DC: NAS.

National Research Council. 2008. Water implications of biofuels production in the United States. Washington, DC: Academic Press.

Nolan, S., and Goddard, T. 2002. Measurements of water erosion and infiltration in Alberta using a rainfall simulator. Edmonton: Alberta Ministry of Agriculture and Rural Development. http://www1.agric.gov.ab.ca/$department/deptdocs.nsf/all/sag5362.

Opie, J. 2000. *Water for dry land*. 2nd ed. Lincoln: University of Nebraska Press.

Pfuderer, S., Davies, G., and Mitchell, I. 2009. *The role of demand for biofuel in the agricultural commodity price spikes of 2007/08*. Annex 5 of *The 2007/08 agricultural price spikes: Causes and policy implications*. London: Department for Environment, Food and Rural Affairs (DEFRA). http://archive.defra.gov.uk/foodfarm/food/pdf/ag-price-annex%205.pdf.

Pimentel, D. 2003. Ethanol fuels: Energy balance, economics, and environmental impacts are negative. *Natural Resources Research*, 12(2): 127–134.

———. 2004. Livestock production and energy use. In *Encyclopedia of energy*, ed. Matsumura, R., 671–676. San Diego, CA: Elsevier.

———. 2006. Soil erosion: A food and environmental threat. *Environment, Development and Sustainability*, 8:119–137.

———. 2008. The ecological and energy integrity of corn ethanol production. In *Reconciling human existence with ecological integrity: Science, ethics, economics and law*, ed. Westra, L., Bosselmann, K., and Westra, R., 245–256. London: Earthscan.

Pimentel, D., Berger, B., Filiberto, D., Newton, M., Wolfe, B., Karabinakis, B., Clark, S., Poon, E., Abbett, E., and Nandagopal, S. 2004. Water resources: Agricultural and environmental issues. *Bioscience* 54(10): 909–918.

Pimentel, D., Doughty, R., Carothers, C., Lamberson, S., Bora N., and Lee, K. 2002. Energy inputs in crop production: Comparison of developed and developing countries. In *Food security and environmental quality in the developing world*, ed. Lal, L., Hansen, D., Uphoff, N., and Slack, S., 129–151. Boca Raton, FL: CRC Press.

Pimentel, D., Marklein, A., Toth, M. A., Karpoff, M., Paul, G. S., McCormack, R., Kyriazis, J., and Krueger, T. 2008. Biofuel impacts on world food supply: Use of fossil fuel, land and water resources. *Energies* 1:41–78. http://www.mdpi.org/energies/papers/en1020041.pdf.

Pimentel, D., and Patzek, T. 2005. Ethanol production using corn, switchgrass, and wood; biodiesel production using soybean and sunflower. *Natural Resources Research*, 14(1): 65–76.

———. 2008. Ethanol production using corn, switchgrass and wood; biodiesel production using soybean. In *Biofuels, solar and wind as renewable energy systems: Benefits and risks*, ed. Pimentel, D., 373–394. Dordrecht, The Netherlands: Springer.

Pimentel, D., and M. Pimentel. 2008. *Food, energy and society.* Boca Raton, FL: CRC Press.

Pimentel, D., and Satkiewicz, P. Forthcoming. World malnutrition. In *Encyclopedia of Sustainability*, Vol. 4. Great Barrington, MA: Berkshire.

Pimentel, D., Whitecraft, M., Scott, Z. R., Zhao, L., Satkiewicz, P., Scott, T. J., Phillips, J., et al. 2010. Will limited land, water, and energy control human population numbers in the future? *Human Ecology*, 38(5): 599–611.

Pitts, D. J., Smajstrla, A. G., Harman, D. Z., and Clark, G. A. 2002. Irrigation costs for tomato production in Florida. http://edis.ifas.ufl.edu/AE010.

Postel, S. 1992. *Last oasis: Facing water scarcity.* New York: Norton.

———. 1993. The politics of water. *World Watch*, 6(4): 10–18.

———. 1999. *Pillar of sand: Can the irrigation miracle last?* New York: Norton.

PRB (Population Reference Bureau). 2010. *2010 world population data sheet.* Washington, DC: PRB. http://www.prb.org/Publications/Datasheets/2010/2010wpds.aspx.

Prial, D. 2009. Dam use in the U.S. has passed its prime. *Fox Business*, February 2. http://www.geiconsultants.com/stuff/contentmgr/files/0/30714b09d670 3a71da8fd027576817c6/misc/feb09_johnson_foxbiz.pdf.

Pugh, C. A. 2001. Improving irrigation operations to restore in-stream flows. Denver: U.S. Bureau of Reclamation, Water Resources Laboratory. http:// www.iahr.org/membersonly/grazproceedings99/pdf/D156.pdf.

Qi, L., and Tishun, Z. 2004. Sustainable agrosilvicultural management techniques in northern China dryland. Jinan, China: Shandong Statistics Bureau.

Rangeland. 1994. *1994 forest, range and recreation resource analysis.* Victoria: British Columbia Forest Service. http://www.for.gov.bc.ca/hfd/library/ frra/1994/.

Rosengrant, M. K., Cai, X. and Cline, S. A. 2002. *World water and food to 2025.* Washington, DC: International Food Policy Research Institute. http://www .ifpri.org/publication/world-water-and-food-2025.

Santa Barbara, J. 2007. The false promise of biofuels. San Francisco: International Forum on Globalization; Washington, DC: Institute for Policy Studies. http://www.ifg.org/pdf/biofuels.pdf.

Schlesinger, W. H. 1997. *Biogeochemistry: An analysis of global change.* San Diego, CA: Academic Press.

Shapouri, H., Duffield, J., McAloon, A., and Wang, M. 2004. The 2001 net energy balance of corn-ethanol. Washington, DC: USDA. http://www.usda.gov/ oce/reports/energy/net_energy_balance.pdf.

Shapouri, H., and Gallagher, P. 2005. USDA's 2002 ethanol cost-of-product survey: Agricultural Economic Report No. 841. Washington, DC: USDA. http:// www.usda.gov/oce/reports/energy/USDA_2002_ETHANOL.pdf.

Shiklomanov, I. A. 1993. World fresh water resources. In *Water in crisis: A guide to the world's fresh water resources*, ed. Gleick, P., 13–24. New York: Oxford University Press.

Shiklomanov, I. A., and Rodda, J. C. 2003. *World water resources at the beginning of the twenty-first century.* Cambridge: Cambridge University Press.

Shock, C. C. 2006. Drip irrigation: An introduction. Ontario: Malheur Agricultural Experiment Station, Oregon State University. http://www.cropinfo.net/ drip.htm.

Smil, V. 2008. *Energy in nature and society.* Cambridge, MA: MIT Press.

Snyder, R. 2000. Measuring crop water use in California rice, 2000. Davis: Department of Land, Air and Water Resources, University of California. http://www.carrb.com/00rpt/WaterUse.htm.

Solley, W. B., Pierce, R. R., and Perlman, H. A. 1998. Estimated use of water in the United States in 1995. USGS Circular 1200. Washington, DC: USGS. http://water.usgs.gov/watuse/pdf1995/html/.

Troeh, F. R., and Thompson, L. M. 2004. *Soils and soil fertility.* 5th ed. New York: Oxford University Press.

UNEP (United Nations Environmental Program). 2003a. General Assembly Resolution 53-21: Water scarcity in the Middle East-North African region. http://www.skmun.freeservers.com/unep/unepres1.htm.

———. 2003b. *Global environment outlook 3: Past, present and future perspectives.* London: Earthscan. http://www.unep.org/geo/geo3/english/pdfs/prelims .pdf.

UNESCO (United Nations Educational, Scientific and Cultural Organization). 2001a. Managing risks. Paris: World Water Assessment Program, UNESCO. http://www.unesco.org/water/wwap/facts_figures/managing_risks .shtml.

———. 2001b. Securing the food supply. Paris World Water Assessment Program, UNESCO. http://www.unesco.org/water/wwap/facts_figures/food_ supply.shtml.

———. 2006. Percentage of undernourished people. Paris: UNESCO. http://www .unesco.org/water/wwap/wwdr/indicators/pdf/CH07-IPSFinalEdits.pdf.

USCB (U.S. Census Bureau). 2007. *Statistical abstract of the United States, 2008.* Washington, DC: USCB.

USDA (U.S. Department of Agriculture). 1997. Farm and ranch irrigation survey, 1998. In *1997 census of agriculture*, Vol. 3, *Special studies*, Part 1. Washington, DC: USDA, National Agricultural Statistics Service. http://www.agcensus .usda.gov/Publications/1997/Farm_and_Ranch_Irrigation_Survey/index .asp

———. 2001. *Agricultural statistics, 2001.* Washington, DC: USDA.

———. 2007. *Agricultural statistics, 2007.* Washington, DC: USDA.

USDI (U.S. Department of the Interior). 2001. *Quality of water: Colorado River basin.* Progress Report No. 20, July. Washington, DC: USDI.

USGS (U.S. Geological Survey). 2003a. Ground water use in the United States. Washington, DC: USGS. http://wwwga.usgs.gov/edu/wugw.html.

———. 2003b. Water Q&A: Water use. Washington, DC: USGS. http://water.usgs .gov/droplet/qausage.html

———. 2009. Where do the salts go? The potential effects and management of salt accumulation in south-central Arizona. USGS Fact Sheet 170-98, June. Washington, DC: National Water Quality Assessment Program, USGS. http://ag.arizona.edu/azwater/awr/janfeb05/awr-janfeb05-usgs-supplement-web.pdf.

Vörösmarty, C. J., Green, P., Salisbury, J., and Lammers, R. B. 2000. Global water resources: Vulnerability from climate change and population growth. *Science,* 289:284–288.

Vörösmarty, C. J., Letternmaier, D., Leveque, C., Maybeck, M., Pahl-Wostl, C., Alcamo, J., Cosgrove, H., et al. 2004. Humans transforming the global water system. *Eos,* 85(48): 509–520.

Wada, Y., Beek, L. P. H., van Kempen, C. M., van Reckman, J. W. T. M., Vasak, S., and Bierkens, M. F. P. 2010. Global depletion of groundwater resources. *Geophysical Research Letters* 37: L20402, 5 pages. doi: 10.1029/2010GL044571, 2010 http://www.agu.org/journals/gl/gl1020/ 2010GL044571/2010GL044571.pdf.

Wall, G. J., Pringle, E. A., and Sheard, R. W. 1991. Intercropping red clover with silage corn for soil erosion control. *Canadian Journal of Soil Science*, 71:137–145.

Wallace, J. S. 2000. Increasing agricultural water use efficiency in meet future food production. *Agriculture, Ecosystems and Environment*, 82:105–119.

Washington State Department of Ecology and West Water Research. 2004. *Analysis of water banks in the western states*. Pub. No. 04-11-011. http://www.ecy.wa.gov/pubs/0411011.pdf.

WHO (World Health Organization). 2000. *Nutrition for health and development: A global agenda for combating malnutrition*. Department of Nutrition for Health and Development. Geneva: WHO. http://whqlibdoc.who.int/hq/2000/WHO_NHD_00.6.pdf.

World Resources Institute. 2007. Nutrition: Grain fed to livestock as a percent of total grain consumed. [Online database accessed April 27, 2011.] http://earthtrends.wri.org/searchable_db/index.php?action=select_countries&theme=8&variable_ID=348.

Youngquist, W. 1997. *GeoDestinies: The inevitable control of earth resources over nations and individuals*. Portland, OR: National Book.

Zhi, M. 2000. Water-efficient irrigation and environmentally sustainable irrigated rice production in China. International Commission on Irrigation and Drainage. Department of Irrigation, Wuhan University, China. http://www.icid.org/wat_mao.pdf.

Zhu, X. G., Long, S. P., and Ort, D. R. 2008. What is the maximum efficiency with which photosynthesis can convert solar energy into biomass? *Science*, 19:153–159.

chapter nine

The potential of Onondaga County to feed its own population and that of Syracuse, New York: Past, present, and future

Stephen B. Balogh, Charles A. S. Hall, Aileen Maria Guzman, Darcy Elizabeth Balcarce, and Abbe Hamilton

Contents

9.1   Introduction

9.1.1   The changing state of fossil fuel availability and its potential to impact food production systems

The world today faces enormous problems related to population and resources. These ideas have been discussed intelligently and, for the most part, accurately in many papers from the middle of the last century (e.g., Ehrlich, 1968; Meadows et al., 1972; Odum, 1971; Hubbert, 1969, 1974). However, these concepts largely disappeared from scientific and public discussion, in part because of an inaccurate understanding of both what those earlier papers said and the validity of many of their predictions (Hall and Day, 2009). Today, most environmental science textbooks focus far more on the adverse environmental impacts of fossil fuels than on the implications of our overwhelming economic and even nutritional dependence on basic resources and the implications of their depletion. This failure to bring the potential reality and implications of peak oil—indeed, the peak of most major economic and agricultural commodities—into scientific discourse and teaching is a grave threat to industrial society.

The possibility of a huge, multifaceted failure of a substantial part of industrial civilization is so completely beyond the understanding of our leaders that we are almost totally unprepared for it. One reason is that general public acceptance and policy actions have rarely occurred for large environmental and health issues—from smoking to flooding in New Orleans—before several decades of evidence of negative impacts have confronted decision makers. The increasing availability of inexpensive petroleum, at least until the middle of this past decade, allowed the "papering over" of energy supply issues by, for example, replacing lost soil nutrients due to erosion with cheap fossil-fuel derived fertilizers, allowing fishing boats to fish ever more distant regions, concentrating animals away from their feed production, and encouraging the mass migration of people from problem-rife cities to suburbs.

Everything we do displays an astonishing dependence on oil. Beyond shoe leather and bicycles, virtually no extant forms of transportation exist for ourselves or our food that are not based on oil—and even our shoes and bicycle tires are now generally made of petroleum products. Food production, transportation, and processing are extremely energy intensive, to the degree that they use nearly 20% of the energy consumed in the United States. Clothes, furniture, and most pharmaceuticals are made from and with petroleum, and most existing jobs would cease to exist without petroleum.

The very large volume of fossil fuels used in the United States means that each of us has the equivalent of 60–80 hardworking laborers to "hew our wood and haul our water," as well as to grow, transport, and cook our food; make, transport, and import our consumer goods; provide sophisticated medical and health services; and so forth. A North American taking a hot shower in the morning has already used far more energy than probably two-thirds of the Earth's human population will use in an entire day. Oil is especially important for the transportation of ourselves, our goods, and our services; for the production of gas for heating, cooking, and some industries; and as a feedstock for fertilizers and plastics. But one would be hard-pressed to have any sense of this extreme oil dependence in our public debates and on our university campuses beyond complaints about the increasing price of gasoline: This is despite a situation similar to the summer of 2008 and five years of flat oil production, assuaged only by the subsequent, and many would say ensuing, financial collapse that decreased demand for oil.

No substitutes for oil have been developed on anything like the scale required, and most of those that exist are very poor net energy performers. Despite considerable potential, renewable sources (other than hydropower or traditional wood) currently provide less than 1% of the energy used in the United States and the world. Until 2008, the annual increase in the use of each fossil fuel was generally much greater than the total production (let alone increase) in electricity from wind turbines and photovoltaics (EIA, 2010a). Our new sources of "green" energy simply increase along with (rather than displacing) the traditional ones.

Petroleum possesses important and unique qualitative attributes, including a very high energy density and transportability (Cleveland, 2005), that lead to high economic utility and the magnitude of current use that makes its future supply prospects worrisome, including its role in food production. The relation between petroleum supply and potential demand will be the continuing issue, rather than the point at which oil actually runs out. Barring the continuation of our present worldwide recession, demand will continue to increase as human populations, petroleum-based agriculture, and economies (especially Asian) continue to grow. Petroleum supplies had been growing most years since 1900 at 2–3% per year, a trend that most investigators think cannot continue and in fact ceased in 2005 (e.g., Campbell and Laherrere, 1998; Heinberg, 2003).

How much oil and gas are left in the world? A lot remains, although probably not a lot relative to our increasing needs, and maybe not a lot that we can afford to exploit with a large financial and, especially, energy profit. We will probably always have enough oil to oil our bicycle chains. But will we have anything like the quantity that we use now at prices that allow for the things we are used to having? The issue of how much oil remains usually does not develop from the perspective of "When will we run out?" but rather "When will we reach 'peak oil' globally?" Worldwide, we have consumed a little more than a trillion barrels of oil. The current debate centers on whether 1, 2, or even 3.5 trillion barrels of economically extractable oil remain to be consumed. Fundamental to this debate, yet mostly ignored, is an understanding of the capital, operating, and environmental costs to find, extract, and use whatever new sources of oil remain to be discovered and to generate whatever alternatives we might choose to develop. Thus, the investment issues, in terms of both money and energy, will become ever more important.

Several prominent geologists have suggested that the peak production rate of conventional oil occurred for the world in 2004–2005 (e.g., Deffeyes, 2005; Campbell, 2010). A preliminary peak of all petroleum liquids (including unconventional oil, natural gas liquids, and so on) occurred in 2008. This peak was surpassed in 2010–2011, mainly due to increased production in biofuels. If global demand regains its prerecession growth, regardless of technology or price, global petroleum supplies will not be

likely to continue to increase or even to maintain current levels. The arguments of these geologists and their organization, the Association for the
Study of Peak Oil (ASPO), spearheaded by the analyses and writings of
geologists Colin Campbell and Jean Laherrere, have the support of many
other geologists who more or less agree that dozens of oil-producing
countries have already reached production peaks. These investigators
also believe that essentially all regions of the Earth favorable for oil production have been well explored for oil, so there are few surprises left,
except perhaps in regions that will be nearly impossible to exploit. The
costs of exploiting lower-grade or harder-to-reach oil deposits have been
made clear to the world by the Gulf of Mexico oil spill of 2010.

Thus, we have entered the second half of the age of oil (Campbell,
2006). In the first half of the age, oil supply grew year by year; in the second half, the importance of oil will continue, but will be coupled with
a year-by-year decline in supply. The impacts of the peak and decline
appear to be modulated somewhat by the continuation of an "undulating
plateau" at the peak, the general decline in economic growth for much
of the world, and some help from still-abundant natural gas. We are of
the opinion that it will not be possible to fill in the growing gap between
supply and demand of conventional oil with, for example, liquid biomass
alternatives on the scale required (Hall et al., 2008), and even were that
possible, the investments and time required to do so would mean that we
needed to get started some decades ago (Hirsch et al., 2005).

Clearly, we are at a "bumpy plateau" in oil production, as predicted
by Campbell (2004). As petroleum extraction declines, it causes prices to
increase, but then the increased oil price constrains economic growth,
which decreases oil use and prices to the point that demand rises again,
and so on—the chickens and eggs can keep going for some time. That is
why many "peak oilers" speak of a bumpy plateau. However, if potential
demand keeps growing, then the difference between a steady or declining
supply and an increasing demand presumably would continue upward
pressures on price. When the decline in global oil production begins in
earnest, we will see the end of cheap oil and an economic climate very
different from even the difficult times now.

9.1.2   Historical energy costs of production and transport

Two important agricultural transitions increased productivity and
changed the way America's farmers grew and supplied food. The first
was the transition from draft animals to mechanized tractors that took
place in the early 20th century (EIA, 2002). The second took place after
World War II when the military-industrial complex shifted from producing tanks, planes, and munitions to mass-producing tractors, combines,
and fertilizers. This shift from draft animals to machines, combined

with advances in fertilizer production, led to a radical restructuring of agriculture in the United States and an enormous increase in output per hectare and per unit of human labor. The era of solar-powered agriculture came to an end, and our dependency on fossil-fuel-powered farming began.

Modern agriculture is extremely fossil-fuel intensive. It is dependent on large tractors and combines and the production of the fertilizers and pesticides needed to sustain high yields on continually degrading land. Food is transported long distances by ship, rail, and truck to be processed, and then longer distances to reach American dinner tables. Pirog and Benjamin (2003) have estimated that fresh produce travels 1,500 miles on average from production to consumption.

A decline in the production rate of petroleum has troublesome implications for our current agricultural system. According to Pimentel and Pimentel (1979), one manpower is equal to roughly 1/10 horsepower, so a 10-man-hour workday produces one horsepower-hour of work. A gallon of gasoline used in a 20% efficient engine can perform the equivalent of about 10 horsepower-hours, or 100 manpower-hours, of work (Pimentel and Pimentel, 2008). Therefore, a gallon of gasoline (about $3 in 2010) would generate about 100 times the daily work output of a hardworking human, at less than 0.5% of the cost.

Early U.S. grain farms, based primarily on human labor, required about 373 man-hours per 100 bushels of wheat and 344 man-hours per 100 bushels of corn. By 1900, with draft animals and steel plows now an integral part of farming, the man-hours were reduced by more than half for corn and nearly 70% for wheat, though during that period yields remained steady. After World War II, agriculture efficiency rose dramatically, with man-hours per 100 bushels in 1955 reduced to 18 for wheat and 22 for corn. By 2000, further efficiency gains and increased fossil energy inputs had reduced the human labor inputs to 8 hours per acre for wheat and 3.3 for corn (see Figures 9.1 and 9.2).

The transition from draft animals to tractors did more than increase the efficiency of human labor. Farms grew larger, and although the frontier officially closed in 1890, a new frontier of land became available to farmers: land previously dedicated to growing fodder for their animals (Olmstead and Rhode, 2001). The horse and mule population for the United States peaked in 1920 at about 26 million, with about 20 million of those being dedicated work animals. These populations had fallen to less than 4 million by 1960. The percentage of farms reporting only tractors (no horses or mules) grew from 8% in 1940 to 55.8% just 14 years later (see Table 9.1). By 1999, the number of equine animals (horses, ponies, mules, and donkeys) on farms totaled just 3.2 million, though the total U.S. equine population, including recreational horses, was up to 5.25 million (USDA, 1999).

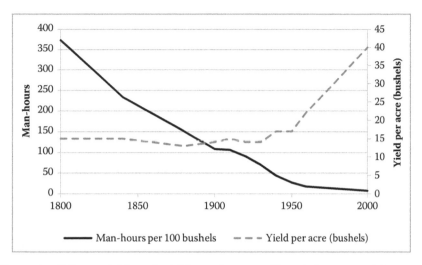

Figure 9.1 Wheat, yield, and yield per human effort. 1 hectare = 2.54 acres. (1800–1960 data from Rasmussen, W. D., *Journal of Economic History*, 22, 578–591, 1962; Year 2000 data from Pimentel, D., and Pimentel, M., *Food, energy and society*, 3rd ed., Boca Raton, FL: CRC Press, 2008.)

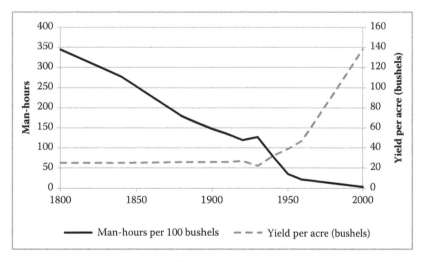

Figure 9.2 Corn, yield, and yield per human effort. (1800–1960 data from Rasmussen, W. D., *Journal of Economic History*, 22, 578–591, 1962; Year 2000 data from Pimentel, D., and Pimentel, M., *Food, energy and society*, 3rd ed., Boca Raton, FL: CRC Press, 2008.)

Table 9.1 Use of Draft Animals and Tractors on Mid-Atlantic States Farms, 1940–1954

Percentage of mid-Atlantic states farms reporting:	Year			
	1940	1945	1950	1954
Horses/mules and no tractors	41.5	24.3	15.8	6.6
Horses/mules and tractors	24.9	31.7	25.3	17.2
Tractors only	8.1	18.4	38.3	55.8
No tractors or horses/mules	25.4	25.6	20.6	20.4

Source: U.S. Census Bureau, *United States Census of Agriculture*, 1954, Washington, DC: USCB, 1956/1957.

The general trend of agriculture in the United States has been supplementing human labor, initially with animal power and later with fossil fuels. The net result has been a very large increase in the amount of grain produced per farmer per year. Less commonly understood is that there has been a great *decrease* in the productivity of each unit of *energy* invested. Perhaps the single most important way in which fossil fuel contributed to agricultural productivity was the widespread application of the Haber-Bosch process to fertilizer production following World War II when large-scale ammonium production facilities were no longer needed to produce ordnance. In addition, there was a widespread application of fossil fuels in the extraction of phosphorus in Florida and Morocco. Finally, a whole suite of additional agrochemicals, pesticides, and herbicides were produced that require fossil fuels as a feedstock.

Although draft animals and fossil-fuel-powered machines improved the efficiency of human labor on farms, yields did not rise substantially until the advent of chemical fertilizers and hybrid seeds, especially in the 1960s. Steinhart and Steinhart (1974) studied the rising energy inputs into U.S. agricultural production and found that by 1970 the energy inputs into the food system were nearly 10 times that of the food energy consumed by Americans. Fertilizer use grew quickly after the 1950s and then leveled out after concerns arose about eutrophication of wetlands and waterways and improved application techniques became available (see Figure 9.3). Cleveland (1995a) found improving agricultural efficiency after the low was reached in 1978 for the ratio of food energy output per unit of energy input, reversing the near-linear increase in energy inputs.

The mechanization of farming also reduced the number of crop types grown on each farm. On average, farms in 1900 produced five different crop types. By 2000, this number was reduced to one (Dimitri et al., 2005). In 2010, a single farmer driving a 60-foot-wide field cultivator tractor was able to work 43 acres in one hour, burning only 0.32 gallons of diesel per acre (University of Minnesota Extension, 2009). It appears that both labor and energy efficiency have increased recently (Cleveland, 1995a, 1995b).

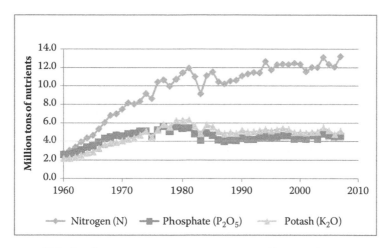

Figure 9.3 U.S. fertilizer consumption, 1960–2008. (From U.S. Department of Agriculture, 2010a, Fertilizer use and price [online database accessed July 25, 2010], http://www.ers.usda.gov/Data/FertilizerUse/.)

9.1.3   Current national patterns of food production and distribution

The availability of cheap oil has led to the changes in the agriculture system that we now practice. We use massive, but decreasing, areas of land to produce the agricultural products we need, plus a large quantity for export. We use genetically modified (but highly fertilizer dependent) high-yield seed varieties, large quantities of petroleum-requiring pesticides, and fertilizers on a single species to generate massive output per acre. The 2007 U.S. Agriculture Census indicates that the total land area devoted to agricultural production is 922 million acres. This is a decline of more than 6 million acres since 2002. At the same time, farm expenses used to purchase feeds, fertilizers, chemicals, and gasoline and other fuels are estimated to have increased from about $40 billion in 2002 to around $90 billion in 2007.

Likewise, the availability of cheap oil has also changed consumption patterns. For instance, based on the 2007 U.S. Agriculture Census, an average American consumes 2,775 kcal per day, 28% more than in 1970 (USDA, 2002). More than 3,700 kcal are available per capita per day in the United States, but 27% of food is wasted by final consumers (Kantor et al., 1997). American consumers are also eating more meat compared to the past (USDA, 2002). The U.S. livestock population consumes more than seven times as much grain as the U.S. human population consumes directly; if this grain were consumed in a plant-based diet, it could feed 840 million people (Pimentel and Pimentel, 2003).

As the tastes of American consumers change and the demand increases for produce not traditionally available year-round and not grown in a particular area, the distance produce travels has increased, and the energy consumed by this transport has also increased. Pirog and Benjamin (2003) at Iowa State University, as noted earlier, indicate that on average fresh produce travels 1,500 miles. Weber and Matthews (2008) calculate that the total freight required for food and food products in 1997 amounted to approximately 1.2 trillion tonne-km (0.82 trillion short ton-mi), or 12,000 tonne-km (8,200 short ton-mi) per household. They estimate the average final delivery distance of food to be 1,640 km (1,020 mi), with total supply chain transportation of 6,760 km (4,200 mi).

In Onondaga County, New York, the 2007 U.S. agricultural census indicates that the number of farms declined from 725 in 2002 to 692 in 2007, with a 1% reduction in the land area devoted to farms. Additionally, the county grew more grains for feeds than in the past (see below). A quick walk through the top four grocery chains operating in Syracuse and Onondaga County shows that many of the fruits, vegetables, grains, and other food items now come from other states such as California and Florida, as well as from other countries such as China, Chile, and Mexico.

The same liberalization of markets that helps boost U.S. exports also increases competition for American farmers and growers. Now, Florida orange groves compete for market share with not only California but also Brazil. Upstate New York apple farmers find competition from Washington State and Chinese fruit shipped halfway around the world. This global specialization and competition is predicated on large amounts of cheap fossil fuels, mostly petroleum products.

9.1.4 Overview and history of the city of Syracuse and Onondaga County

The Syracuse Metropolitan Area is located within Onondaga County in central New York State (see Figure 9.4). Syracuse is the subject of an Urban Long-Term Research Area (ULTRA-Ex) study through a grant from the National Science Foundation. Some of the questions that drive the research include: What is the socioecological metabolism (SEM) of a Rust Belt city? How has it changed over time? How might it be vulnerable to future external factors such as restrictions in oil availability? How might city revitalization emphasizing natural ecosystem processes via green infrastructure affect the SEM at both the city/regional and the household/ neighborhood levels in the future?

In order to begin to answer these questions, one must first have a proper understanding of the metabolism of the city and its surroundings, meaning the stocks, flows, and investments of energy used now to feed and employ the residents of the city and the surrounding population in Onondaga County.

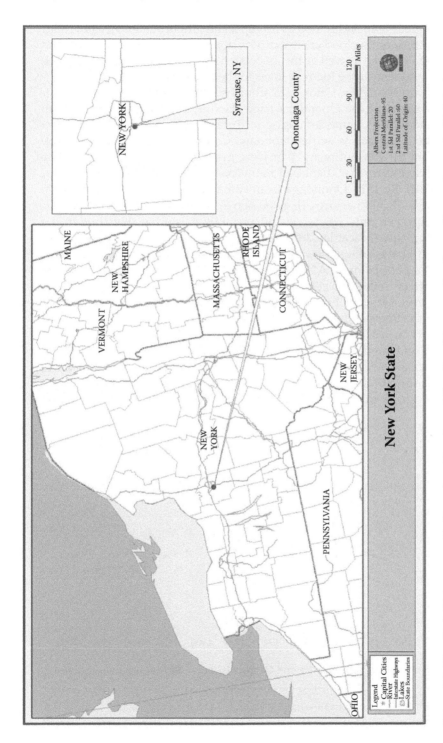

Figure 9.4 New York State map, with Onondaga County and the city of Syracuse *(inset)* as well as major roads. (Original map, created in ESRI Arcview 9.2 by Aileen Guzman, Syracuse University College of Geography, community geographer.)

This perspective includes food as an especially important energy flow. Our study focuses on the food energy production and demand in Syracuse and Onondaga County, as well as the associated fossil energy required to grow, process, and deliver the food products to the consumer.

Prior to the 1700s, Onondaga County was home to some two thousand members of the Onondaga tribe (Hodge, 1907), who lived well on the abundant fish, game, and agriculture. Ephraim Webster was the first European-American to settle Onondaga County, in 1786. Disbanded soldiers from the Revolutionary War were granted tracts of land in Upstate New York and further displaced many of the Onondaga (Anderson and Flick, 1902). By 1825, Onondaga County had a population of over 48,000 and nearly 78,500 hectares (194,000 acres) under cultivation (Macauley, 1829). However, only 100 homes purportedly existed at that time in the city of Syracuse (Macauley, 1829). Syracuse grew in population and prominence after the completion of the Erie Canal in 1825 and by 1835 had reached a population of 4,100 (Anderson and Flick, 1902). The chief economic driver in the 19th century was salt production.

By 1855, Syracuse had grown from a small village to a city of more than 25,000. The salt industry fueled the economic and population growth of the city. It was later replaced by the chemical, automotive, electrical devices, and metals industries. By 1902, over 150 separate industries employed the 130,000 residents living in Syracuse (Anderson and Flick, 1902). The population of the city peaked in 1950 at 218,830, while the county reached a peak population of 472,746 in 1970 (see Figure 9.5). Industrial

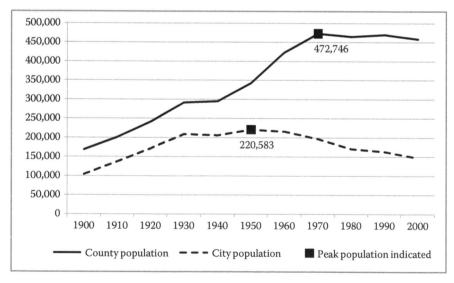

Figure 9.5 Syracuse and Onondaga County population, 1900–2000. (From U.S. Census Bureau, 2000.)

Table 9.2 Population and Economic Statistics for Onondaga County, 1990–2008

	1990	2000	2006–2008 (est.)	Change 1990–2000 (%)
Total population	468,973	458,336	452,633	–2.2
Housing units	190,878	196,633	—	+3.0
Persons per unit	2.46	2.33	—	–5.3
Registered passenger vehicles[a]	208,515	264,831	278,200	+27
Mean travel time to work (min)	18.3	19.3	18.7	+5.5
Persons per square mile	601	587	581	–2.3
Persons in commercial employment	178,212	177,360	182,964	–0.4
Persons in industrial employment	49,968	38,354	35,417	–23
Roadway mileage[b]	3,032.4	3,052.0	3,081.5	+0.6

[a] No data available for 1990; earliest available data at county level from 1996 (New York State Department of Motor Vehicles).
[b] No data available for 1990; earliest available data at county level from 1998 (New York State Department of Transportation).

production began declining in the 1970s and continues to decline to this day. As industries left the area, the population began to decline. The main economic activity in Onondaga County today is in the commercial sector—namely, health care, education, and retail—although agriculture continues to be an important source of economic production.

The post–World War II trend in land development in Onondaga County, like in a majority of the United States, has been a migration from the city and inner suburbs to new suburban developments built on formerly productive farmland (Kunstler, 1993). This urban-to-suburban migration continues today, as the city population continues to decline at a rate faster than the relatively stable county population (see Figure 9.5) The number of housing units increased by 3% from 1990 to 2000, a time when the population declined by 2.2% (see Table 9.2).

9.1.4.1   A brief history of inputs and outputs in Onondaga County
Solar energy and especially biomass were the major energy inputs into early Onondaga County. Forests provided wood for heat. Firewood was abundant. In the United States, some 5 million acres per year were cleared for new farmland (Schurr, 1960). In 1850, the per capita consumption of firewood was estimated to be 4.5 cords,* with more than 90% of it

* A cord is a measure of wood, comprising a stack 4 feet wide, 4 feet high, and 8 feet long, estimated to contain 20 million BTU (or 21 GJ) when burned (USDA, 2004).

consumed in the household (Schurr, 1960). To keep comfortably warm, an American family would use 17.5 cords of firewood a year. Firewood remained the primary fuel for home heating until coal overtook it in the late 19th century (EIA, 2002).

Energy inputs into the commercial and industrial sectors also relied heavily on solar inputs. The salt industry used wood to boil saltwater in kettles initially, then solar salt works were incorporated in 1821 reducing the biomass fuel need (Anderson and Flick, 1902). The remainder of energy inputs came from mills that tapped into the region's hydrological power (also solar) and small amounts of imported oils for light. Fossil fuels became available initially in the form of small amounts of coal imported from Pennsylvania. The importance of coal grew with the advent of the coal-fired boiler used in most trains and ships by the 19th century.

Syracuse lies at a natural intersection of east-west and a north-south low elevation transportation corridors. Draft animals such as mules and horses provided the propulsion to barges on the Erie Canal and local transportation by carriage. Locomotive service began in 1839 when the rail line from Albany to Utica was extended to Syracuse (Anderson and Flick, 1902). In 1926, the city boasted that the area was serviced by six interurban electric lines and that 181 passenger trains arrived and departed each day (City of Syracuse, 1926). Personal automobiles replaced the streetcars in the early 20th century. The first Franklin automobile, which was produced in Syracuse, was sold in 1902. Over the next few decades, the light and heavy rail systems in Syracuse were dismantled or rerouted, and the automobile became the transportation vehicle of choice. Home furnaces, previously fueled with coal, were switched over to "cleaner" home heating oil. Eventually, a majority of furnaces were replaced when natural gas service was installed in the city and later the surrounding areas.

9.1.4.2   Trends in Onondaga County energy consumption, 1990–2000

Onondaga County experienced a 2.2% decline in population from 1990 to 2000 (see Table 9.2), and the population of the city of Syracuse declined at an even faster rate—from 163,860 in 1990 to 147,306 in 2000, a decline of 10.1%. Rolf Pendall (2003) has termed the phenomenon in central New York as "sprawl without growth." Between 1982 and 1997, central New York (which includes Syracuse, Utica, and Rome) urbanized more than 160,000 acres despite a population decline (Pendall, 2003). The trend in relocating from urban neighborhoods to suburban locations requires increased energy consumption to power larger suburban homes and for transportation during longer commutes and travel from suburban neighborhoods to schools and shopping centers. Many times, these suburban developments are built on the highest quality farmland. From 1990 to 2000 in Onondaga County, the total anthropogenic energy consumption increased 14%. Per capita energy

Table 9.3 Estimated Energy Use in Onondaga County in 1990 and 2000

	Total energy in Petajoules (J × 10¹⁵)			Gigajoules per capita		
	1990	2000	Change	1990	2000	Change
Residential energy use	24.9	29.8	20%	53	65	23%
Commercial energy use	25.2	32.3	28%	54	70	30%
Industrial energy use	13.4	13.2	–1%	29	29	—
Transportation energy use	34.2	35.6	4%	73	78	7%
Total	*97.7*	*110.9*	*14%*	*209*	*242*	*16%*

Onondaga County energy consumption was estimated by Ngo and Pataki (2008): New York State level historical energy consumption (EIA, 2010b) was prorated based on population and employment. Individual sectors were adjusted based on percent of statewide gasoline consumption, home heating fuel choice, heating degree days.

Source: Balogh, S., Simulating the potential effects of plug-in hybrid electric vehicles on the energy budget and tax revenues for Onondaga County, New York, M.S. thesis, State University of New York, College of Environmental Science and Forestry.

consumption increased 16% (see Table 9.3). Residential and commercial sector energy use increased greater than 20%, and transportation energy consumption increased by 4%. Industrial energy use declined slightly.

9.1.5   Objectives

Our objectives in this chapter are to answer the following questions:

1. Can we feed the current population of the city of Syracuse and of Onondaga County from local agricultural production?
2. Has Onondaga County been calorically self-sufficient in agricultural production over the last century?
3. What changes to the current production and distribution systems would be necessary to meet the nutritional needs of the present citizens of Onondaga County?
4. How would a decline in fossil fuel availability affect local food production?

9.2   Methods

We estimated food production and use by constructing:
1. A *demand* model that calculates the farm-derived food demand from humans in Onondaga County
2. A *supply* model that calculates historical caloric production and crop yields from U.S. Department of Agriculture (USDA) published statistics on the area and production of major crops in Onondaga County

and also a spatial analysis of historical farmland in Onondaga County

3. A *foodshed* model that calculates the land needed to support the county's population over time

4. An *energy-flow* model that quantifies the current fossil energy costs for agricultural inputs and transportation and examines the change in energy consumption associated with a potential shift from national/global food supply chains to local food production and delivery.

We then synthesize and compare the outputs of these models to examine historical food security and self-sustainability of the county. Our detailed methods are described below.

9.2.1 Demand model: Food requirements of Onondaga County residents

For the demand model, we used historical U.S. per capita food availability (USDA, 2005; FAO, 2010) to estimate the daily amount of calories available to feed the average Onondaga County resident. Though differences in diet and food availability may have existed regionally, we assume that the diet of the average citizen in Syracuse and Onondaga County is no different than that of the average American. This seems reasonable, given that Syracuse is considered an "average" American city for marketing purposes, with gross domestic product, age, ethnic makeup, and so on thought to be similar to the national mean.

We calculated annual food demand in kcal for Onondaga County and the city of Syracuse by multiplying the daily per capita food availability (in kcal) for a given year by 365 days, and then by the respective population. Results are reported in Tcal (1×10^9 kcal). Similarly, to calculate the food requirement by weight to feed Syracuse and Onondaga County residents, we used FAO and USDA food availability statistics by food type and by individual crop. The recommended USDA diet includes seven major categories: sweeteners, grains, dairy, meats and other proteins, fats and oils, fruit, and vegetables. We assigned individual foods and food subgroups to these major categories (for example, see Table 9.4). We then calculated the annual food supply quantity needed, by weight (in Mt), to meet demand from Syracuse and Onondaga County by multiplying the food supply quantity in kg/person/yr by the respective populations.

9.2.2 Supply model: Historical production

Our objective with the supply model was to determine crop yields in metric tons per hectare and the total caloric output of major crops of

Table 9.4 U.S. per Capita Food Availability by Weight, Calories, and Protein Supply in 2007

Item	Food supply quantity (kg/capita/yr)	Food supply (kcal/capita/day)	Protein supply quantity (g/capita/day)
Grains	170	925	26.9
Cereals	111.6	830	24.4
Sweeteners	67.6	629	0.2
Fats and oils	40.2	830	2.8
Oil crops	5.4	61	2.5
Vegetable oils	29.1	661	0.2
Animal fats	5.7	108	0.1
Fruits	111	116	1.3
Vegetables (incl. potatoes)	186	80	3.6
Meat and other proteins	168	602	53.4
Bovine meat	41.2	114	14
Pig meat	29.7	131	8
Poultry meat	50.7	199	18.2
Eggs	14.3	54	4.2
Tree nuts	3.8	27	0.8
Pulses (legumes)	4.2	39	2.6
Fish, Seafood	24.1	38	5.6
Dairy	253.8	373	21.9
Grand total	*938.2*	*3555*	*110.1*

Sources: U.S. Department of Agriculture, 2005, Nutrient availability data set, [online database accessed July 25, 2010], http://www.ers.usda.gov/Data/FoodConsumption/Nutrient Avail Index.htm; Food and Agriculture Organization, 2010, FAOSTAT food balance sheets [online database accessed July 25, 2010], http://faostat.fao.org/site/368/default.aspx.

Onondaga County from historical data appearing in the U.S. Agriculture and Market Census from 1910 to 2007. The U.S. agricultural census publishes production data and the number of acres in production for each state at the county level. The USDA has also estimated crop production in Onondaga County for the years between agricultural censuses, starting in 2002.

We selected the crops with the highest total annual production, based on the relevance to our study and availability of data: oats, wheat, grain from corn, silage from corn, alfalfa hay, other hay, beans, potatoes, rye, soybeans, and apples. There are several important vegetable crops grown in Onondaga County—for example, onions, garlic, and, in lesser amounts, tomatoes, squash, zucchini, and others—but the data series for these crops are less contiguous, and the caloric output from these vegetables is much smaller than the major crops listed above.

Agricultural production is measured in bushels (most grains), cwt (hundred lb.; e.g., potatoes), or short tons (e.g., hay). Farmland harvested is reported in acres. To calculate agricultural production in calories, yields are first expressed by weight, using a pounds-per-bushel conversion factor (University of Missouri Extension, 2010) and then converted to kilograms (0.454 kg = 1 lb.). Finally, we multiply by 2.47 acres/hectare and divide by 1,000 (1,000 kg = 1 metric ton) to determine yield in metric tons/hectare for each crop. We used caloric values for each crop (in raw form) from the USDA Nutrient Data Laboratory (USDA, 2010d) to calculate total output in calories and correct for water weight differences between harvest weight and percent water in the USDA database.

We used data obtained from the U.S. agricultural census from 1910 to 2002 to determine the area devoted to farmlands in Onondaga County historically. The current area devoted to farmlands was obtained from the Onondaga County Planning Agency. For the city of Syracuse, we obtained historical land use from 1950 to 2000 from the Onondaga County Planning Agency. The most current land use maps and data (2008) were obtained through the Syracuse University community geographer. These data are based on the 2008 tax parcel map from the Syracuse City Assessor's Office.

We analyzed the distance from farms to the two major farmers' markets located in Syracuse, namely, the Central New York Regional Farmers Market and the Downtown Syracuse Farmers Market. We attempted to obtain a current list of farms participating in those farmers' markets, but that information was not available publicly. We therefore relied on data from the Farmers Market Federation of New York (2010) to estimate the average distance from farms to the farmer's markets using the distance calculator of ArcGIS.

9.2.3   Foodshed model: Land area requirements calculated from top-down and bottom-up analyses

The term *foodshed* originated with Walter Hedden (1929) and was reintroduced in 1991 by Arthur Getz (1991; see Peters et al., 2009 for further history of the concept). The foodshed model is used to calculate the land area required to produce the nutritional requirements for a given population. The analogy of a watershed is apt, in that the foodshed is defined by a structure of supply. Though not strictly defined, Kloppenburg et al. (1996) noted that foodsheds have no fixed or determinate boundaries. They are a function of multiple and overlapping features such as plant communities, soil types, ethnicities, cultural traditions, and culinary patterns (Kloppenberg et al., 1996).

For the purposes of this chapter, we will use the term *foodshed* to denote the agricultural land area requirement to meet the nutritional needs of a defined population, specifically, Onondaga County. Using the historical local yields calculated from the production analysis and the demand in tons calculated above, we were able to determine the historical foodshed (in hectares) for Onondaga County and the city of Syracuse. We substituted proxy yields for components of the representative diet where specific production yields for Onondaga County were unavailable (e.g., fruits, sugars, fats, etc.).

We calculated the hectares of food production required to meet the nutritional demands of Onondaga County residents for each of the seven major categories of the USDA recommended diet—sweeteners, grains, dairy, meats and other proteins, fats and oils, fruit, and vegetables—by dividing demand in tons by the local yield (in tonnes/ha). Next, we compared it with the land in agricultural production and the total land area in the county to determine the relative food security. Our detailed methods for determining land area requirements for each food category follow.

9.2.3.1   Sweeteners

An average American eats 69 kilograms of sweeteners per year (FAO, 2010). The amount of sweeteners in the American diet has increased 19% from 1970 to 2005 (Wells and Buzby, 2008). We assumed sweeteners could be generated locally from sugar beets. They are adaptable to cold weather climates and are grown currently in 12 U.S. states (Asadi, 2007). The average yields for U.S. sugar beets are about 40 metric tons (wet) per hectare, of which 17% by weight is extractable sugars—meaning an average yield of 7 tonnes of sugar per hectare (Asadi, 2007). To estimate the historical yields, we assumed that early 20th-century yields would follow a trend similar to corn production, with 1910 yields being approximately a third of current yields and a linear trend in increasing production levels.

9.2.3.2   Grains

An average American eats 111 kilograms of grains per year. We assumed that wheat is the grain of choice for Syracuse and Onondaga County residents and represents 80% of grain demand. Corn and corn products meet remainder of grain demand (20%).

9.2.3.3   Dairy

An average American consumes 250 kilograms of dairy products per year. We used milk consumption to represent dairy demand. The demand for milk before losses was divided by the average production of dairy cows

Table 9.5 Yearly Ration for Lactating Dairy Cows and Assumed Equivalent Food Demand

	% of food (by weight)	Tons/yr/head	Assumed for model calculations
Corn silage	60%	14.42	13.1 metric tons
Alfalfa hay	13%	3.05	Hay: 4.5 metric tons
Orchard grass	7%	1.79	
SBOM (48%)	5%	1.12	Soybeans: 2 metric tons
Soybeans	2%	0.53	
Corn grains	10%	2.44	Corn: 2.2 metric tons
Distillers grain	3%	0.61	

for the year to determine the number of lactating cows needed to support dairy demand in Onondaga County. To estimate the hectares needed to produce the feed for these cows, we used the average yearly ration (see Table 9.5) for dairy cows from the Virginia Cooperative Extension (2007).

9.2.3.4   Meats and other proteins

We took the USDA Food and Nutrient Intakes by Region data and determined percentages of meats consumed in the Northeast (see Table 9.6). We multiplied these percentages by the annual demand for meat products to determine how much of each commodity was required. We then calculated the hectares of land to support production of each type of meat. We assumed fish were imported and could not be harvested locally for this study. We also assumed that the food production in the "nuts" category need not be produced by large monoculture crops, but could be successfully grown in urban, suburban, and rural areas.

- *Beef:* An average beef animal weighs 1,200 pounds and yields approximately 456 pounds of beef (Thiboumery and Jepson, 2009). We assumed that beef cattle's diet is similar to that of dairy cows.

Table 9.6 Proportion of Meat/Other Protein by Type Consumed on Average in the Northeastern US

Meat product	Percent consumed
Beef	32%
Pork	20%
Chicken	20%
Eggs	12%
Fish	10%
Nuts	6%

- *Pork:* The average hog weighs 250 pounds and produces 133 pounds of pork (Thiboumery and Jepson, 2009). Hogs require an average 5.9 kg of grain for every net kg of animal product (Pimentel and Pimentel, 2003).
- *Chicken:* The average broiler weighs 5.5 pounds and contains 3.75 pounds of meat (USDA, 2010c; PoultryHub, 2010). Chickens require 2.3 kg of feed (grain) for every kg of animal product (Pimentel and Pimentel, 2003).
- *Eggs:* The average chicken produces 290 eggs per year. Each kg of eggs produced requires 11 kg of grain feed (Pimentel and Pimentel, 2003).
- *Fish:* This source of protein is excluded from this study because, other than recreational fishing, all seafood is imported into Onondaga County.

9.2.3.5   Fats and oils

Soybean oil is the most commonly used oil in the United States for food consumption. A 60-lb. bushel of soybeans yields 11 lb. (18%) oil and 47 lb. soybean meal (North Carolina Soybean Producers Association, 2007). We assumed that 80% of the fats and oils category could be met through soybean oil, and the remaining demand could be met by butter, as a coproduct of milk production. We multiplied oil consumption by 5.55 to approximate tons of total soybeans that need to be grown. Then, we divided that value by the yearly estimated soybean yield to calculate the hectares needed to meet oil demand. Pre-1960 yields were not available and were assumed to be 1.0 tonnes/ha.

9.2.3.6   Fruit

We used New York State data to determine the yield per acre of common fruit crops. We weighted the different yields per hectare by multiplying pounds per acre for 1992 for each commodity by the total acres in production to determine the total pounds produced. We then determined for each commodity the percentage of total production. We multiplied the average yield per acre by the percentage produced to calculate a weighted average yield per acre for all fruit in New York State (see Table 9.7). We then converted the yield from lb./acre to tonnes/ha: 15,250 lb./acre = 6.919 tonnes/acre = 2.8 tonnes/ha. We used this value to estimate fruit yields in Onondaga County.

9.2.3.7   Vegetables

We used the average of bean yields (1.3 tonnes/ha) and yields for potatoes* (varies by year) for Onondaga County to estimate the land area needed to meet vegetable demand.

---

* Potatoes are included in vegetable category by the USDA.

Table 9.7 Fruit Yield Data for New York State

	1992 yield (million lb.)	Percent of total production	Average yield (lb./acre)
Apples	1,170	72%	17,500
Sweet cherries	2.2	<1%	3,953
Tart cherries	31	1.9%	5,904
Peaches	14	<1%	6,998
Pears	33	2%	11,750
Grapes	360	20%	9,792
Blueberries	1.4	<1%	1,910
Strawberries	8.7	<1%	3,450
Total	1,620	100%	

These figures were used to estimate average fruit yield for Onondaga County.
Source: U.S. Department of Agriculture, 2010, Fruit historic data, [online database accessed July 25, 2010], http://www.nass.usda.gov/Statistics_by_State/New_York/Historical_Data/Fruit/Fruitindex.htm.

9.2.4 Energy requirements of food production and transport

9.2.4.1 Food production

Besides solar insolation, energy inputs into agriculture come from three major sources:

1. Human labor
2. Fossil fuel inputs used in mechanical equipment
3. Fossil fuels used in the fabrication of chemical fertilizers and pesticides

Irrigation, when required, sharply increases energy inputs. Irrigated corn requires about three times the energy of rain-fed corn to produce similar yields (Pimentel et al., 2004).

Prior studies have examined the embodied energy of agricultural products (see Steinhart and Steinhart, 1974; Pimentel and Pimentel, 1979, 2003). We used the most recent energy input/output data available (Pimentel and Pimentel, 2003, 2008) to estimate the total energy inputs required to produce the agricultural outputs in Onondaga County in 2007. For example, to calculate the total input energy to produce corn in Onondaga County, we took the actual yield in kcal and divided by 4.23 (a food to input energy ratio of 4.23:1 from Pimentel and Patzek, 2008). This method was used also to estimate the current energy required to produce the food needed to meet demand in Onondaga County in 2000. We broke down food production and demand into the seven USDA

Table 9.8 Energy Inputs into Food Production

Food type	Farm inputs (petroleum)	Energy outputs (food)	Protein kcal output/ input ratio	Total kcal output/ input ratio
Grains (average of corn and wheat)	6,532,000 (kcal/ha)	19,149,000 (kcal/ha)	n/a	3:1
Sweeteners (sugar beet)	27.39 (GJ/ha)	99.1 × 17% sugar (GJ/ha)	n/a	0.62:1
Silage	6,284,000 (kcal/ha)	25,284,000 (kcal/ha)	n/a	4.02:1
Vegetables (dry beans)	2,740,000 (kcal/ha)	4,954,000 (kcal/ha)	n/a	1.81:1
Fruits (apples)	18,000,000 (kcal/ ha)	9,587,000 (kcal/ha)	n/a	0.53:1
Dairy	Feed: 30 (kcal); fossil Fuels: 36 (kcal)	1 (kcal milk protein)	0.015:1	0.073:1
Chicken	Feed: 19 (kcal); fossil fuels: 22 (kcal)	1 (kcal chicken protein)	0.024:1	0.136:1
Beef	Feed: 122 (kcal); fossil fuels: 78 (kcal)	1 (kcal beef protein)	0.005:1	0.021:1
Pork	Feed: 65 (kcal) × 0.5 (waste); fossil Fuels: 35 (kcal)	1 (kcal pork protein)	0.015:1	0.10:1
Eggs	Feed: 20 (kcal); fossil fuels: 13 (kcal)	1 (kcal egg protein)	0.030:1	0.085:1
Oils (soybeans)	1,827,000 (kcal/ha)	7,584,000 × 18% oil (kcal/ha)	n/a	0.75:1

Source: Pimentel, D., and Pimentel, M., *Food, energy and society*, 3rd ed. (Boca Raton, FL: CRC Press, 2008); U.S. Department of Agriculture, Nutrient Data Laboratory, 2010, USDA national nutrient database [online database accessed April 4, 2010], http://www.nal.usda.gov/fnic/foodcomp/search/ (used to calculate per total kcal for dairy, chicken, beef, and pork).

categories used above and used the energy input estimations described below (see Table 9.8).

Using the energy input/output data shown in Table 9.8 and the caloric output from 2007 derived from the supply model above, we estimated the agricultural energy inputs. Likewise, we used the output from the demand model to estimate the agricultural energy inputs needed to satisfy the total food demand for the city of Syracuse and Onondaga County.

Table 9.9 Efficiency of Agricultural Transport Modes

	BTU per ton mile	kcal per tonne-km[d]
Tractor-trailer[a]	n/a	340
Box truck[b]	5,346	3,663
1-ton pickup truck[c]	9,928	6,802

[a] Pimentel, 1980.
[b] 13 mpg * 2 tons = 26 ton mi per gallon. At 139,000 BTU/gal of diesel, BTU per ton mile = 5,346
[c] 14 mpg * 1 ton = 14 ton mi per gallon. At 139,000 BTU/gal of diesel, BTU per ton mile = 9,928
[d] 3.9683 kcal/BTU

9.2.4.2   Transportation

Energy consumption for food transportation is a function of the distance from farm to processing and distribution centers, the distance from the distribution centers to retail grocery stores, and the efficiency of the mode of transportation. Transport by tractor-trailer makes up the majority of long-haul transportation for agricultural products, followed by transportation by rail. To compare the short-distance transportation by smaller vehicles (i.e., box-style delivery trucks and 1-ton pickups), with the efficiencies from long-haul transport, we tabulated the energy used (converting from BTU per ton-mile to kcal per tonne-km) for each mode (see Table 9.9).

We assumed that 10% of the current food demand is currently met by local food production, and half of that local food is transported by delivery box-trucks and pickups. We assumed the remaining food products are transported by tractor-trailer. We used the estimated food demand for Onondaga County for 2007 from the demand model above. From these assumptions, we calculated a rough estimate of food transport energy consumption. Our goal was to examine the energy savings or cost associated with increased local food production. We then calculated the transportation energy inputs for two alternate scenarios:

1. 50% of food is produced locally and 50% of this food is transported by box-truck and 1-ton pickup to market.
2. 100% of the food is produced locally, with half of the food transported by box-truck and half by 1-ton pickup trucks.

9.3   Results

The human food demand in Onondaga County reached a peak in 1990 of 610 Tcal and then fell slightly to 589 Tcal by 2006.

Onondaga County farms have increased their caloric output since the 1930s despite a consistent decline in the area dedicated to farming. This can be attributed to rising yields and the shift to more productive crops such as hybrid corn. However, given the rising population and dietary choices, both the demand and foodshed models estimate that at no time in the 20th and 21st centuries has Onondaga County had the potential to be self-sufficient in agricultural production. The land in production in Onondaga County in 2006 is only 15% of the land needed to satisfy food demand according to the demand model. Today, the land in production would not be sufficient for even a low meat or vegetarian diet. Hence, we conclude that Onondaga County must receive a net energy subsidy from elsewhere for its people to be fed.

We estimate that the actual agricultural production in Onondaga County requires 1.16 million barrels (7.1 Petajoules) of oil-equivalent, with dairy requiring the largest percentage of energy inputs. The consumption of the equivalent of 2.5 million barrels of oil (15.1 Petajoules) would be required to feed the population in Onondaga County.

Transporting the food consumed in Onondaga County represents 11% of the total energy inputs. If the county were able to produce half of the food required to feed its population, the transportation energy requirements would be cut by 43%.

Our results are detailed below.

9.3.1   Human food requirements for Syracuse and Onondaga County

The population in Onondaga County peaked in the 1970s. The estimated caloric and metric tons of food demand, however, continued to grow through the 1990s before falling slightly by 2006 (see Table 9.10 and Figure 9.6).

9.3.2   Food production in Onondaga County

Yield increased for most field crops from 1900 through 2008, though some crops, such as beans, alfalfa, and hay, did not have a clear trend (see Figure 9.8). Total food production (in kcal) was relatively high in the early 20th century, declined until reaching a low in the 1930s, then rose again (see Figure 9.7), despite the fact that the total area farmed in Onondaga County has declined since at least the 1940s (see Figure 9.9).

There has been a large shift in crops grown over time. Initially, a large portion of local crops consisted of oats and potatoes, presumably for horses and human consumption. Agriculture in Onondaga County since the 1960s has been dominated by corn and dairy production. By 2007, dairy production accounted for 22% of total production by energy content, and corn was nearly two-thirds of gross output (see Table 9.11).

Table 9.10 Per Capita kcal, Population, Caloric Demand, and Demand by Weight for the City of Syracuse and Onondaga County Residents, 1910–2006

Year	Per capita food consumption (kcal)	Population		Demand (Tcal)		Demand (kt)	
		Syracuse	Onondaga County	Syracuse	Onondaga County	Syracuse	Onondaga County
1910	2,868	137,249	200,298	144	210	109.2	159.3
1920	2,868	171,717	241,465	180	253	136.6	192.1
1930	2,868	209,326	291,606	219	305	166.5	232.0
1940	2,868	205,967	295,108	216	309	163.8	234.8
1950	2,749	220,583	341,719	221	343	172.2	266.7
1960	2,749	216,038	423,028	217	424	168.6	330.2
1970	2,868	197,208	472,746	206	495	156.9	376.1
1980	2,982	170,105	463,920	185	505	159.8	435.8
1990	3,561	163,860	468,973	213	610	153.9	440.6
2000	3,557	147,306	458,336	191	595	140.7	437.8
2006	3,565	140,658	452,978	183	589	130.6	420.7

Tcal (teracalories) = 1 x 10⁹ kcal; kt (kilotonnes) = 1,000 metric tons.

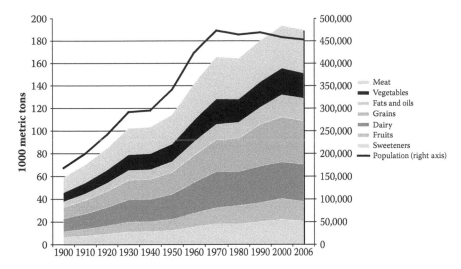

Figure 9.6 Food demand for Onondaga County by category (in thousand metric tons) and population, 1910–2006.

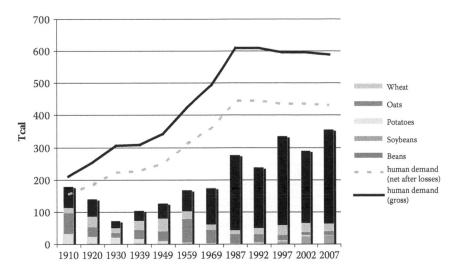

Figure 9.7 Production of major crops in Onondaga County, 1910–2007. The total energy required by humans is also included. The dashed line indicates gross demand, and the dotted line indicates net demand after losses. Smaller crops, e.g., potatoes, rye, beans, apples, were included but are not visible at this scale.

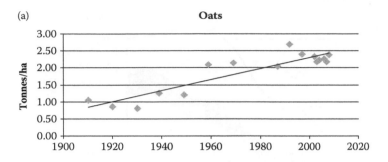

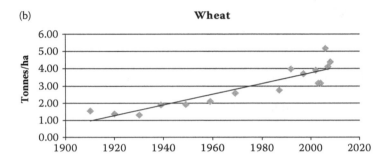

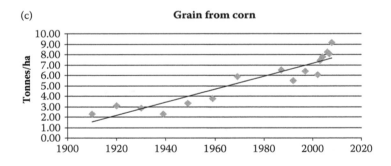

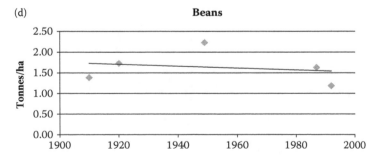

Figure 9.8 Agricultural yield in metric tons per hectare from 1910 to 2008 in Onondaga County.

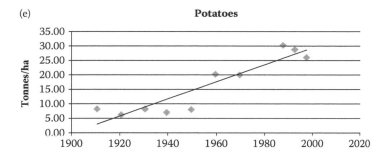

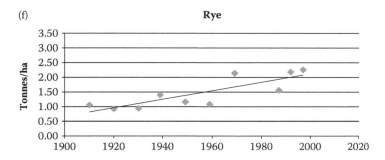

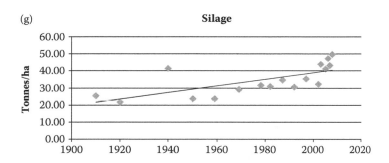

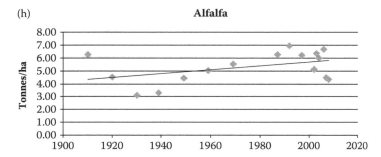

Figure 9.8 Continued

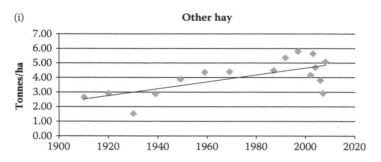

Figure 9.8 Continued

9.3.2.1 Land under agricultural production in Onondaga County and Syracuse

The total agricultural land in Onondaga County has declined from 177,251 hectares in 1900 (91% of the county) to 61,763 hectares (30%) in 2006 (see Figure 9.10). At the same time, the number of farms in the county decreased from 4,564 in 1990 to 692 in 2007 (see Figure 9.11). Aside from the decrease in the total land area and number of farms in Onondaga County, the total land area allotted for crops has declined, from 97,824 hectares in 1924 to 43,020 hectares in 2007. Likewise, the area of cropland harvested has also declined from 91,114 hectares in 1924 to 37,238 hectares in 2007 (see Figure 9.9).

Not all lands devoted to farms are harvested. Based on these results, about 50% of the total farmland is harvested. The rest of the land is used

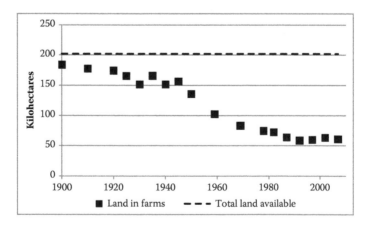

Figure 9.9 Total cropland and harvested cropland (in thousand hectares) for Onondaga County, 1900–2007. Using linear best fit lines. The difference between the two lines appears to be fallow land or unharvested crops. (From U.S. Census Bureau, *United States Census of Agriculture* [various editions], Washington, DC: USCB, 1900–2007.)

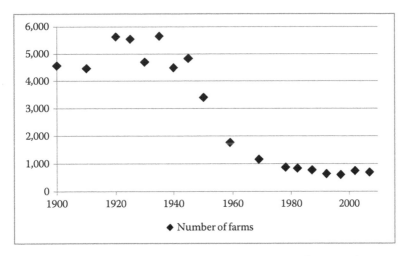

Figure 9.10 Total land area (in thousand hectares) in farms for Onondaga County, 1900–2007. (From U.S. Census Bureau, *United States Census of Agriculture* [various editions], Washington, DC: USCB, 1900–2007.)

either as pasture, storage, living quarters, or forested land or simply remains idle. At the same time, based on Figure 9.11, we can observe that not all land devoted to crops is utilized. The area of total cropland not in use ranges from 7% to 33%.

Table 9.11 Production of Major Crops in Onondaga County in 2007

Product	Amount produced (kt)	Calories produced (Tcal)
Oats	3.3	12.7
Wheat	6.2	22.7
Corn	80.2	293
Rye	2.39	0.8
Beans	0.4 (est.)	1.4
Potatoes	0.4	0.3
Soybeans	7.5	25.3
Silage	265	n/a
Hay	19.1	n/a
Alfalfa	32.1	n/a
Dairy	198	99.2
Eggs	2.6	3.3
Beef	0.04	2.5
Chicken	*	*
Total	617 (*without forage, 300*)	461

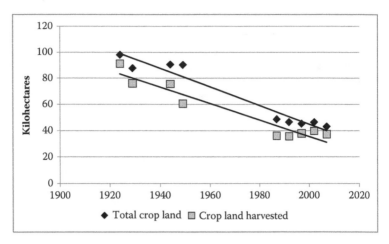

Figure 9.11 Number of farms in Onondaga County, 1900–2007. (From U.S. Census Bureau, *United States Census of Agriculture* [various editions], Washington, DC: USCB, 1900–2007.)

9.3.2.2 Distance from local farms to farmers' markets

The average distance of farms participating in the downtown Syracuse Farmers Market is 13 miles, while that for the regional farmers' market was 17 miles. These two were chosen for the analysis because they are the two most popular farmers' markets in Syracuse. Aside from these two markets, there are six other farmers' markets located throughout Onondaga County that are open during the summertime (New York State Department of Agriculture and Markets, 2010). The closest farm that participates in the two farmers' market is located approximately 7 miles away.

9.3.3 Foodshed model

The foodshed model combines population data, per capita demand for specific categories of food, and historical yields for crops in Onondaga County to estimate the number of hectares required to grow food locally to feed the residents of Onondaga County. Given the assumptions in the foodshed model, at no time in the 20th century or since could Onondaga County been self-sufficient in food production (see Figure 9.12).

9.3.4 Energy cost of feeding today's population

The Onondaga County caloric human food requirement for 2000 was estimated to be 559 Tcal (2.3 Petajoules), and the energy inputs required to produce this much food was estimated at more than 3,600 Tcal (15.1 Petajoules), equivalent to approximately 2.5 million barrels of oil (see

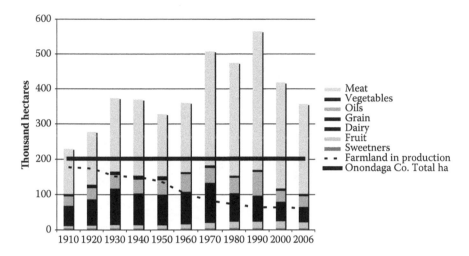

Figure 9.12 Agricultural footprint to feed the population of Onondaga County as calculated from the demand model and historical yields in Onondaga County vs. farmland in production and total available land in Onondaga County.

Table 9.12). Transportation energy adds an additional 447 Tcal of energy inputs (see Table 9.13a), for a total of 4,047 Tcal to grow and transport food to Onondaga County grocery stores. The energy required to transport goods to the area is approximately 11% of the total energy inputs.

Table 9.12 Estimated Fossil Energy Inputs Required to Meet Onondaga County Food Demand in 2000

Category	Demand (Tcal)[a]	Energy output/ input ratio[b]	Estimated input requirements (Tcal)
Sweeteners	105.7	0.62	170
Fruits	21.6	0.53	41
Dairy	64.2	0.073	879
Grains	143.2	2.93	49
Fats and oils	131.5	0.41	321
Meat	97.9	—	—
Chicken	20	0.136	147
Pork	20	0.10	200
Beef	31	0.021	1,476
Eggs	11.7	0.085	364
Vegetables	30.9	1.81	17
Total			*3,664*

[a] Calculated as: Onondaga population × FAO per capita caloric availability for 2000 × 365 (see Table 9.10 for total caloric demand by year).
[b] See Table 9.8.

Table 9.13 Transportation Energy Required for Agricultural Products to Onondaga County

	Mode	Kilotonnes delivered	Average distance (km)	Kilotonne-km	Efficiency (kcal/tonne-km)	Energy consumed (Tcal)
(a) 2000 demand, assuming 10% local production and transportation with an average 44 km from farm to market; imported food assumed to travel an average distance of 3,300 km						
Imported food	Tractor-trailer	394 (90%)	3,300	1,300,000	340	442 (99%)
Local food	Tractor-trailer	22 (5%)	44	963	340	0.3
	Box truck	11 (2.5%)	44	482	3,663	1.8
	1-ton pickup	11 (2.5%)	44	482	6,802	3.3
Total						447
(b) Scenario: 50% local food production and transportation						
Imported food	Tractor-trailer	219 (90%)	3,300	722,300	340	246 (90%)
Local food	Tractor-trailer	109 (5%)	44	4,816	340	1.6
	Box truck	55 (2.5%)	44	2,408	3,663	8.8
	1-ton pickup	55 (2.5%)	44	2,408	6,802	16.4
Total						272
(c) Scenario: 100% local food production and transportation by small trucks only						
Local food	Box truck	104,542 (50%)	44	2,090,840	3,663	16.0
	1-ton pickup	104,542 (50%)	44	2,090,840	6,802	29.8
Total						45.8

Table 9.14 Estimated Energy Inputs into Specific Crops Important in
Agricultural Production in Onondaga County in 2007

Product	Metric tons produced[a]	Gcal/metric ton[b]	Calories produced (Tcal)[c]	EROI ratio used to derive energy cost[d]	Energy inputs (Tcal)
Oats			12.7	3.1	4.1
Wheat			22.7	2.4	9.4
Corn			293	2.9	100
Rye			0.8	2.4	0.3
Beans			1.4	1.8	0.8
Potatoes			0.3	1.2	0.2
Soybeans			25.3	4.2	6.1
Silage	240,718	0.8	196	4.0	48.8
Hay	19,183	1.7	0.03	5.0	6.6
Alfalfa	30,112	2.3	0.07	6.2	11.0
Dairy (2002)			99.2	0.1	1,360
Eggs			3.3	0.1	39.0
Beef			2.5	0.02	119.5
Chicken	1,339	6,842.3	0.003	0.1	0.03
Total					*1,706*

[a] New York Agricultural Statistics Service, *Onondaga County farm statistics, August 2009* (Albany: New York Agricultural Statistics Service, 2009)
[b] Pimentel, D., and Pimentel, M., *Food, energy and society*, 3rd ed. (Boca Raton, FL: CRC Press, 2008)
[c] As calculated in Table 9.11
[d] Pimentel, D., and Pimentel, M., *Food, energy and society*, 3rd ed. (Boca Raton, FL: CRC Press, 2008); also as calculated in Table 9.8

Therefore, food production and transport to meet the requirements of Onondaga County residents requires approximately 6.4 (3,600/559) kcal of embodied energy (fossil fuel and other inputs) to produce 1 kcal of food energy. We estimate that actual agricultural production in 2007 in Onondaga County required 1.7 Tcal (7.1 Petajoules) of energy inputs, with dairy requiring the largest percentage of energy inputs (see Table 9.14).

Despite the decreased efficiency of small trucks when compared to large tractor-trailers, shifting half of the food production to the local level would reduce agricultural transportation energy costs by about 39% (see Table 9.13b). If all of the food were grown locally, the transportation energy demand could by reduced by 90% (see Table 9.13c). This is because of the large distances we assume for imported food. Were we to grow food 100–500 miles away and ship it by tractor trailer this might require less energy for shipping than local food.

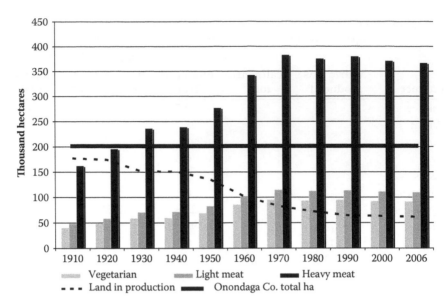

Figure 9.13 Agricultural footprint to feed the population of Onondaga County as calculated with Peters et al. (2007) assumptions.

9.4   Discussion

Our analysis indicates that, despite being situated in a moderately productive agricultural county with only moderate population density, and despite recent increases in the productivity of crops, the total food production for the county is insufficient to meet the demand for food of the county or even the city of Syracuse. Only 30% of the county is in agricultural production at this time. If all possible agricultural land (i.e., that proportion of the county in production in 1920) were utilized, it would be possible to feed the city and county a reduced-protein diet.

The most important implications of this study relate to the probability that oil, now heavily used for growing and transporting food in the developed world, may be much less available in the future. While this study is important in determining the degree to which local food production could feed Syracuse, it also shows that transportation energy appears less important than the energy necessary to grow the food itself, especially for a protein-intensive diet.

9.4.1   Land area requirements

Peters et al. (2007, 2008) calculated the area in acres required to feed a population center in Upstate New York, basing their study on three diets with increasing land requirements: vegetarian, light meat, and heavy meat diets. They assumed that a vegetarian diet requires 0.5 acres (0.2 ha) per capita, while a diet with small amounts of meat and dairy per day (2 oz.) requires 0.6 acres (0.24 ha) per capita in New York State. A diet more consistent with the current average diet, with a large amount of meat, requires more than 2 acres (0.81 ha). To compare our results to those of Peters et al., we multiplied these land area requirements by the city and county population to determine the foodshed for the region over time. Next, we compared the foodshed calculation outputs against the land in production and all available land in Onondaga County to determine the proportion of food needs that could be met by local production.

The total number of hectares needed exceeds the farmland in production (see Figure 9.13). Using the required hectares per capita for the range of diets from Peters (vegetarian, light meat, and heavy meat) to estimate the historical foodshed for Onondaga shows that the land in production from 1910 through 1960 could support the county's population at a lower level of meat consumption. By the 1970s, however, hectares required for the population exceeded the available land, even for a vegetarian diet. As early as 1920, a diet rich in meat required a foodshed that exceeded the land in production, and after 1930, it exceeded the total land within the county borders (see Figure 9.13).

The land in production in 2000 satisfies only a fraction (17%) of the heavy meat diet's land requirement described in Peters et al. (2007, 2008) and our demand model and only 56% and 67% of the land needed to sustain a light meat or vegetarian diet, respectively (see Table 9.15). Interestingly, according to Peters et al.'s estimates, the land area requirement for the city of Syracuse could be met for a vegetarian or light meat diet by the production in the surrounding area of Onondaga County.

We compared the output from our demand analysis and the high meat foodshed from Peters et al. (see Figure 9.14). Both models calculate a similar number of hectares to support the population in 2006 (356 vs. 366 kilohectares), even though their methods differ. By the 2000s, the demand analysis estimates a decreasing land area needed to support the population. In earlier decades, however, the demand model estimates a much higher land area is required to support the population, a reflection of poorer yields in the past.

9.4.2 "Local" food

Onondaga County is bordered by four counties that do not have large cities and also produce a large amount of agricultural products. Peters, Bills, Wilkins, et al. (2009) calculate that the food demand of the city of Syracuse and the surrounding population centers could be met by the surrounding area. They discuss the arbitrary nature of "local" food production and the varying definitions offered other studies. It seems

Table 9.15 Farmland Needed for Food Self-Sufficiency for the Residents of the City of Syracuse and of Onondaga County

Diet	Land area requirement (ha/capita)	Population (2000)		Land area requirement (ha)		Land in production (2000)	Percent satisfied
		Syracuse	Onondaga County	Syracuse	Onondaga County		
Our demand model	0.91	147,306	458,336	134,235	417,668	63,301	15%
Heavy meat[1]	0.81	147,306	458,336	119,318	371,252	63,301	17%
Light meat[1]	0.24	147,306	458,336	35,353	110,000	63,301	56%
Vegetarian[1]	0.20	147,306	458,336	29,461	91,667	63,301	67%

[1] Peters foodprint acreage (2007).

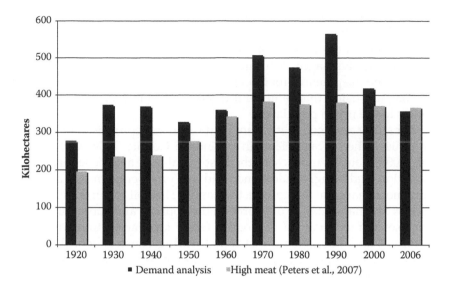

Figure 9.14 Comparison of the hectares required to feed human population of Onondaga County based on (a) our demand model and (b) Peters et al. (2007) high-meat-diet land requirements.

reasonable, given the relatively low population density in the surrounding counties—112 people per square mile versus the relatively higher 588 people/sq. mi. in Onondaga County (USCB, 2010)—that the production in the five-county region could meet the food demand of the combined population. Of course, transporting food from longer distances would mean an increase in transportation energy costs.

9.4.3 Land quality/urban farming potential

Vacant land in Syracuse has tested positive for high levels of lead and other contaminants (Johnson and Bretsch, 2002). Urban gardeners in other major cities have demonstrated the potential for producing large amounts of food in small areas using intensive farming methods and increased manual labor (see Dervaes Family, 2010; Allen, 2010). Cubans were able to produce copious amounts of organic food on previously contaminated land during the "Special Period" after the Soviet Union cut off their subsidized oil, using a composting system and raised beds to ensure soil quality (Altieri et al., 1999).

Will Allen, a leader in urban agriculture, trains community members to grow nutritious, affordable food, assuring the urban population a source of fresh food despite the prevailing political or economic forces that shape their area (Allen, 2010). His systems approach to urban farming captures compostable materials from the waste stream, which allows his gardens to create new soil and use the heat from the composting process to heat small greenhouses. More importantly, the nonprofit Growing Power improves the health, self-esteem, and self-sufficiency of the economically disadvantaged. A future study under the Syracuse ULTRA-Ex will examine the potential for food production on vacant urban lands and will compare using that same land to grow trees to increase ecosystem services to the urban population.

Our study assumed that vacant land could be repurposed into farms that achieve the average yields in Onondaga County. However, as Hall et al. (2000) note, in most cases the best farmlands are those already in production, and bringing additional land into production means incorporating more marginal, less productive lands. The soil in currently vacant or fallow farmland may be of poorer quality. Sprawl and suburbanization have also claimed many of the former fertile farmlands. In many cases, the topsoil and vegetation are stripped during construction. After the foundations are built, infill and a thin layer of topsoil are added. Until the new grass and plantings take root, the entire area is subject to soil erosion. On the other hand, suburban lawns might be considered as good farmland close to farmers, with the forest mostly cleared, irrigation systems installed, and fertile soil.

9.4.4 Diet

This study assumed that the local population would shift their current diet of highly processed and convenient food to a diet high in fresh food that is available only on a seasonal basis. A local food diet might require increased food preparation time and a change in menu and shopping habits, and foods that were previously available year-round would become scarce when out of season.

The results of this study and others (Pimentel and Pimentel, 2003, 2008; Peters et al., 2007, 2008) demonstrate that the percentage of meat and dairy products in a region's diet strongly affect the land area required to support that population. Our data show that Onondaga County is more likely to become self-sufficient if residents adopt a vegetarian or low meat diet. Reducing the meat intake to a few small portions per week, as in the Peters et al. scenario, would decrease the additional area needed to feed the population in Onondaga County from more than 350,000 to around 100,000 hectares.

9.4.5 Energy inputs

Our boundaries for the energy inputs into agricultural production end at the distribution point. To be sure, further energy is consumed to package, process, store, and display the food, as well as the energy required by the consumer to travel to and from the store, refrigerate, store, and then cook the food. It was our assumption that the distribution, storage, and cooking energy costs would be comparable in a locally based food system to the current American system, but this may not be the case. Local food production would require the construction or addition of processing capacity to handle the large volume of locally produced fruit and vegetables. Meat products pose a similar dilemma. Local butchering and meat-processing facilities would need to be encouraged or rules relaxed to allow farmers to sell directly to the public. Also, consumers themselves would have to do more of the processing, for instance, canning and freezing to extend seasonal foods. These practices may increase energy consumption, as individual consumers are less efficient than large processing and canning facilities.

The energy subsidies needed to produce a kilocalorie of food energy do not increase along the trend line as suggested by data published by Steinhart and Steinhart (1974) and in fact indicate an increasing efficacy. One average kcal of food energy requires a minimum input of approximately 6 kcal of fossil energy, down from 9 or 10 in 1970. This decline in inputs is consistent with the results of Cleveland (1995a), who calculated that agricultural efficiency had improved after reaching a low in 1978 and by 1990 the ratio of food output per unit of energy had increased to levels comparable with the 1950s.

Although some 3,700 kcal of food is available per capita per day (FAO, 2010), Kantor et al. (1997) have estimated that on average 27% of food produced on American farms is not consumed by the population. They determined that the largest proportions of food loss are in the fresh produce, dairy, and grains categories. Less waste occurs in categories such as meat, oils, and sweeteners. The USDA estimates that this ratio has remained fairly consistent since the 1970s (Kantor et al., 1997). For our demand model, which calculates caloric demand based on population, available calories, and eating habits, we assumed net consumption of food is 27% less than the gross food availability across all food categories. However, it is difficult to say whether a local food system might incur more or less waste than in the current system.

9.4.6 Potential implications of a low-energy future on agricultural self-sufficiency in Onondaga County

Pimentel estimates that the energy spent on growing, processing, transporting, storing, and preparing food is approximately 19% of U.S. energy consumption (pers. comm.). We found that just over 4,000 Tcal (17 Petajoules) was required to grow and transport the food demand of Onondaga County. This is equivalent to 15% of the 111 Petajoules of energy estimated to be consumed each year in the county. Our calculation does not include the energy to process, store, or prepare food. Our just-in-time system of food production and delivery, and our reliance on long-distance shipping by tractor-trailer, however, renders us vulnerable to any future energy supply disruptions.

The price of fertilizers increased rapidly in the summer of 2008 as energy prices reached an all-time high, before falling back by the end of the year (Huang, 2009). Increased competition for grain exports and falling per capita production led to a "run on rice" later that year, as developing nations held tightly to their excess supply and other nations scrambled to secure imports (Christiaensen, 2009). In a future with decreasing energy availability and increased variability of crop yield due to climate change, this type of event may continue to plague developing nations and could raise food commodity prices for consumers living in exporting nations such as the United States. For these reasons, communities may need to assess their food security and encourage more local and sustainably grown food.

If the production rate of oil, the economy, and fuel prices maintain an undulating plateau as some energy analysts predict, food prices could also destabilize and fluctuate rapidly. Farmers' future profits could be reduced during periods of increasing fuel prices; during economic downturns, collapsing food prices could be equally devastating.

It is also important to note that over time, land under agricultural production degrades. A highly productive piece of farmland requires increased fertilizer inputs to maintain high yields. Similarly a more marginal farm can raise yields by increasing the amount of fertilizer used. However, this relationship is nonlinear, and at high levels of fertilizer use, the response becomes asymptotic (Hall et al. 2000). Hall et al. (2000) believe that as the farmlands degrade over time due to nutrient depletion and erosion, any disruption in inputs would reduce yields to a level below the original yields obtained on the site without inputs (see Figure 9.15). Locally produced soil nutrient additives, such as compost, require human and fossil energy inputs to separate, collect, compost, and then apply to the soil.

9.4.7   Possible impact of biofuel production on food production

A decline in the availability of petroleum products, especially gasoline and natural gas, could increase the demand for liquid biofuels and biomass to heat homes during the cold Northeast winters. Subsidies and mandates by the U.S. government have already led to increased corn production for ethanol (Vedenov and Wetzstein, 2008). In 2006, 27% of U.S. corn crop was converted to ethanol (USDA, 2010e). Biofuels cannot be transported through petroleum pipelines (water is soluble in ethanol, and pipes would corrode) and therefore must be shipped by tanker truck or rail car. Since the energy return on energy invested for ethanol is already precariously low (by some accounts, negative), increasing the shipping distance of ethanol can turn a marginal net energy positive fuel into a net

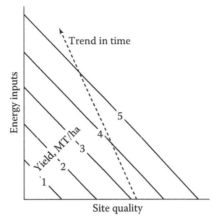

Figure 9.15 Isopleths of agricultural yield. (From Hall, C. A. S., Perez, C. L., and Leclerc, G., *Quantifying sustainable development: The future of tropical economies*, San Diego, CA: Academic Press, 2000.)

energy loser (Murphy et al., 2010). Therefore the pressure may be to produce these fuels locally. Home heating fuel costs rose in 2008, prompting some to switch to wood and other biomass heating systems, just as consumers did during the energy crises during the late 1970s and early 1980s. Both the demand for liquid biofuels for transportation and the potential shift toward biomass for home heating may put the future of local food production at odds with local energy production.

9.5   Conclusions

Our study shows the vulnerability of a medium-size city to oil supply disruptions because of the energy intensity of its food production systems. Our study also shows the potential of, but also the difficulty of, adjusting to a world with less oil. Examination of the historical, current, and future potential for agricultural self-sufficiency can provide insight for local, regional, and state leaders. Even though it might be possible to meet the caloric requirements of the people of the city of Syracuse and of Onondaga County from food production within the county (and certainly within the local five counties) by adapting our diet and increasing land in farms, this assumes that the oil and petrochemical inputs will be available to maintain current levels of agricultural production.

Acknowledgments

This material is based upon work supported by the National Science Foundation under Grant No. BCS-0948952.

References

Allen, W. 2010. Growing power. http://www.growingpower.org/about_us.htm.

Altieri, M. A., et al. 1999. The greening of the "barrios": Urban agriculture for food security in Cuba. *Agriculture and Human Values*, 16 (June): 131–140.

Anderson, J. J., and Flick, A. C. 1902. *A short history of the state of New York*. New York: Maynard, Merrill.

Asadi, M. 2007. *Beet-sugar handbook*. Hoboken, NJ: Wiley-Interscience.

Balogh, S. 2010. Simulating the potential effects of plug-in hybrid electric vehicles on the energy budget and tax revenues for Onondaga County, New York. M.S. thesis, State University of New York, College of Environmental Science and Forestry.

Campbell, C.J. 2004. Presentation at the Association for the Study of Peak Oil conference, Lisbon.

———. 2006. The Rimini Protocol, an oil depletion protocol: Heading off economic chaos and political conflict during the second half of the age of oil. *Energy Policy*, 34 (August): 1319–1325.

———. 2010. The post-peak world. London: Institute for Policy Research and Development. http://iprd.org.uk/wp-content/plugins/downloads-man ager/upload/The%20Post-Peak%20World.pdf.

Campbell, C. J., and Laherrere, J. 1998. The end of cheap oil. *Scientific American* (March): 78–83.

Christiaensen, L. 2009. Revisiting the global food architecture: Lessons from the 2008 food crisis. WIDER Discussion Paper 2009/04. Helsinki: UNU-WIDER. http://www.wider.unu.edu/publicatiions/working-papers/discussion papers/2009/en_GB/dp2009-04/_files/82162024434499621/default/dp2009-04.pdf.

City of Syracuse. 1926. Syracuse—Convention City. [Brochure.] http://syracuse thenandnow.org/ History/SyrConventionCity1926.pdf.

Cleveland, C. J. 1995a. The direct and indirect use of fossil fuels and electricity in USA agriculture, 1910–1990. *Agriculture, Ecosystems & Environment*, 55 (September): 111–121.

———. 1995b. Resource degradation, technical change, and the productivity of energy use in U.S. agriculture. *Ecological Economics*, 13 (June): 185–201.

———. 2005. Net energy from the extraction of oil and gas in the United States. *Energy*, 30(5): 769–782.

Deffeyes, K. 2005. *Beyond oil: The view from Hubbert's Peak*. New York: Farrar, Straus and Giroux.

Dervaes Family. 2010. Urban homesteading. Accessed July 25, 2010. http://urban homestead.org.

Dimitri, C., Effland, A., and Conklin, N. 2005. The 20th-century transformation of U.S. agriculture and farm policy. Economic Information Bulletin No. EIB3, June. Washington, DC: USDA, Economic Research Service.

Ehrlich, P. 1968. *The population bomb*. New York: Ballantine Books.

EIA (U.S. Energy Information Administration). 2002. History of energy in the United States, 1635–2000. http://www.eia.doe.gov/aer/eh/frame.html.

———. 2010a. International energy statistics. [Online database accessed July 9, 2010.] http://tonto.eia.doe.gov/cfapps/ipdbproject/IEDIndex3.cfm.

———. 2010b. State Energy Data System: New York. [Online database accessed July 14, 2010.] http://www.eia.gov/state/seds/seds-states.cfm?q_state_ a=NY&q_state=New York.

Farmers Market Federation of New York. 2010. Markets and farmers. Accessed July 25, 2010. http://www.nyfarmersmarket.com/index.htm.

FAO (Food and Agriculture Organization). 2010. FAOSTAT food balance sheets. [Online database accessed July 25, 2010.] http://faostat.fao.org/site/368/default.aspx.

Getz, A. 1991. Urban foodsheds. *Permaculture Activist*, 24:26–27.

Hall, C. A. S., and Day, J. W., Jr. 2009. Revisiting the limits to growth after peak oil. *American Scientist*, 97:230–237.

Hall, C. A. S., Perez, C. L., and Leclerc, G. 2000. *Quantifying sustainable development: The future of tropical economies*. San Diego, CA: Academic Press.

Hall, C. A. S., Powers, R., and Schoenberg, W. 2008. Peak oil, EROI, investments and the economy in an uncertain future. In *Biofuels, solar and wind as renewable energy systems: Benefits and risks*, ed. Pimentel, D., 109–132. Dordrecht, The Netherlands: Springer.

Hedden, W. P. 1929. *How great cities are fed*. Boston: D. C. Heath.

Heinberg, R. 2003. *The party's over: Oil, war and the fate of industrial societies*. Gabriella Island, BC: New Society.

Hirsch, R., Bezdec R., and Wending, W. 2005. Peaking of world oil production: Impacts, mitigation and risk management. Washington, DC: U.S. Department of Energy, National Energy Technology Laboratory.

Hodge, F. W. 1907. *Handbook of American Indians north of Mexico*. Washington, DC: Government Printing Office.

Huang, Wen-yuan. 2009. Factors contributing to the recent increase in U.S. fertilizer prices, 2002–2008. Economic Research Service Report AR-33. Washington, DC: USDA. http://www.kfb.org/commodities/commodities images/Fertpriceincreases.pdf.

Hubbert, M. K. 1969. Energy resources. In *Resources and man: A study and recommendations*, ed. Cloud, P., Committee on Resources and Man of the National Research Council, and National Research Council Division of Biology and Agriculture, 157–242. San Francisco: W. H. Freeman.

———. 1974. Testimony before the Subcommittee on the Environment of the Committee on Interior and Insular Affairs, House of Representatives, 93rd Cong., Serial no. 93-55. Washington, DC: Government Printing Office.

Johnson, D., and Bretsch, J. 2002. Soil lead and children's BLL levels in Syracuse, NY, USA. *Environ Geochem Health*, 24(4): 375–385.

Kantor, L. S., Lipton, K., Manchester, A., and Oliviera, V. 1997. Estimating and addressing America's food losses. *Food Review* (January–April): 2–12.

Kloppenberg, J., Jr., Hendrickson, J., and Stevenson, G. W. 1996. Coming in to the foodshed. *Agriculture and Human Values*, 13(3): 33–41.

Kunstler, J. H. 1993. *The geography of nowhere: The rise and decline of America's manmade landscape*. New York: Simon & Schuster.

Macauley, J. 1829. *The natural, statistical, and civil history of the state of New-York*. New York: Gould & Banks.

Meadows, D. H., Meadows, D. L., Rander, J., and Behrens, W. W., III. 1972. *The limits to growth*. Washington, DC: Potomac Associates.

Murphy, D. J., Hall, C. A. S., and Powers, B. 2010. New perspectives on the energy return on (energy) investment (EROI) of corn ethanol. *Environment, Development and Sustainability*, 13 (February): 179–202.

New York Agricultural Statistics Service. 2009. *Onondaga County farm statistics, August 2009*. Albany: New York Agricultural Statistics Service.

New York State Department of Agriculture and Markets. 2010. New York State farmers' markets. [Online database accessed July 25, 2010.] http://www.agmkt.state.ny.us/AP/CommunityFarmersMarkets.asp#OnondagaCounty.

Ngo, N., and Pataki, D. 2008. The energy and mass balance of Los Angeles County. *Urban Ecosystems*, 11(2): 121–139.

North Carolina Soybean Producers Association. 2007. How soybeans are used. http://www.ncsoy.org/ABOUT-SOYBEANS/Uses-of-Soybeans.aspx.

Odum, H. T. 1971. *Environment, power and society*. New York: Wiley-Interscience.

Olmstead, A. L., and Rhode, P. W. 2001. Reshaping the landscape: The impact and diffusion of the tractor in American agriculture, 1910–1960. *Journal of Economic History*, 61:663–698.

Pendall, R. 2003. Sprawl without growth: The Upstate paradox. Washington, DC: Brookings Institution. http://www.brookings.edu/~/media/Files/rc/reports/2003/10demographics_pendall/200310_Pendall.pdf.

Peters, C., Bills, N. L., Lembo, A. J., Wilkins, J. L., and Fick, G. W. 2009. Mapping potential foodsheds in New York State: A spatial model for evaluating the capacity to localize food production. *Renewable Agriculture and Food Systems*, 24:72–84

Peters, C., Bills, N. L., Wilkins, J. L., and Fick, G. W. 2009. Foodshed analysis and its relevance to sustainability. *Renewable Agriculture and Food Systems*, 24:1–7.

Peters, C., Wilkins, J. L., and Fick, G. W. 2007. Testing a complete-diet model for estimating the land resource requirements of food consumption and agricultural carrying capacity: The New York State example. *Renewable Agriculture and Food Systems*, 22:145–153.

———. 2008. Land and diet: What's the most land efficient diet for New York State? *Rural New York Minute*. Ithaca, NY: Community and Rural Development Institute, Cornell University.

Pimentel, D. 1980. Energy used for transporting supplies to the farm. In *Handbook of energy utilization in agriculture*. ed. Pimentel, D. 55. Boca Raton, Florida: CRC Press.

Pimentel, D., Berger, B., Filiberto, D., Newton, M., Wolfe, B., Karabinakis, B., Clark, S., Poon, E., Albert, E., and Nandagopal, S. 2004. Water resources: Agricultural and environmental issues. *BioScience*, 54(10): 909–918.

Pimentel, D., and Patzek, T. 2008. Ethanol production using corn, switchgrass, and wood; biodiesel production using soybean. In *Biofuels, solar and wind as renewable energy systems: Benefits and risks*, ed. Pimentel, D., 373–394. Dordrecht, The Netherlands: Springer.

Pimentel, D., and Pimentel, M. 1979. *Food, energy, and society*. London: Arnold.

———. 2003. Sustainability of meat-based and plant-based diets and the environment. *American Journal of Clinical Nutrition*, 78:660S–663S.

———. 2008. *Food, energy and society*. 3rd ed. Boca Raton, FL: CRC Press.

Pirog, R., and Benjamin, A. 2003. Checking the food odometer: Comparing food miles for local versus conventional produce sales to Iowa institutions. Ames: Leopold Center for Sustainable Agriculture, Iowa State University. http://www.leopold.iastate.edu/pubs/staff/files/food_travel072103.pdf.

PoultryHub. 2010. Chicken meat (broiler) industry. Accessed July 25, 2010. http://www.poultryhub.org/index.php/Chicken_meat_(broiler)_industry.

Rasmussen, W. D. 1962. The impact of technological change on American agriculture, 1862–1962. *Journal of Economic History*, 22 (December): 578–591.

Schurr, S. H. 1960. *Energy in the American economy, 1850–1975: An economic study of its history and prospects.* Baltimore, MD: Johns Hopkins University Press.

Steinhart, J. S., and Steinhart, C. E. 1974. Energy use in the US food system. *Science,* 184:307–314.

Thiboumery, A., and Jepson, K. 2009. Beef and pork whole animal buying guide. PM 2076. Ames: Small Meat Processors Working Group, Iowa State University, University Extension. http://www.extension.iastate.edu/Publications/PM2076.pdf.

University of Minnesota Extension. 2009. Machinery cost estimates. http://www.extension.umn.edu/distribution/businessmanagement/df6696.pdf.

University of Missouri Extension. 2010. Tables for weights and measurement: Crops. http://extension.missouri.edu/publications/DisplayPub.aspx?P=G4020.

USCB (U.S. Census Bureau). 1956/1957. *United States Census of Agriculture, 1954.* Washington, DC: USCB.

———. 2010. Onondaga County. *State and county quick facts.* Washington, DC: USCB. http://quickfacts.census.gov/qfd/states/36/36067.html.

USDA (U.S. Department of Agriculture). 1999. National Agricultural Statistics Service (NASS). Equine report. http://usda.mannlib.cornell.edu/usda/nass/Equine/equi1999.txt.

———. 2002. Profiling food consumption in America. In *Agriculture factbook, 2001–2002,* Chap. 2. http://www.usda.gov/factbook/chapter2.htm.

———. 2004. Forest Products Laboratory. Fuel value calculator. *Techline.* http://www.fpl.fs.fed.us/documnts/techline/fuel-value-calculator.pdf.

———. 2005. Nutrient availability data set. [Online database accessed July 25, 2010.] http://www.ers.usda.gov/Data/FoodConsumption/NutrientAvailIndex.htm.

———. 2010a. Fertilizer use and price. [Online database accessed July 25, 2010.] http://www.ers.usda.gov/data/fertilizeruse/.

———. 2010b. Fruit historic data. [Online database accessed July 25, 2010.] http://www.nass.usda.gov/Statistics_by_State/New_York/Historical_Data/Fruit/Fruitindex.htm.

———. 2010c. Livestock, dairy and poultry outlook. Washington, DC: USDA, Economic Research Service. http://www.ers.usda.gov/Publications/LDP/2010/08Aug/ldpm194.pdf.

———. 2010d. Nutrient Data Laboratory. USDA national nutrient database. [Online database accessed April 4, 2010.] http://www.nal.usda.gov/fnic/foodcomp/search/.

———. 2010e. World supply and demand estimates. [Online database accessed July 25, 2010.] http://www.usda.gov/oce/commodity/wasde/latest.txt.

Vedenov, D., and Wetzstein, M. 2008. Toward an optimal U.S. ethanol fuel subsidy. *Energy Economics,* 30(5): 2073–2090.

Virginia Cooperative Extension. 2007. Dairy cow—27000 Lb production—corn silage/alfalfa haylage ration. Pub. 446-047. http://pubs.ext.vt.edu/446/446-048/PDF_DairyCow27000Lb_ProductionCornSilageAlfalfaHaylage.pdf.

Weber, C. L., and Matthews, H. S. 2008. Food-miles and the relative climate impacts of food choices in the United States. *Environmental Science & Technology,* 42(10): 3508–3513.

Wells, H. F., and Buzby, J. C. 2008. Dietary assessment of major trends in U.S. food consumption, 1970–2005. Washington, DC: USDA Economic Research Service. http://www.ers.usda.gov/Publications/EIB33/EIB33.pdf.

chapter ten

Energy production from corn, cellulosic, and algae biomass

David Pimentel, Jason Trager, Sarah Palmer, Jessica Zhang, Bari Greenfield, Emily Nash, Kate Hartman, Danielle Kirshenblat, and Alan Kroeger

Contents

10.1   Introduction

Each year, the United States and other nations import more than 60% of their oil at a tremendous cost to themselves (USCB, 2009). Oil represents nearly 40% of U.S. energy consumption, and this level of oil consumption leads the International Energy Agency (IEA/OECD, 2008) and other organizations to estimate that cheap world oil supplies will be depleted by 2040 (Green et al., 2006; Hodge, 2008; W. Youngquist, pers. comm., December 8, 2009). This forecast has created an urgent need for alternate liquid fuels and has stimulated many nations to seek diverse ways to produce liquid fuels. As a consequence, corn has become a popular feedstock for ethanol production. Unfortunately, the production of ethanol from corn grain has proven to be energetically and environmentally costly, even taking into account the subsidies, which now total $12 billion per year (Koplow and Steenblik, 2008). In addition, converting corn into ethanol has increased U.S. food prices (Pimentel, Marklein et al., 2009) and the price of corn on world commodity markets. Clearly, using food as a source of ethanol presents important ethical problems.

Increasing food costs and shrinking food supplies worldwide have both the director general of the United Nations and president of the World Bank warning that using grains and other human foods to produce fuel is leading to increased malnutrition and starvation worldwide (Spillius, 2008). A total of 2.3 billion tons of grains is produced annually in the world, and about 20% of this total is used for ethanol production. Another important food product—vegetable oils, including soybean, canola, rapeseed, and palm oil—is being used for biofuel. Currently in Europe, 60% of the rapeseed oil is being used for biodiesel, producing about 1.5 billion gallons (6 billion liters; FAO, 2009).

Using food products in the production of biofuels is particularly troublesome because of the limited supply of biofuel energy that can be produced from foods. For example, the United States currently produces 34 billion liters of ethanol, consuming 33% of all U.S. corn production, but this 34 billion liters of ethanol replaces only 1.7% of total oil consumption in the United States, assuming no fossil energy inputs (USCB, 2009). In fact, converting 100% of U.S. corn into ethanol would provide the United States with only 5% of its needed oil fuel, assuming again zero fossil energy inputs.

Other countries also produce bioethanol. Brazil produces about 27 million liters, but its feedstock for ethanol is sugarcane (CONAB, 2010). Again, the 27 million liters of ethanol is not enough to meet consumption needs, as Brazil's oil consumption during the past 10 years has increased 42% (CONAB, 2010). Additional costs to consumers in Brazil include subsidies that total several billion dollars per year just for ethanol (Coelho, 2005; Green et al., 2006; Hodge, 2008; Berg, 2004; FEE, 2009; Schmitz et al., 2009), although some claim that there are no subsidies for Brazilian ethanol (UNICA, 2010; Walter, 2009). Nevertheless, the subsidies for ethanol are contributing to deforestation and other environmental problems in Brazil (*Pacific Ecologist*, 2009).

Can green plants such as corn, switchgrass, willow, and all other kinds of biomass be suitable sources of liquid fuels? Unfortunately, these

Table 10.1 Total Amount of Biomass and Solar Energy Captured Each Year
in the United States

	Million ha	Tons/ha	$\times 10^6$ tons	Total energy collected $\times 10^{15}$ BTU
Crops	160	5.5	901	14.4
Pasture	300	1.1	333	9.6
Forests	264	2.0	527	8.4
Total	*724*		*1,758*	*27.8*

An estimated 27.8×10^{18} BTU of sunlight is collected by green plants in the United States per year. The green plants (crops, grasses, and forests) in the United States are collecting 0.1% of the solar energy reaching the United States (Pimentel et al., 2009).

green plants in the United States convert only about 0.1% solar energy into plant material (see Table 10.1; Pimentel, Marklein et al., 2009). The use of grain and other biomass for liquid fuels at present results in a greater contribution of CO_2 to the atmosphere (Pimentel, Marklein et al., 2009) than simply burning fossil fuel and *not* producing liquid biofuels. In contrast, photovoltaic cells collect more than 150 times the solar energy that green plants collect and add relatively little CO_2 to the atmosphere (Patzek, 2010).

In this chapter, we examine the potential for improving the efficiency of converting corn grain and cellulosic biomass into ethanol. Also, we examine the production of biodiesel using algae. In addition, we attempt to define the impact of biofuel production on greenhouse gas emissions and the prevention of malnutrition and hunger.

10.2   Ethanol production from corn grain

10.2.1   Energy inputs in corn ethanol production

In this analysis, the most recent scientific data for corn fermentation/distillation are used. All current fossil energy inputs are included in corn production and for fermentation and distillation to determine the entire energy cost of ethanol production. Additional costs to consumers include federal and state subsidies (Koplow and Steenblik, 2008), plus costs associated with environmental pollution and/or degradation that occur during the entire production process.

In a large ethanol conversion plant, the ethanol yield from 2.69 kg of corn grain is 1 liter (approximately 9.5 liters of pure ethanol per bushel of corn). The production of corn in the United States requires a significant energy and monetary investment for an average of 14 inputs, including labor, farm machinery, fertilizers, irrigation, pesticides, and electricity (see Table 10.2). As listed in Table 10.2, an average corn yield

Table 10.2 Energy Inputs and Costs of Corn Production per Hectare
in the United States

Inputs	Quantity	kcal × 1,000	Costs
Labor	11.4 hr[a]	520[b]	$148.20[c]
Machinery	55 kg[d]	1,018[e]	110.00[f]
Diesel	62 L[g]	620[h]	46.42
Gasoline	9 L[f]	90[i]	7.14
Nitrogen	150 kg[j]	2,475[k]	85.25[l]
Phosphorus	55 kg[j]	228[m]	48.98[n]
Potassium	62 kg[j]	202[o]	26.04[p]
Lime	1,120 kg[f]	315[q]	28.64
Seeds	21 kg[d]	520[d]	74.81[f]
Irrigation	8.1 cm[r]	320[s]	123.00[t]
Herbicides	2.3 kg[a]	230[w]	35.29
Insecticides	0.7 kg[u]	70[w]	32.55
Electricity	13.2 kWh[v]	34[x]	7.22
Transport	107 kg[y]	122[z]	61.20
Total		*7,438*	*$834.74*
Corn yield 9,500 kg/ha[aa]	34,200	kcal input:output 1:4.60	

[a] USDA, 2005.
[b] It is assumed that a person works 2,000 hours per year and utilizes an average of 9,000 liters of oil equivalents per year.
[c] It is assumed that labor is paid $20 an hour.
[d] Pimentel and Pimentel, 2008.
[e] Prorated per hectare and 10-year life of the machinery. Tractors weigh from 6 to 7 tons and harvesters 8 to 10 tons, plus plows, sprayers, and other equipment.
[f] Estimated.
[g] William McBride, USDA, personal communication, 2010.
[h] Input 11,400 kcal per liter.
[i] Input 10,125 kcal per liter.
[j] NASS, 2003.
[k] Cost $0.55 per kg.
[l] Patzek, 2004.
[m] Input 4,154 kcal per kg.
[n] Cost $0.62 per kg.
[o] Input 3,260 kcal per kg.
[p] Cost $0.31 per kg.
[q] Input 281 kcal per kg.
[r] USDA, 1997.
[s] Batty and Keller, 1980.
[t] Irrigation for 100 cm of water per hectare costs $1,000 (Larsen et al., 2002).
[u] Personal communication with William McBride, USDA-ERS.
[v] USDA, 1991.
[w] Input 100,000 kcal per kg of herbicide and insecticide.
[x] Input 860 kcal per kWh and requires 3 kWh thermal energy to produce 1 kWh electricity.
[y] Goods transported include machinery, fuels, and seeds that were shipped an estimated 1,000 km.
[z] Input 0.34 kcal per kg per km transported.
[aa] Average from USDA, 2008, and USCB, 2007.

of 9,500 kg/ha (151 bu/ac) using up-to-date production technologies requires the expenditure of about 7.4 million kcal of energy inputs (mostly natural gas and oil). This is the equivalent of about 743 liters of oil equivalents expended per hectare of corn. The production costs for corn total $835/ha for the 9,500 kg/ha yield, or approximately 11¢/kg or $2.20/bushel (see Table 10.1).

Full irrigation (when there is insufficient or no rainfall) requires about 100 cm/ha of water per growing season. Because 15–19% of U.S. corn production is irrigated (USDA, 1997; Supalla, 2007), only 8.1 cm/ha of irrigation was included for the growing season. On average, irrigation water is pumped from a depth of 100 m (USDA, 1997). On this basis, the average energy input associated with irrigation is 320,000 kcal per hectare (Table 10.2).

10.2.2 Energy inputs in corn fermentation/distillation

The average costs in terms of energy and dollars for a large, modern, dry-grind ethanol plant are significant (see Table 10.3). In the fermentation/distillation process, the corn is finely ground and approximately 8 liters of water is added per 2.69 kg of ground corn; some of this water may be recycled. After fermentation, the mixture is distilled to obtain 95% pure ethanol from the 8–12% ethanol beer. A liter of ethanol must be extracted from approximately 11 liters of the ethanol-water mixture. Although ethanol boils at 78°C and water boils at 100°C, all the water is not extracted from the ethanol in the first distillation, which obtains 95% ethanol (Maiorella, 1985; Wereko-Brobby and Hagan, 1996; S. Lamberson, pers. comm., 2000). To be mixed with gasoline, the 95% ethanol must be further processed and more water removed, requiring additional fossil energy inputs to achieve 99.5% pure ethanol (Table 10.3). Thus, a total of 8 liters of wastewater is produced in the production of 1 liter of ethanol, and the disposal/treatment of this relatively large amount of sewage effluent comes at an energetic, economic, and environmental cost.

In this analysis, the total cost, including the energy inputs for the fermentation and distillation and the apportioned energy costs of steam, electricity, and stainless steel tanks and other industrial materials, is substantial (see Table 10.3).

The largest energy inputs in corn-ethanol production are for corn feedstock production, steam energy, and the electricity used in the fermentation and distillation process. The corn feedstock requires more than 26% of the total energy input.

The total energy input to produce a liter of 99.5% ethanol is 8,143 kcal (Table 10.3). However, a liter of ethanol has an energy value of only 5,130 kcal. Based on a net energy loss of 3,013 kcal of ethanol produced, 58%

Table 10.3 Inputs per 1,000 Liters of 99.5% Ethanol Produced from Corn

Inputs	Quantity	kcal × 1,000	Cost
Corn grain	2,690 kg[a]	2,106[b]	$634.14
Corn transport	2,690 kg[b]	264[c]	27.63[d]
Water	7,721 L[e]	46[d]	3.86[d]
Stainless steel	3 kg[f]	42[g]	8.52[h]
Steel	4 kg[f]	40[i]	2.39[h]
Cement	8 kg[f]	11[i]	1.86
Steam	2,564,764 kcal[j]	2,362[j]	59.94[k]
Electricity	395 kWh[j]	2,863[j]	26.38[l]
95% ethanol to 99.5%	9 kcal/L[m]	9[m]	40.00
Sewage effluent	20 kg BOD[n]	69[o]	6.00
Distribution	331 kcal/L[p] 331	375.00	
Total		8,143	$1,185.72

[a] Output: 1 liter of ethanol = 5,130 kcal (low heating value). The mean yield of 9.5 L pure EtOH per bushel has been obtained from the industry-reported ethanol sales minus ethanol imports from Brazil, both multiplied by 0.95 to account for 5% by volume of the #14 gasoline denaturant, and the result was divided by the industry-reported bushels of corn inputs to ethanol plants (see Patzek, 2006).
[b] Data from Table 10.2.
[c] Calculated for a 144-km round trip.
[d] Pimentel et al., 2009.
[e] 7.7 liters of water mixed with each kg of grain.
[f] Estimated from the industry reported costs of $85 million per 65 million gallons/yr dry grain plant amortized over 30 years. The total amortized cost is $43.6/1000L EtOH, of which an estimated $32 go to steel and cement.
[g] Johnson et al., 2008.
[h] Pimentel, 2003.
[i] Venkatarama and Jagadish, 2003.
[j] Patzek, 2004.
[k] Calculated based on coal fuel. Below the 1.95 kWh/gal of denatured EtOH in South Dakota, see Patzek, 2010.
[l] $0.07 per kWh (USCB, 2004).
[m] 95% ethanol converted to 99.5% ethanol for addition to gasoline (T. Patzek, University of California, Berkeley, pers. comm., 2004).
[n] 20 kg of BOD per 1,000 liters of ethanol produced (Kuby et al., 1984).
[o] 4 kWh of energy required to process 1 kg of BOD (Blais et al., 1995).
[p] DOE, 2002.

more fossil energy is expended than is produced as ethanol and costs $1.19 per liter ($4.48 per gallon; Table 10.3).

10.2.3　Economic costs

Current corn-ethanol production technology uses more fossil fuel and costs substantially more in dollars than its produced energy value is worth on the market. Without the more than $12 billion annual federal and state government subsidies, U.S. ethanol production would be reduced or would cease, confirming the basic fact that ethanol production is uneconomical and does not provide the United States with any net energy benefit (Koplow and Steenblik, 2008). Government subsidies for ethanol production are mainly paid to large corporations (Koplow and Steenblik, 2008), while corn farmers receive a minimum profit per bushel for their corn (Pimentel and Patzek, 2008). Sen. John McCain reports that direct subsidies for ethanol, plus the subsidies for corn grain, amount to 79¢ per liter (McCain, 2003).

About 80% of the ethanol from sugarcane in Brazil is also heavily subsidized (Berg, 2004). Even with heavy subsidies, only about half of the fuel burned in autos in Brazil is ethanol; about 50% is gasoline (Berg, 2004). Sugar subsidies have a major impact on ethanol production from sugarcane.

If the production cost of a liter of ethanol were added to the tax subsidy cost, then the total cost for a liter of ethanol would be $1.54. The mean wholesale price of ethanol was almost $1.00 per liter without subsidies. Because of the relatively low energy content of ethanol, 1.6 liters of ethanol has the energy equivalent of 1 liter of gasoline. Thus, the cost of producing an amount of ethanol equivalent to a liter of gasoline is about $2.33 ($8.82 per gallon of gasoline). This is considerably more than the 53¢ per liter cost of producing a liter of gasoline. The subsidy per liter of ethanol is 60 times greater than the subsidy per liter of gasoline! This is the reason why ethanol is so attractive to large corporations.

10.2.4 Cornland use

In 2008, about 34 billion liters of ethanol (9 billion gallons) was being produced in the United States (EIA, 2008). The total amount of petroleum fuels used in the United States is about 1,270 billion liters (USCB, 2009). Therefore, 34 billion liters of ethanol (energy equivalent to 22 billion liters of petroleum fuel) provided only 1.7% of the petroleum utilized. To produce this 34 billion liters of ethanol, about 9.6 million ha or 34% of U.S. cornland was used. Expanding corn-ethanol production to 100% of U.S. corn production would provide just 4% of the petroleum needs of the United States while removing cropland needed for food production.

However, U.S. corn cultivation may continue to increase because of the ethanol targets set by the most recent energy legislation (Donner and Kucharik, 2008): 36 billion additional gallons, of which 15 billion gallons (60 billion liters) are to be produced from corn grain.

The nitrogen fertilizer requirement for corn production is the prime cause of the "dead zone"—areas just below the surface water with little or no dissolved oxygen—in the Gulf of Mexico (NAS, 2003). Increased corn ethanol production will increase the need for more nitrogen fertilizer and increase the nitrogenous water pollution in the Gulf of Mexico, which promotes the algal blooms that die and decompose to produce the dead zone (Donner and Kucharik, 2008).

10.2.5   By-products

The energy and dollar costs of producing ethanol can be offset partially through by-products, for example, the dry distillers grains (DDG) made from dry-milling of corn. From about 10 kg of corn feedstock, about 3.3 kg of DDG with 27% protein content can be harvested (Stanton and LeValley, 1999). The DDG is suitable for feeding cattle, which are ruminants, but has only limited value for feeding hogs and chickens. In practice, this DDG is generally used as a substitute for soybean feed that contains 49% protein (Stanton and LeValley, 1999). However, soybean production for livestock feed is more energy efficient than corn production because little or no nitrogen fertilizer is needed for the production of this legume feed (Pimentel et al., 2002).

In practice, only 2.1 kg of soybean protein provides the equivalent nutrient value of 3.3 kg of DDG (or nearly 60% more DDG is required to equal the soybean meal protein). Thus, the credit fossil energy per liter of ethanol produced is about 445 kcal. Factoring this credit for a nonfuel source into the production of ethanol reduces the negative energy balance for ethanol production from 158% to 151% (see Table 10.3). The high energy credits for DDG given by some researchers are unrealistic because the production of livestock feed from ethanol production is uneconomical due to the high costs of fossil energy, plus the costs of soil depletion to the farmer (Patzek, 2004). The resulting overall energy output/input comparison remains negative even with the large credits for the DDG by-product.

10.2.6   Environmental impacts

Some of the economic and energy contributions of the by-products are negated by the widespread environmental pollution problems associated with corn-ethanol production. First, U.S. corn production causes more soil erosion than any other U.S. crop (Pimentel et al., 1995; NAS, 2003). In addition, corn production uses more herbicides, insecticides, and nitrogen fertilizer than any other crop produced in the United States. Consequently, corn causes more water pollution than any other crop since a large quantity of these chemicals invades ground and surface waters and the acreage in corn production covers more land area than any other grain (NAS, 2003).

Another environmental impact of biomass crop production is the land-use changes they demand. Nabuurs et al. (2007) report that the limit factor for biomass crops is the availability of arable land, and the massive scale of biomass crop production necessary will require deforestation. However, an important consideration when evaluating the environmental effects of biofuels is whether CO_2 emissions decrease because of biofuel production or not; if emissions do not decrease, this would favor forest preservation and expansion (Righelato, 2007). According to the IEA (2004), forests converted to cropland have a negative environmental impact because this land change destroys the carbon sink that the forest provided.

Renton Righelato of the World Land Trust investigated the impacts of land-use changes from forest to biofuel cropland and found that the amount of carbon sequestered—emissions avoided—by tropical forests is three to four times more than the emissions avoided by bioethanol production (Righelato, 2007). Only after the forest area reaches maturity, in 50 to 100 years, would the carbon sequestered from cropland conversion be able to surpass the amount of carbon stock that is accumulated and calculated according to models for the power of age in a forest structure (Righelato, 2007; Alexandrov, 2007; Sylvester-Bradley, 2008).

As mentioned, the production of 1 liter of ethanol requires 1,700 liters of freshwater both for corn production and for the fermentation and distillation processing of ethanol (Pimentel and Patzek, 2008). In some Western irrigated cornlands, as in some regions of Arizona, ground water is being pumped 10 times faster than the natural recharge of the aquifers (Pimentel et al., 2004). Ethanol production using sugarcane requires slightly more water per ethanol liter than corn ethanol, or about 2,000 liters of water.

In addition, 1.59 liters more fossil fuel is required to produce 1 liter of ethanol, which confirms that ethanol production is significantly contributing to the global warming problems (Pimentel and Pimentel, 2008). All these factors—soil erosion, water pollution, ground water depletion, increased use of agriculturally valuable land to produce corn—confirm that the environmental and agricultural system that produces U.S. corn is experiencing major degradation. Further, it substantiates the conclusion that the U.S. corn production system, and indeed the entire corn-ethanol production system, is not environmentally sustainable now or for the future unless major changes are made in the cultivation of this major food/feed crop. Corn as a raw material for ethanol production in the face of what has been presented cannot be considered a renewable energy source.

Pollution problems associated with the production of ethanol at the ethanol production plant sites are also emerging. The EPA already has issued warnings to ethanol plants to reduce their air pollution emissions or be shut down (FoodProductiondaily.com, 2002). Another pollution problem concerns the large amounts of wastewater produced by each ethanol plant. As noted, producing 1 liter of corn ethanol produces 6–12 liters of wastewater. This wastewater has a biological oxygen demand (BOD) of 18,000–37,000 mg/liter, depending on the type of plant (Kuby et al., 1984; Patzek, 2004). The cost of processing this sewage in terms of energy (4 kWh/kg of BOD) was included in the cost of producing ethanol in Table 10.3.

The major problem with corn and all other biomass crops is that they collect, on average, only 0.1% of the solar energy per year (Pimentel, Marklein et al., 2009). At a fairly typical gross yield of 3,000 liters of ethanol per hectare per year, the power density achieved is only 2.1 kW/ha. That is tiny compared with the gross power density achieved via oil, after delivery for use, on the order of 2,000 kW/ha (Ferguson, 2007).

10.3 World malnutrition and use of food for biofuel

The Food and Agriculture Organization (FAO, 2009) of the United Nations has estimated that there were 1.02 billion undernourished people worldwide in 2009, representing approximately a sixth of the entire population. In its 2009 report, *The State of Food Insecurity in the World*, the FAO defined undernourishment as being "when caloric intake is below the minimum dietary energy requirement (MDER)," where MDER "is the amount of energy needed for light activity and a minimum acceptable weight for attained height" (FAO, 2009, p. 8).

Caloric intake is certainly not the only measurement of malnourishment; micronutrient deficiencies can also have severe health impacts. In 2000, the World Health Organization (WHO, 2000) estimated the number of people who have iron deficiency anemia at around two billion. Anemia can result in extreme fatigue, impairment of physical and mental development in children, and higher maternal deaths. WHO also estimated that 740 million people have iodine deficiency disorder, which can have severe impacts on children's brain development. In addition, up to 140 million children are vitamin A deficient, putting them at higher risk for blindness and making them more susceptible to illnesses.

As more land and crops are devoted to the production of biofuels, rather than to human consumption, concerns have been raised that these statistics will worsen. Jacques Diouf, head of the FAO, stated in 2007 that he feared that a number of factors, including the production of crops for biofuels, create "a very serious risk that fewer people will be able to get food" (Rosenthal, 2007). The president of the World Bank, Robert Zoellick, shared a similar apprehension, asserting that demand for biofuels has been a "significant contributor" to ballooning food prices. According to Zoellick, "It is clearly the case that programs in Europe and the United States that have increased biofuel production have contributed to the added demand for food" (Zoellick, 2008). Jean Ziegler, the UN special rapporteur on the right to food, has taken a more extreme stance. In 2007,

Ziegler claimed biofuels were a "crime against humanity" and called for a five-year moratorium on their production (Ferrett, 2007).

10.4   Cellulosic biomass for ethanol production

10.4.1   Biomass

Due to the inherent problems with corn ethanol, there is growing interest in using cellulosic biomass from nonfood biological material to produce ethanol. However, such cellulosic biomass materials have fewer carbohydrates and more complex matrices of lignin and hemicellulose, thus complicating the ethanol conversion processes. But in terms of biomass energy produced (not liquid fuel), switchgrass and willow are more efficient than corn in terms of fossil energy inputs versus biomass energy output per hectare (Pimentel and Patzek, 2008).

This analysis focuses on the potential of cellulosic biomass to serve as feedstock for the production of liquid fuel. Since cellulosic biomass, like straw and wood, has very few of the simple starches found in corn, two to three times more cellulosic material must be produced and processed to obtain similar amounts of cellulosic ethanol (Patzek, 2010).

10.4.2   Biomass resources

Biomass resources have been receiving increasing attention because of their flexibility as a renewable resource (heat or liquid fuels). All growing biomass material requires solar energy, water, land, and nutrients in appropriate proportions. Balancing these inputs is a key factor limiting the feasibility of biomass production. For instance, assuming maximum efficiency, corn on a sunny, warm, July day can fix 5% of the solar energy reaching the plant. For the total year, however, the corn plant collects only 0.2% of the solar energy. On average, most green plants in the United States collect annual average of only 0.1% of the solar energy (Pimentel, Marklein et al., 2009).

10.5   Cellulosic biomass for liquid fuel

Two substantial questions remain: How much cellulosic biomass resource is available? And what is the annual potential for ethanol production? Perlack et al. (2005) estimate that 1.3 billion tons of biomass could be collected to replace 30% of current U.S. liquid fuel consumption per year by 2030. Assuming a conversion rate of 11 kg of biomass per liter of ethanol, this yields about 118 million liters (31 million gallons) per year by 2030. However, no one has yet achieved any positive net liquid energy in the conversion of cellulosic biomass into liquid fuel (Patzek, 2010).

The analysis of available biomass by Perlack et al. (2005) included forest biomass, crop residues, perennial crops such as switchgrass and willow, grain crops, processing wastes, and manure. In their report, no assessment of the conversion rate between dry biomass and ethanol was attempted. Furthermore, they assumed a 50% increase in the yields of corn, wheat, and other small grains and improvements in harvest technology so that 75% of crop residues could be harvested. As pointed out by Lal and Pimentel (2009), however, the harvest of crop residues would be a disaster for agriculture, because crop residues are vital in protecting vital topsoil and maintaining soil productivity.

President George W. Bush, in both his 2006 and 2007 State of the Union addresses, cited the need to reduce the nation's reliance on foreign oil. In 2007, he set a goal for the United States to produce 35 billion gallons of ethanol by 2017. However, he did not explain how this ethanol would be produced. The Energy Independence and Security Act of 2007 called for 36 billion gallons of ethanol by 2022, 21 billion gallons of which must be derived from cellulosic biomass and 15 billion gallons from "advance biofuels" corn. Currently, there are no commercial plants producing liquid net energy from any cellulosic biomass (Patzek, 2010).

Sandia National Laboratories has reported that 90 billion gallons of ethanol could be produced by 2030 based on simulations (West et al., 2009). A total of 15 billion gallons would come from current corn ethanol and the remaining 75 billion gallons would be produced from cellulosic biomass materials. This Sandia study assumed a minimum conversion rate of 74 gallons of ethanol per dry metric ton of cellulosic biomass, which is significantly above any current attainable yields. Sandia is suggesting the extraordinary yield of 1 liter of ethanol from only 3.6 kg of cellulosic biomass!

The best data available report about 1.8 billion tons of biomass produced annually in the United States, including all agricultural food crops, all forests, and all grasses (see Table 10.1). This 1.8 billion tons is larger than the Perlack et al. (2005) value of 1.3 billion tons, but the data for each source are quite different.

Note, however, that this net annual primary productivity of about 1.8 billion tons per year for the United States includes all the human food produced and consumed each year. The total food consumed annually in the United States is about 400 million tons (USDA, 2008). This leaves net primary productivity of only 1.4 billion tons of all other cellulosic biomass produced each year (including crop residues).

After harvesting of the 901 million tons of crops, there will be about 500 million tons of crop residues remaining on the land (see Table 10.1; Lal et al., 1999). In contrast to our estimate of 500 million tons of crop residues left on U.S. cropland (Table 10.1), Perlack et al. (2005) estimate that 73% of the 1.3 billion tons, or about 950 million tons, of cellulosic biomass is from agricultural residues. This is nearly twice the estimate of our reported 500 million tons: 901 million tons of crops minus 400 million tons of food removed from the cropland (Table 10.1).

In addition to being a potential fuel source, cellulosic biomass, such as switchgrass and forests, are important resources for wildlife. The removal of forest biomass for fuel has immediate impacts for that formerly forested land, but also negative impacts on adjacent forests and other ecosystems (Righelato, 2007).

Moving beyond the United States, our estimate for global biomass produced annually worldwide is $1,764 \times 10^9$ tons (see Table 10.4). This is about 1,000 times greater than the biomass produced in the United States. Most of the biomass is crop biomass, totaling 750 billion tons, with forest biomass adding 629 billion tons (Xu and Wang, 2009; Table 10.4). According to the IEA (2009), the best estimate for total world biomass conversion for energy is 15×10^{15} kcal. Most of the forest biomass was already being used for fuel and some for building (Nabuurs et al., 2007).

Crop residues or agricultural residues left on the land after crop production are vital to protecting the soil from serious erosion. After harvesting, as much as 70% of the crop remains as residue (Lal, 2005). Removing corn residues will increase soil erosion 100-fold and totally devastate agricultural production (Lal et al., 1999; Pimentel and Lal, 2007). Crop residues enable the soil to resist rain and wind erosion. Already, U.S. corn land is losing soil about 15 times faster than sustainability (Troeh et al., 2004). In addition, greater soil erosion due the removal of crop residues increases the oxidation of soil organic matter, which releases CO_2 and may increase global climate change (Lal et al., 1999). The forest soils, too, are a critical part of forest carbon, sequestering carbon in the forest soil organic matter (Lal et al., 1999). According to Righelato (2007), 600 tons of CO_2 is lost per hectare when secondary forests are burned for crop production. Of the 75 billion tons of soil eroded worldwide each year, about two-thirds comes from agricultural land. If we assume a cost

Table 10.4 Total Amount of Biomass and Solar Energy Captured Each Year in the World

	Billion ha	Tons/ha	$\times 10^9$ tons	Total energy collected $\times 10^{18}$ BTU
Crops	1.5	5.0	750	12.0
Grasses	3.5	1.1	385	6.2
Forests	3.0	1.6	629	7.8
Total	*8.0*		*1,764*	*26.0*

An estimated 26×10^{18} BTU of sunlight reaching the Earth is captured as biomass by green plants (crops, grasses, and forests) each year in the world, suggesting that green plants are collecting about 0.1% of the solar energy reaching these plants (Pimentel et al., 2009).

of $3 per ton of soil for nutrients, $2/t for water loss, and $3/t for off-site impacts, this massive soil loss costs the world about $400 billion per year (Pimentel et al., 1995).

Some pro-ethanol investigators have suggested that all U.S. grasses could be harvested and converted to ethanol to produce an estimated 12% of U.S. liquid fuels (Tilman et al., 2006). However, these calculations fail to take into account the implications of removing this grass biomass from use. Current estimates indicate that there are 97 million head of cattle eating this grass, plus 6 million sheep and goats, 7.3 million horses, and millions of wild animals such as white-tailed deer, mule deer, elk, mice, woodchucks, and numerous other wild mammals that depend on these grasses for food (USDA, 2008; AVMA, 2011). Also, it has been reported that grasshoppers and various other insects that depend on grasses consume as much or more grass than some cattle (Moseley, 2009). In New York State pastures, it was reported that pasture insects consume twice as much forage as cows in the pasture per day (Wolcott, 1937). The cows ate about 180 kg during a study period, whereas the insects, for the same period, consumed 364 kg per hectare.

According to the U.S. Forest Service and others, there is already a serious overgrazing problem in the United States (Brown, 2009). Thus, it appears that few, if any, grasses should be harvested for use as cellulosic ethanol, contrary to the suggestions of Tilman et al (2006).

The forest ecosystem appears to have several million tons of forest biomass that could be harvested for cellulosic ethanol if this technology is perfected. At present, an estimated 225 million metric tons of forest biomass is harvested for use as lumber, paper, furniture, and fuelwood (USCB, 2009). Once the conversion technique is perfected, an estimated 300 million tons might be harvested for cellulosic ethanol. However, this means only 18% of the total biomass production per year in the United States would be available for harvest for ethanol production, not 1.3 to 1.6 billion tons—five times this amount—as suggested by Perlack et al (2005). Sandia National Laboratories (2009) suggests that we could harvest more than three times what Perlack and Bush are suggesting—a production level of 90 billion gallons per year. However, it is impossible that three times the biomass suggested by Perlack and Bush could be harvested from the plant biomass groups in the United States.

10.6   Cellulosic biomass conversion technologies

Biomass materials differ greatly in the structure and makeup of their carbohydrates, sugars, and oils. These differences influence their potential as an energy source. Serious problems exist with using cellulosic biomass materials such as wood or switchgrass because they contain very little simple sugars but contain cellulose and 12–25% lignin. In contrast, corn grain contains 77% carbohydrates that can be easily fermented by microbes and converted into ethanol (USDA, 1975). Wood or wheat straw contains only about 30–35% hemicellulosic materials and carbohydrates that can be fermented.

Not only does the hemicellulosic material make up less than half of carbohydrates in cellulosic biomass, but they are tightly enmeshed in the lignin in woody or straw biomass and cannot be freed unless the biomass is treated with a strong acid or alkali or with an enzyme (Ververis et al., 2004; Patzek and Pimentel, 2005). Pretreatment is required to break down the lignin and hemicellulose matrix to increase the porosity and surface area of the carbohydrates, making fermentation more effective. Releasing these sugars for fermentation into ethanol requires treatment to physically degrade the complex matrices of lignin and cellulose as well as to break down the cellulose polymers into their glucose monomer subunits, which can be fermented by microbes (De Wit et al., 2010). Traditional treatments involve strong acids or alkalis and newer mechanisms include steam explosions and various enzymes and genetically engineered microbes.

Research in lignin degradation for ethanol production purposes has yet to provide promising results. Although there has been progress in using white fungus, *Echinodontium taxodii* (Yu et al., 2009), on enzymatic hydrolysis of woody biomass, there is no commercial procedure for lignin degradation. Most of the techniques considered for lignin degradation— lignin modification through plant genetics, use of supercritical fluid pretreatment—also fall into the category of speculation more than into a sound commercial method. Pretreatment methods are judged on their ability to separate carbohydrate material from the lignin without inhibiting future hydrolysis and fermentation (Sun and Cheng, 2002). Of these techniques, enzymatic hydrolysis tends to be the preferable technology, because it is milder and less corrosive than the other treatments. To date, the yields of simple starches are low, and the enzymatic treatments are expensive (Patzek, 2010).

The fermentation stage can be used with any of the aforementioned processes to produce simultaneous saccharification and fermentation. This includes a two-stage dilute sulfuric acid hydrolysis and biomass fractionation processes for ethanol production. Thermochemical conversion techniques rely on gasification with two distinct techniques: fermentation based on ethanol production, and catalyst-based ethanol production (Dwivedi et al., 2009).

The fermentation process is complicated by the fact that the yeasts and bacteria can convert only six-carbon sugars into ethanol, leaving a large portion of the lignocellulosic material, hemicellulose sugars, unutilized. The fermentation process is similar to that of corn ethanol.

After fermentation, the residue has to be dewatered and dried to be used as a fuel for the distillation process or as feed for livestock. Of course, dewatering and drying the residue requires a significant energy input.

The implications of these factors for the use of cellulosic biomass for ethanol production are fourfold:

1. More than twice as much cellulosic biomass has to be grown, harvested, and processed as cellulosic biomass, compared with corn grain.
2. The cellulose biomass conversion technology for ethanol production is not sufficiently developed for economic production of ethanol.
3. The conversion process currently has several inherent limiting factors that may be overcome with future research.
4. The lignin by-product could be burned as a fuel if it were possible to separate it from the water in its diluted form.

10.7 Cellulosic ethanol thermodynamics (switchgrass)

Current energy efficiency of producing cellulosic ethanol is extremely low compared with other investigated paths to liquid fuels like corn ethanol (Patzek and Pimentel, 2005; Patzek, 2010; Howarth and Bringezu, 2009). At present, not a *single* commercial ethanol plant that uses cellulosic biomass exists in the world. The average energy input for switchgrass is only about 3.9 million kcal per hectare per year (see Table 10.5). With an exceptional average yield of 10 t/ha/yr, this suggests that for each kcal invested as fossil energy, the return in switchgrass thermal energy is 9–11 kcal (Samson et al., 2000)—an excellent return. This return, however, is impossible to realize for more than a year without the addition of significant amounts of fertilizers and some herbicides. Nonetheless, massive industrial monocultures of switchgrass are proposed and are being investigated. These large monocultures of switchgrass would result in a major increase in CO_2 releases due to fossil energy inputs and changes in land use (Patzek and Pimentel, 2005; Kumar and Sokhansanj, 2007).

If switchgrass is pelletized for use as a biomass fuel in stoves, the reported 1:14.6 kcal return as thermal energy is excellent (Samson et al., 2004). The 14.6 kcal return is higher than the 11 kcal return reported in Table 10.5 because a few additional inputs are included in the table that were not considered in the Samson et al. report. If the realistic sustained yield of switchgrass were 4 t/ha/yr, rather than 10 t/ha/yr, the return of 14.6 kcal would decline to about 4–5 kcal, similar to corn (Pimentel and Patzek, 2008). The cost of switchgrass pellets ranges from $125 to $175 per ton (Samson and Bailey Stamler, 2009). This is clearly a reasonable price.

Table 10.5 Average Inputs and Energy Inputs per Hectare per Year
for Switchgrass Production

Inputs	Quantity	kcal × 1,000	Cost
Labor	2.5 hr[a]	114[b]	$24.77[c]
Machinery	30 kg[d]	555	82.85
Diesel	64.3 L[e]	643	42.56
Nitrogen	85 kg[e]	1,403	45.60
Phosphorus	18 kg[e]	75	32.00
Potassium	36 kg[e]	59	60.00
Seeds	10 kg[e]	200	6.58
Herbicides	2.6 kg[g]	260	35.00
Total	*10,000 kg yield[hi]*	*3,935*	*$329.36*
	40 million kcal yield input/output ratio 1:2.21[j]		

[a] Estimated.
[b] It is assumed that an average person works 2,000 hours per year and uses about 9,000 liters of oil equivalents. Prorated, this works out to be 114,000 kcal.
[c] The agricultural labor is paid $9.90 per hour.
[d] The machinery estimate also includes 25% more for repairs.
[e] Data from Samson, 1991.
[f] Calculated based on data from Henning, 1993.
[g] Calculated based on data from Brummer et al., 2000.
[h] Samson et al., 2000.
[i] Brummer et al. (2000) estimated a cost of about $400/ha for switchgrass production. Thus, the $268 total cost is about 49% lower than that estimate, and this includes several inputs not included by Brummer et al.
[j] Samson et al. (2000) estimated an input per output return of 1:14.9, but we have added several inputs not included in their estimate. Still, the input/output return of 1:11 would be excellent if the sustained yield of 10 t/ha/yr were possible.

Fuelwood in Ithaca, New York, is selling for $250 per ton (Anne Wilson, 2009, pers. comm.), but the energy input:output ratio to produce the wood fuel is only 1:22 (energy returned on energy invested) without the pelletizing energy input (Pimentel, 2008).

However, converting switchgrass into ethanol would result in a negative energy return (see Table 10.6). The energy required to produce 1 liter of ethanol using switchgrass is about 2.13 liters of oil equivalents. This is significantly more negative than producing corn ethanol, which requires 1.59 liters of oil equivalents (see Table 10.3). The cost of producing a liter of ethanol using switchgrass is 64¢ ($2.40 per gallon; see Table 10.6). The four major inputs into ethanol production using cellulosic biomass are the switchgrass, steam, sulfuric acid, and lime.

If switchgrass were carefully managed so that the birds and other wildlife in the grassland were minimally disturbed, the many bird species and mammals could be maintained in the switchgrass pasturelands (McCoy et al., 2001; Roth et al., 2005). This would require special timing of harvest and other manipulations.

Table 10.6 Inputs per 1,000 Liters of 99.5% Ethanol Produced
from U.S. Switchgrass

Inputs	Quantity	kcal × 1,000	Cost
Switchgrass	6,500 kg[a]	2,558	$214
Switchgrass transport	6,500 kg[a]	600[b]	35[c]
Water	97,500 L[d]	55[e]	16[e]
Stainless steel	3 kg[f]	165[f]	11[f]
Steel	4 kg[f]	92[f]	11[f]
Cement	8 kg[f]	384[f]	11[f]
Grind switchgrass	6,500 kg	200[g]	16[g]
Sulfuric acid	240 kg[h]	0	168[i]
Steam	8.1 tons[h]	4,404	36
Lignin	780 kg[j]	0	0
Electricity	666 kWh[h]	1,703	46
95% ethanol to 99.5%	9 kcal/L[k]	9	40
Sewage effluent	40 kg BOD[l]	138[m]	12
Distribution	331 kcal/L[n]	331	20
Total		*10,819*	*$636*

Output: 1 liter of ethanol = 5,130 kcal. The ethanol yield here is 154 L/t dry biomass (dbm).
[a] Data from Table 10.5 and Patzek, 2010.
[b] Calculated for a 144-km round trip.
[c] Pimentel, 2003.
[d] 15 liters of water mixed with each kg of biomass.
[e] Pimentel et al., 2004.
[f] Newton, 2001.
[g] Calculated based on grinder information (Naval Facilities Engineering Service Center, 2003).
[h] Estimated based on cellulose conversion (Arkenol, 2004).
[i] Sulfuric acid sells for $7 per kg.
[j] Switchgrass contains 12% lignin.
[k] 95% ethanol converted to 99.5% ethanol for addition to gasoline (T. Patzek, University of California, Berkeley, pers. comm., 2004).
[l] 20 kg of BOD per 1,000 liters of ethanol produced (Kuby et al., 1984).
[m] 4 kWh of energy required to process 1 kg of BOD (Blais et al., 1995).
[n] DOE, 2002.

10.8   Cellulosic ethanol thermodynamics (willow trees)

Because the cellulosic conversion technologies and problems for willow are similar to those of switchgrass, no details like those in Tables 10.5 and 10.6 will be calculated for willow. Instead, we will focus on willow culture and production, which are quite different from switchgrass. Willow, a fast-growing tree that coppices, like switchgrass, is native to the United States and has an important advantage in that it can fix nitrogen with the help of microbes.

Employing a short-rotation scheme of cutting willow plantings every three years, the average yield is reported to be about 10 t/ha/yr, or about the same as switchgrass (Samson et al., 2000). Since willow fixes nitrogen and would require about 80 kg/ha/yr of nitrogen for a yield of 10 t/ha/yr, about 33 kg of additional nitrogen has to be applied annually (Keoleian and Volk, 2005) to achieve the same yields as switchgrass. Nitrogen and herbicides are the two major inputs for production (Keoleian and Volk, 2005). Thus, the inputs per hectare are about 4 million kcal per year (Clinch et al., 2009).

Willow plantings number from 10,000 to 20,000 per hectare (Keoleian and Volk, 2005). This density of planting eliminates soil erosion as a problem. With an average willow yield of 10 t/ha/yr, this suggests that for each kcal invested as fossil energy, the energy return in willow as thermal energy is 10 kcal. With care, this return can be maintained for 10 years (Kopp et al., 1997).

Willow wood can also be pelletized for use as a biomass fuel. Wood pellet production typically requires three processes: drying, grinding, and densification. Depending on the type of fuel used in drying the wood pellets, wood pellets require anywhere from 700,000 to 900,000 kcal per ton of energy to produce (Magelli et al., 2009). For wood with high moisture content, like willow, the energy requirement is higher. However, the final energy return of willow pellets is about 1:10 kcal when burned as pellets. The cost of willow wood pellets range from $125 to $170 per ton (Samson et al., 2009).

As with switchgrass, there is no effective technology to convert willow cellulosic biomass into ethanol at present. Thus, like switchgrass, an estimated 2 liters of oil equivalents is needed to produce 1 liter of ethanol. A comparison of ethanol fermentation techniques from willow biomass shows that the most efficient method of ethanol conversion is through steam pretreatment of willow biomass after SO_2 impregnation (Sassner et al., 2005).

10.9   Costs of woody biomass

Unfortunately, the current cost of producing ethanol using willow appears to be too expensive because the biomass itself is expensive and commercial ethanol conversion technologies tend to produce relatively low yields. Since the willow biomass industry is still in its infancy, there is a dearth of reliable data on willow prices. However, estimates have put the cost of willow biomass production and delivery to end users at $49.66–57.30 per oven-dried metric ton (odmt), which is consistent with the Volk et al. (2006) estimate of $50–57/odmt in New York State. Industry reports put wood chips in the Northeast at an average of $75/odmt, with prices ranging from $48 to $90 per ton (Wood Resources International, LLC, 2007). A technical brief by Cornell University (2008) sets the costs of willow biomass at $71/odmt.

Another avenue for obtaining woody biomass is through international trade. The United States, for example, imports 100,000 odmt of wood chips, a large majority of which comes from Brazil and Canada. Canada reports average prices of $4.03 per 24,000 kcal (a gallon of gasoline contains 30,000 kcal) (Ribeiro et al., 2008).

Most biomass is burned for cooking and heating; however, it can also be converted into electricity and liquid fuel. In both temperate and tropical forest ecosystems, approximately 3 dry metric tons per hectare per year of woody biomass can be harvested sustainably (Birdsey, 1992; Ferguson, 2001). This amount of woody biomass has a gross energy yield of 13.5 million kcal/ha, but it requires an energy expense of approximately 33 liters/ha of diesel fuel, plus the embodied energy for cutting and collecting wood for transport to an electric power plant. Thus, the energy input/output ratio for a woody biomass system is calculated to be 1:22 (Pimentel, 2008).

The cost of producing 1 kWh of electricity from woody biomass is about 6¢, which is competitive with other electricity production systems that average 7¢/kWh in the United States (USCB, 2009). Between 3.0 and 3.5 kWh of thermal energy is expended to produce 1 kWh of electricity, which, when combined with the 1:22 ratio for producing woody biomass, gives an energy input/output ratio of 1:7. Per capita consumption of woody biomass for heat in the United States amounts to 625 kg per year (USCB, 2009). In developing nations, use of diverse biomass resources (wood, crop residues, and dung) averages about 630 kg per capita (Kitani, 1999). Developing countries use only about 500 liters of oil equivalents of fossil energy per capita compared with nearly 9,000 liters per capita in the United States.

Woody biomass could supply the United States with about 5 quads thermal (1.5×10^{12} kWh) of its total gross energy supply by 2050, provided there is adequate forest available. For 5 quads thermal, we would need a yield of forest biomass of 3 t/ha/yr. A city of 100,000 people using the biomass from a sustainable forest (3 t/ha/yr) for electricity would require approximately 200,000 ha of forest area, based on an average electrical demand of slightly more than 1 billion kWh (860 kcal = 1 kWh; Pimentel, 2008).

The environmental impacts of burning biomass are less harmful than those associated with coal, but more harmful than those of natural gas (Pimentel, 2001). Biomass combustion releases more than 200 different chemical pollutants, including 14 carcinogens and 4 co-carcinogens, into the atmosphere (Pimentel, 2001). Globally, primarily in developing nations where people cook with fuelwood over open fires, some 4 million people per year die from exposure to wood smoke (Pimentel et al., 2007). In the United States, wood smoke kills 30,000 people each year (Burning Issues, 2003). However, the pollutants from electric plants that use wood and other biomass can be controlled.

Gasification of willow poses some particular environmental risks. Willow chips have 20–60% moisture, but have to be dried to 10–15% before they can be gasified (Keoleian and Volk, 2005). Often the driers emit volatile organic compounds, mostly terpenes, that require air pollution control systems (McKendry, 2002). Tars may also build up in the gasifiers and require removal.

10.10   Algae production of biodiesel

Microalgae encompass a diverse array of aquatic and marine photosynthetic microbes, all with the capacity for rapid growth and having a considerable range of growing conditions. However, there are only a few species of algae that consist of 30–50% oil (Dimitrov, 2007; Mata et al., 2009). This potential for high oil yield in a few species has produced great interest in culturing algae to provide a portion of the U.S. oil needs. Some theoretical claims suggest that as much as 934,000 liters of biodiesel oil could be produced by algae per hectare per year (or 100,000 gallons per acre per year; Walton, 2008). Based on these and other data, the calculated cost of a liter of algae oil would be only 5–10¢ ($8–16 per barrel; see Table 10.7) (Global Green Solutions, 2007). Currently, a barrel of oil in the U.S. market is selling for about $100, but has been as high as $147 per barrel. If the predicted production yield from algae proves to be true, some of current U.S. annual oil needs of 2.8 × 10^{12} liters could be met, and the land required to provide the United States with its current transport oil needs would be 5–10% of U.S. land area.

Despite all the research dating from the early 1970s, the projected algae and oil yields have yet to be achieved (Dimitrov, 2007). One calculated estimate based on all the included economic costs of using algae for oil production suggests that it would cost about $800 per barrel, not $20 per barrel (Dimitrov, 2007). The $800 per barrel translates into nearly $20 a gallon ($5.29 per liter; see Table 10.10). However, even at $20

Table 10.7 Dollar Costs of Ethanol Production per Liter from Switchgrass

Source of estimate	Cost per liter
Hay & Forage Grower, 2008	$0.15–0.16
Mongabay.com, 2008	$0.26
USDA (2002)	$0.26–0.75
Our investigation	$0.64
Khanna, 2009	$3.15

per gallon, this might be acceptable when oil is a critical resource in the future.

Microalgae grow rapidly and can double their biomass in 24 hours under optimal conditions (Chisti, 2008; Mata et al., 2009; Rosenburg et al., 2008; Tsukahara and Sawayama, 2005). Keeping in mind the laws of thermodynamics (Rodolfi, 2009), algae have the capacity to fix an estimated 5% of the incoming solar energy, much like corn on a bright sunny day in July. Suppose we optimistically assume that algae can collect 5% of the solar energy all year long under favorable climate culture conditions. At this optimal 5% rate, algae could collect 520×10^6 kcal per hectare per year and convert the sunlight into algae biomass. Reports suggest that a few algae species can contain up to 40% oil, which would equate to about 48 million kcal of energy per hectare in three months.

While some species of microalgae can adapt to a wide range of conditions, all photosynthetic growth requires sunlight, CO_2, water, nitrogen, and several other nutrients (Chisti, 2007). Inorganic elements such as nitrogen and phosphorus are essential to algae and must be present in adequate quantities. Algae also need large quantities of water, about 300 liters of water per kilogram of algae (Pimentel, 2008a). If algae cultures were located in the Southwest where it is warm with ample sunlight, they would grow rapidly. However, one essential limiting resource in this dry region is water.

Development of algal growth technology has progressed through the last two decades (Chisti, 2007). Algae require about 183 kg of CO_2 per 100 kg of algae (Chisti, 2007; Table 10.8). There is abundant CO_2 being released from coal-fired power plants that could be effectively utilized in algae production. An average power plant emitting about 1.0×10^8 m^3 of CO_2 could fuel an algae facility producing about 1,000 kg of algae (dry weight per day) while having minimal impact on electricity production. It should be pointed out, however, that when oil and other parts of the algae are utilized, the CO_2 is again released back into the atmosphere, and thus algae production for biofuel production is not a sink for CO_2.

Another beneficial use of algae could be utilization of some wastewater. Algae have been successfully used as a tertiary treatment for sewage, as they require both nitrogen and phosphorus for growth (Greer, 2009). Several species of algae have been utilized in wastewater systems (Tsukahara and Sawayama, 2005). However, the use of wastewater also has associated risks related to the thousands of different chemicals in sewage. While some species of algae tolerate heavy metals, other chemical compounds may be highly toxic to the algae. Thus, before any sewage is utilized, it would have to be carefully tested with algae cultures to make sure that it is not toxic.

Table 10.8 Estimated Inputs to Produce 30,000 kg Algae for Oil Production in a Bioreactor for 3 Months

Inputs	Quantity	kcal × 1,000	Cost
Labor	10 hr	450[a]	$130
Fuel	100 L	100[b]	53
Nitrogen	150 kg[c]	2,400[d]	85
Phosphorus	50 kg[c]	200[e]	50
Potassium	50 kg[c]	200[e]	26
Water	90 ML[f]	11,000[f]	2,000
Electricity (pumps)	12,000 kWh	10,000[g]	840
Bioreactor tubes	1,500	600[h]	400
Steel supports for the tubes		1,200[i]	300
Steel tank for heat and CO_2		200[j]	200
Total		*26,350*	*3,984*
Algae yield	30,000 kg/ha[h]	48,000,000 kcal of oil	

[a] It is assumed that a person works 2,000 hours per year and utilizes an average 9,000 liters of oil equivalents per year.
[b] 10,000 kcal/kg.
[c] Estimated.
[d] 16,000 kcal/kg.
[e] 4,000 kcal/kg.
[f] 1 kg of algae requires 300 liters of water for production, and 11 million kcal is required to supply this amount of water.
[g] Electricity for the pumps moving the water through the plastic tubes for the algae growth for three months.
[h] Estimated energy prorated for the 1,500 tubes 10 cm × 80 meter plastic tubes.
[i] Steel supports energy prorated for support of the plastic tubes.
[j] Steel tank for heat and CO_2 from the power plant for three months (0.8 m³ steel, longevity = 100 years).

The extraction of the oil from algae is estimated to provide 2,500 kcal of energy per 10,000 kcal of algae oil (PetroAlgae, 2008). One laboratory study with algae reported that the algae required large quantities of nitrogen and water, plus significant fossil energy inputs for the functioning of the system (Goldman and Ryther, 1977). This system was not self-sustaining even under high light conditions in the laboratory.

We examined a model saltwater algae oil production system based on minimal inputs for an approximately 1-ha culture system in Arizona. Table 10.8 provides a nonexhaustive list of inputs, including only 10 inputs in this system, compared to the 14 energy inputs for a corn production system (Pimentel and Patzek, 2008). The largest energy input was for the water pumped from underground, assuming a depth of only 33 m. The second largest was for electricity to run the pumps to keep the culture stirred. This estimate of 11,630 kWh (Table 10.8) may well be much too little for a bioreactor system. Even with such stirring action, this would not prevent some shading of the algae in the tubes as the population of

Table 10.9 Processing the Algae to Extract the 48 Million kcal of Oil from 30,000 kg of Algae

Items	Quantity	kcal × 1,000
Algae	30,000 kg	26,350
Transport of the algae	20 km	408
Extraction of the oil		12,000[a]
Total		*38,956*
	1.30 kcal return per 1 kcal input	

[a] Extraction of the oil from algae requires 2,500 kcal per 10,000 kcal of oil produced (PetroAlgae, 2008).

Cost of 1 liter is $0.83 ($3.14 per gallon).

algae increases. The total energy expended for the three-month culture of algae on 1 hectare totaled 26.4 million kcal.

The energy inputs to process the 30,000 kg/ha of algae produced and to extract an estimated 12,000 kg of oil was estimated to be 12.5 million kcal (see Table 10.9). The largest energy input for processing is the extraction of the oil from the algae, requiring about 0.25 kcal per 1 kcal extracted (PetroAlgae, 2008).

For the algae oil to be converted into diesel fuel, methanol must be added at a rate of 0.125 liter per liter of algae oil; however, this methanol is replaced by heat from the coal-fired electric plant. The total fossil-energy inputs for the system to produce 12,000 kg of oil was 39.5 million kcal (see Table 10.9). Thus, the net algae oil produced in this model system per 1 kcal invested is 1.29 kcal oil energy. This is clearly more efficient than corn ethanol, which results in a *negative* 1.59 kcal of ethanol energy (see Tables 10.3 and 10.4). However, these oil yields and inputs were only a model system based on optimistic assumptions, and there may be other as-yet undetermined costs and savings that will need to be included in the future, and the algae system needs to be carefully field-tested.

The conversion of algae oil into diesel fuel takes the form of a transesterification reaction. This reaction can take place under several conditions, with two primary modes being considered for mass production of biodiesel: supercritical transesterification and base-catalyzed transesterification. These are superior to other modes due to the conversion efficiency of fatty acids to biodiesel (Fukuda et al., 2001).

Supercritical methanolysis is a directly energy-intensive process involving heating the oils to a temperature of 350°C and exposing them to supercritical methanol for 240 seconds, converting the oils to methyl esters. The ideal ratio of methanol to oil in this reaction is 42:1, which is a large increase over the amount required for catalyzed transesterification.

This process allows for increased conversion of free fatty acids as well, which is critical for any commercial-scale process (Fukuda et al., 2001). Simulation of this technique for 10,000 tons of biodiesel suggested an energy cost of 1,839 kW for production (Glišić et al., 2009).

The 1.29 kcal oil return per 1 kcal invested is much lower than the model algae oil production system investigated by Chisti (2008). In that system, the algae oil produced was greater than 7 kcal of oil per 1 kcal invested. The Chisti (2008) system was also sealed in a tank to prevent contamination with less productive algae, while our system was in a sealed bioreactor.

David Pimentel consulted with Dr. Hoyt Islom of SunEco Energy in Chino, California, in 2009 concerning a model system that Dr. Islom is investigating. That system has 200 acres of small ponds, naturally infested with algae and other microbes that had been used for fish aquaculture. The algae biomass growing in these ponds is 40–48% oil. The cost of oil produced from this algal system is reported to be 21¢ per liter (80¢/gallon). The fossil energy input is 0.19 kcal to produce 1 kcal of oil, or about 5 kcal of oil for 1 kcal input. The design and operation of this system appears to be quite encouraging. What is most encouraging is employing natural algae and other microbes.

Other investigators are not reporting as economical production of algae as Islom. For an example, a European study reported that it takes $1925 to produce 1 metric ton dry algae (New Energy and Fuel, 2009).

Another investigation calculated the cost per gallon of biodiesel using electricity in a bioreactor, with electricity costing only 7¢ per kWh, to be $32.81 per gallon ($8.68 per liter; see Table 10.10). This means that 38.6 million kcal is required to produce 50 million kcal of algae oil. Other studies report that a kilogram of microalgae oil costs from $0.05 to $8.68 per liter (Cleantech Forum, 2010; Table 10.10). These data suggest some cost problems in producing algae for biodiesel.

Table 10.10 Estimated Costs of Producing Algae Biodiesel Fuel

Source of estimate	Biodiesel cost/liter
Al Fin, 2010	$0.05
Sass et al., 2005	0.44
Oilgae, 2010	0.53
Our investigation	0.83
Williams and Laurens, 2010	1.70
Conger, 2009	2.12
Chisti, 2007	2.80
Associated Press, 2007	5.29
Rapier, 2009	8.68

10.11   Conclusions

Several physical and chemical factors limit the production of biofuels, such as the complex process required for the conversion of plant biomass into ethanol and biodiesel. For example, fossil energy inputs needed for the production of ethanol from corn or cellulosic biomass are, respectively, 1.59 times and 2.13 times greater than the ethanol energy output from either system. For biodiesel produced from algae, the fossil energy inputs are 77% less than the biodiesel oil produced.

One of the many factors limiting energy output from biomass is the extremely low fraction of sunlight captured by the plants. On average, only about 0.1% of the sunlight is captured by green plants (mostly crop plants) per year. This is in sharp contrast to photovoltaics (electricity, not crops) that capture about 15% of sunlight—150 times more sunlight than green plant biomass. At peak sunlight capture by green plants, they may capture 4.5–6.0% of sunlight (Walker, 2010).

The environmental impacts of producing ethanol from either corn, cellulosic, or soybean biomass are enormous. These include severe soil erosion, heavy use of fertilizers often with considerable runoff into surface waters or contaminating ground waters, the loss of soybeans and corn to the food system, the use of prime cropland acreage for nonfood and non-fiber purposes, the loss of biomass production in forests and grasslands for wildlife and livestock maintenance, and the use of large quantities of pesticides. In addition to the significant contribution to global climate change because of the heavy use of fossil energy inputs, there is the large requirement of 1,000–2,000 liters of water for the production of each liter of ethanol (Trainer, 2010). Furthermore, for each liter of ethanol produced, there are more than 10 liters of sewage effluent produced.

The problems with cellulosic biomass for ethanol production are threefold:

1. More than twice as much cellulosic biomass has to be grown, harvested, and processed as cellulosic biomass compared with corn grain.
2. The cellulose biomass conversion technology for ethanol production is not sufficiently developed for economic and energetic production of ethanol.
3. The conversion process currently has several limiting factors, although these may be overcome with future research.

Fossil-energy conservation strategies, combined with the active development of renewable energy sources such as wind power, photovoltaics, and hydropower, should be given priority (Pimentel, 2008). The use of food crops, such as corn and soybeans, to produce biofuels creates major

ethical concerns, as about 60% of the world population is seriously malnourished and the need for food is critical. Other methods of obtaining
biofuels have proved thus far to be inefficient and costly, which leaves the
development of renewable energy sources as the best possibility.

Acknowledgments

This research was supported in part by the Podell Emeriti Award at
Cornell University.

References

AVMA. 2007. *U.S. pet ownership & demographics sourcebook (2007 edition)*.
Schaumburg, Illinois: American Veterinary Medical Association, Member
and Field Services.

Alexandrov, G. A. 2007. Carbon stock growth in a forest stand: The power of
age. *Carbon Balance and Management*, 2:4. http://www.fao.org/DOCREP/
ARTICLE/WFC/XII/MS14-E.HTM.

Al Fin. 2010. Algae biofuels for $0.20 a gallon? Posted March 30, 2009. http://
alfin2100.blogspot.com/2009/03/algae-biofuels-for-020-gallon-peak-oil
.html.

Arkenol. 2004. Our technology: Concentrated acid hydrolysis. http://www
.arkenol.com/Arkenol%20Inc/tech01.html.

Associated Press. 2007. Oil from algae? Scientists seek green gold. November 29.
http://www.msnbc.msn.com/id/22027663/.

Batty, J. C., and Keller, J. 1980. Energy requirements for irrigation. In *Handbook of
energy utilization in agriculture*, ed. Pimentel, D., 35–44. Boca Raton, FL: CRC
Press.

Berg, C. 2004. World fuel ethanol analysis and outlook. *Online Distillery Network
for Distilleries and Fuel Ethanol Plants Worldwide*. http://www.distill.com/
World-Fuel-Ethanol-A&O-2004.html.

Birdsey, R. A. 1992. Carbon storage and accumulation in United States forest ecosystems. Washington, DC: U.S. Forest Service. http://www.treesearch.fs.fed
.us/pubs/15215.

Blais, J. F., Mamouny, K., Nlombi, K., Sasseville, J. L., and Letourneau, M. 1995. Les
mesure deficacite energetique dans le secteur de leau. In *Les mesures deficacite
energetique pour lepuration des eaux usees municipales*, ed. Sasseville, J. L., and
Blais, J. F. Scientific Report 405. Vol. 3. Quebec: INRS-Eau.

Brown, L. R. 2009. *Plan B 4.0: Mobilizing to save civilization*. New York: Norton.

Brummer, E. C., Burras, C. L., Duffy, M. D., and Moore, K. J. 2000. *Switchgrass
production in Iowa: Economic analysis, soil suitability, and varietal performance*.
Ames: Iowa State University.

Burning Issues. 2003. References for wood smoke brochure [from 2003]. http://
burningissues.org/car-www/science/wsbrochure-ref-3-03.htm.

CONAB (Companhia Brasileira de Abastecimento). 2010. 9° levantamento de
cana-de açúcar. Brasilia, D.F. Accessed on 16 December 2010. http://www
.conab.gov.br.

Chisti, Y. 2007. Biodiesel from microalgae. *Biotechnology Advances*, 25:294–306.

————. 2008. Response to Reijnders: Do biofuels from microalgae beat biofuels from terrestrial plants? *Trends in Biotechnology*, 26(7): 351–352.

Cleantech Forum. 2010. Algaeventure Systems announces process to reduce cost of dewatering by 99 percent: Potential breakthrough for algae commercialization. http://www.biofuelsdigest.com/blog2/2009/03/10/algaeventure-systems-announces-process-to-reduce-cost-of-dewatering-by-99-percent-potential-breakthrough-for-algae-commercialization/.

Clinch, R. L., Volk, T. A., Sidders, D., Thevathasan, N.V., and Gordon, A.M. 2009. Biophysical interactions in a short rotation willow intercropping system in southern Ontario, Canada. *Agriculture, Ecosystems & Environment*, 131: 61–69.

Coelho, S. 2005. Brazilian sugarcane ethanol: Lessons learned. Presentation at the IEA Bioenergy T40-T31-T30 Event, Brazil, December 2. http://bioenergy-trade.org/downloads/coelhonovdec05.pdf.

Conger, C. 2009. Could cheap algae oil power our energy future? *Discovery News*, November 18. http://news.discovery.com/earth/algae-biofuel-renewable-energy.html.

Cornell University. 2008. Technical brief for large-scale biomass for direct combustion. http://www.sustainablecampus.cornell.edu/climate/docs/Cornell CAP_FuelMixandRenewables_TB_LargeScaleBiomass.pdf.

De Wit, M., Junginger, M., Lensink, S., Londo, M., and Faaij, A. 2010. Competition between biofuels: Modeling technological learning and cost reductions over time. *Biomass and Bioenergy*, 34(2): 203–217.

Dimitrov, K. 2007. GreenFuel technologies: A case study for industrial photosynthetic energy capture. Brisbane, Australia. http://www.nanostring.net/Algae/CaseStudy.pdf.

DOE (U.S. Department of Energy). 2002. Review of transport issues and comparison of infrastructure costs for a renewable fuels standard. Washington, DC: DOE. http://tonto.eia.doe.gov/FTPROOT/service/question3.pdf.

Donner, S. D., and Kucharik, C. J. 2008. Corn-based ethanol production compromises goal of reducing nitrogen export by the Mississippi River. *Proceedings of the National Academy of Sciences*, 105(11): 4513–4518. http://www.pnas.org/content/105/11/4513.full.pdf+html.

Dwivedi, P., Alavalapati, J. R., and Lal, P. 2009. Cellulosic ethanol production in the United States: Conversion technologies, current production status, economics, and emerging developments. *Energy for Sustainable Development*, 13(3): 174–182.

FAO (Food and Agriculture Organization). 2009. *The state of food insecurity in the world*. Rome: FAO. http://www.fao.org/docrep/012/i0876e/i0876e00.htm.

Ferguson, A. R. B. 2001. Biomass and energy. *Optimum Population Trust Journal*, 4(1): 14–18.

————. 2008. The power density of ethanol from Brazilian sugarcane. In *Biofuels, solar and wind as renewable energy systems: Benefits and risks*, ed. Pimentel, D., 493–498. Dordrecht, The Netherlands: Springer.

Ferrett, Grant. 2007. Biofuels "crime against humanity." *BBC News*, October 27. http://news.bbc.co.uk/2/hi/7065061.stm.

FoodProductionDaily.com. 2002. Ethanol plants releasing toxins, EPA reports. FoodProductionDaily.com, May 7. http://www.foodproductiondaily.com/Supply-Chain/Ethanol-plants-releasing-toxins-EPA-reports.

Fukuda, H., Kondo, A., and Noda, H. 2001. Biodiesel fuel production bytrans-esterification of oils. *Journal of Bioscience and Bioengineering*, 92(5): 405–416.

Glišić, S., Lukic, I., and Skala, D. 2009. Biodiesel synthesis at high pressure and-temperature: Analysis of energy consumption on industrial scale. *Bioresource Technology*, 100(24): 6347–6354.

Global Green Solutions. 2007. Renewable energy. http://www.stockupticks.com/profiles/7-26-07.html.

Goldman, J. C., and Ryther, J. H. 1977. Mass production of algae: Bio-engineering aspects. In *Biological solar energy conversion*, ed. Mitsui, A., et al., 367–378. New York: Academic Press.

Green, D. L., Hopson, J. L., and Li, J. 2006. Have we run out of oil yet? Oil peaking analysis from an optimist's perspective. *Energy Policy*, 34(5): 515–531.

Greer, D. 2009. Cultivating algae in wastewater for biofuel. *Biocycle*, 50(2): 36–39.

Hay & Forage Grower. 2008. Research reveals switchgrass production costs. *Hay & Forage Grower*, March 11. http://hayandforage.com/ehayarchive/research-reveals-swithgrass-production/.

Henning, J. C. 1993. Big bluestem, indiangrass and switchgrass. Columbia: Department of Agronomy, Campus Extension, University of Missouri.

Hodge, N. 2008. Future sources of energy: What's a cubic mile of oil? http://www.energyandcapital.com/articles/future-sources-energy/787.

Howarth, R. W., and S. Bringezu, S., eds. 2009. *Biofuels: Environmental conse-quences and interactions with changing land use*. Proceeedings of the Scientific Committee on Problems of the Environment (SCOPE), International Biofuels Project Rapid Assessment, September 22–25, 2008, Grummersback, Germany.

IEA (International Energy Agency). 2004. Biofuels for transport: An international perspective. Paris: OECD/IEA.

IEA (International Energy Agency)/OECD. 2008. *World energy outlook 2008*. Paris: OECD/IEA. http://www.iea.org/textbase/nppdf/free/2008/weo2008.pdf.

Johnson, J., Reck, B. K., Wang, T., and Graedel, T. E. 2008. The energy benefit of stainless steel recycling. *Energy Policy*, 36(1): 181–192.

Keoleian, G. A., and Volk, T. A. 2005. Renewable energy from willow biomass crops: Life cycle energy, environmental and economic performance. *Critical Reviews in Plant Sciences*, 24:385–406.

Khanna, M. 2009. Cellulosic biofuels: Are they economically viable and environ-mentally sustainable? *Choices* 23(3): 16–21. http://www.choicesmagazine.org/magazine/pdf/article_40.pdf.

Kitani, O. 1999. Biomass resources. In *Energy and biomass engineering*, ed. Kitani, O., Jungbluth, T., Peart, R. M., and Ramdami, A., 6–11. St. Joseph, MI: American Society of Agricultural Engineers.

Koplow, D, and Steenblik, R. 2008. Subsidies to ethanol in the United States. In *Biofuels, solar and wind as renewable energy systems: Benefits and risks*, ed. Pimentel, D., 79–108. Dordrecht, The Netherlands: Springer.

Kopp, R. F., Abrahamson, L. P., White, E. H., Burns, K. F., and Nowak, C. A. 1997. Cutting cycle and spacing effects on biomass production by a willow clone in New York. *Biomass and Bioenergy*, 12(5): 313–319.

Kuby, W. R., Markoja, R., and Nackford, S. 1984. Testing and evaluation of on-farm alcohol production facilities. Cincinnati, OH: EPA, Industrial Environmental Research Laboratory, 100 p. http://nepis.epa.gov/Exe/ZyPURL.cgi?Dockey=2000TTXF.txt.

Kumar, A., and Sokhansanj, S. 2007. Switchgrass (*Panicum vigratum, L.*) delivery to a biorefinery using integrated biomass supply analysis and logistics (IBSAL) model. *Bioresource Technology,* 98(5): 1033–1044.

Lal, R. 2005. World crop residues production and implications of its use as a biofuel. *Environment International,* 31(4): 575–584.

Lal, R., Kimble, J. M. Follet, R. F., and Cole, C. V. 1999. *Potential use of U.S. cropland to sequester carbon and mitigate the greenhouse effect.* Boca Raton, FL: CRC Press.

Lal, R., and Pimentel, D. 2009. Biofuels: Beware crop residues. [Letter to the Editor]. *Science,* 326: 1345–1346.

Larsen, K., Thompson, D., and Harn, A. 2002. Limited and full irrigation comparison for corn and grain sorghum. Walsh: Plainsman Research Center, Colorado State University, Agricultural Experiment Station. http://www.colostate.edu/depts/prc/pubs/LimitedandFullIrrigationComparisonfor Corn.pdf.

Magelli, F., Boucher, K., Bi, H. T., Melin, S., and Bonoli, A. 2009. An environmental impact assessment of exported wood pellets from Canada to Europe. *Biomass and Bioenergy,* 33(3): 434–441.

Maiorella, B. 1985. Ethanol. In *Comprehensive biotechnology,* Vol. 3, ed. Moo–Young, M., Blanch, H. W., Drew, S., and Wang, D. I. C., 861–914. New York: Pergamon Press.

Mata, T. M., Martins, A. A., and Caetano, N. S. 2009. Microalgae for biodiesel production and their applications: A review. *Renewable and Sustainable Energy Reviews,* 14(1): 431–447.

McCain, J. 2003. Statement of Senator McCain on the Energy Bill. Press Release, November 21. http://mccain.senate.gov/public/index.cfm?FuseAction= PressOffice.Speeches&ContentRecord_id=faed0c9b-6d5c-46dd-acd2-3163891b6685&Region_id=&Issue_id=79a48974-2bd4-4888-be97-e8a445366a84.

McCoy, T. D., Ryan, M. R., Burger, L. W., Jr., and Kurzejeski, E. W. 2001. Grassland bird conservation: CP1 vs. CP2 plantings in Conservation Reserve Program fields in Missouri. *American Midland Naturalist,* 145(1): 1–17.

McKendry, P. 2002. Energy production from biomass. Part 2: Conversion technologies. *Bioresource Technology,* 83(1): 47–54.

Mongabay.com. 2006. Switchgrass-based ethanol could cost $1 per gallon, reduce foreign oil dependence. Mongabay.com, December 5. http://news .mongabay.com/2006/1205-switchgrass.html.

Moseley, B. 2009. Profit from forage production strategies. *Land and Livestock Post,* April 1. http://www.landandlivestockpost.com/livestock/040109-cover.

Nabuurs, G. J., Masera, O., Andrasko, K., Benitez-Ponce, P., Boer, R., Dutschke, M., Elsiddig, E., et al. 2007. Forestry. In *Climate change, 2007: Mitigation of climate change: Contribution of Working Group III to the Fourth Assessment Report of the Intergovernmental Panel on Climate Change,* ed. Metz, B., Davidson, O. R., Bosch, P. R., Dave, R., and Meyer, L. A. 541–584. Cambridge: Cambridge University Press. http://www.ipcc.ch/publications_and_data/ar4/wg3/en/ch9.html.

NAS (National Academy of Sciences). 2003. *Frontiers in agricultural research: Food, health, environment, and communities.* Washington, DC: NAS.

NASS. 2003. Agricultural chemical usage: 2002 field crops summary (p. 6). May 2003. United States Department of Agriculture, National Agricultural

Statistics Service. http://usda.mannlib.cornell.edu/usda/nass/AgriChem UsFC//2000s/2003/AgriChemUsFC-05-14-2003.pdf.

Naval Facilities Engineering Service Center. 2003. Wood tub grinders. In *Joint service pollution prevention opportunity handbook*. Accessed July 29, 2010. http://205.153.241.230/P2_Opportunity_Handbook/7_III_13.html.

New Energy and Fuel. 2009. An Algae Production Cost Breakthrough. Posted March 30, 2009. http://newenergyandfuel.com/ http:/newenergyandfuel/com/2009/03/30/an-algae-production-cost-breakthrough/.

Oilgae. 2010. Cost of algae fuel: $2 per gallon. February 18. http://www.oilgae.com/blog/2010/02/cost-of-algae-fuel-2-per-gallon.html.

Pacific Ecologist. 2009. Brazil: Sugar cane plantations devastate vital Cerrado region. *Pacific Ecologist*, no. 17: 25–27. http://www.pacificecologist.org/archive/17/pe17-brazil-sugar-cane-devastation.pdf.

Patzek, T. W. 2004. Thermodynamics of the corn-ethanol biofuel cycle. *Critical Reviews in Plant Sciences*, 23(6): 519–567.

———. 2006. The real corn-ethanol transportation system. http://petroleum.berkeley.edu/patzek/BiofuelQA/Materials/TrueCostofEtOH.pdf.

———. 2010. A probabilistic analysis of the switchgrass-ethanol cycle. *Sustainability*, 2(10): 3158–3194.

Patzek, T. W., and Pimentel, D. 2005. Thermodynamics of energy production from biomass. *Critical Reviews in Plant Sciences*, 24(5–6): 327–364.

Perlack, R. D., Wright, L. L., Turhollow, A. F., Robin, L., Graham, R. L., Stokes, B. J., and Erbach, D. E. 2005. Biomass as feedstock for a bioenergy and bio-products industry: The technical feasibility of a billion-ton annual supply. Washington, DC: DOE and USDA. http://www1.eere.energy.gov/biomass/pdfs/final_billionton_vision_report2.pdf.

PetroAlgae. 2008. PetroAlgae develops algae-oil extraction system. Press release, May 14. http://www.zimbio.com/Biodiesel/article/136/PetroAlgae+Develops+Algae+Oil+Extraction+System.

Pimentel, D. 2001. Biomass utilization, limits of. In *Encyclopedia of physical science and technology*, 3rd ed., ed. Meyers, R. A., 2:159–171. San Diego, CA: Academic Press.

———. 2003. Ethanol fuels: Energy balance, economics, and the environmental aspects are negative. *Natural Resources Research*, 12(2): 127–134.

———. 2008. Renewable and solar energy technologies: Energy and environmental issues. In *Biofuels, solar and wind as renewable energy systems: Benefits and risks*, ed. Pimentel, D., 1–17. Dordrecht, The Netherlands: Springer.

———. 2008a. A brief discussion on algae for oil production: Energy issues. In *Biofuels, solar and wind as renewable energy systems: Benefits and risks*, ed. Pimentel, D., 499–500. Dordrecht, The Netherlands: Springer.

Pimentel, D., Berger, B., Filiberto, D., Newton, M., Wolfe, B., Karabinakis, E., Clark, S., Poon, E., Abbett, E., and Nandagopal, S. 2004. Water resources: Agricultural and environmental issues. *Bioscience*, 54(10): 909–918.

Pimentel, D., Cooperstein, S., Randell, H., Filiberto, D., Sorrentino, S., Kaye, B., Yagi, C. J., et al. 2007. Ecology of increasing diseases: Population growth and environmental degradation. *Human Ecology*, 35:653–668.

Pimentel, D., Doughty, R., Carothers, C., Lamberson, S., Bora, N., and Lee, K. 2002. Energy inputs in crop production in developing and developed countries. In *Food security and environmental quality in the developing world*, ed. Lal, R., Hansen, D., Uphoff, N., and Slack, S., 129–151. Boca Raton, FL: CRC Press.

Pimentel, D., Harvey, C., Resosudarmo, P., Sinclair, K., Kurtz, D., McNair, M., Crist, S., et al. 1995. Environmental and economic costs of soil erosion and conservation benefits. *Science*, 267:1117–1123.

Pimentel, D., and Lal, R. 2007. Letter to the editor: Biofuels and the environment. *Science*, 317:897.

Pimentel, D., Marklein, A., Toth, M. A., Karpoff, M., Paul, G. S., McCormack, R., Kyriazis, J., and Krueger, T. 2009. Food versus biofuels: Environmental and economic costs. *Human Ecology*, 37:1–12.

Pimentel, D., and Patzek, T. 2008. Ethanol production: Energy and economic issues related to U.S. and Brazilian sugarcane. In *Biofuels, solar and wind as renewable energy systems: Benefits and risks*, ed. Pimentel, D., 357–371. Dordrecht, The Netherlands: Springer.

Pimentel, D., and Pimentel, M. 2008. *Food, energy and society*. 3rd ed. Boca Raton, FL: CRC Press.

Rapier, R. 2009. More reality checks for algal biodiesel. Consumer Energy Reports: R Squared Energy Blog by Robert Rapier. Posted February 28, 2009. http://www.consumerenergyreport.com/2009/02/28/more-reality -checks-for-algal-biodiesel/.

Ribeiro, S. K., Kobayashi, S., Beuthe, M., Gasca, J., Greene, D., Lee, D. S., Muromachi, Y., et al. 2007. Transport and its infrastructure. In *Climate change, 2007: Mitigation of climate change: Contribution of Working Group III to the fourth assessment report of the Intergovernmental Panel on Climate Change*, ed. Metz, B., Davidson, O. R., Bosch, P. R., Dave, R., and Meyer, L. A. 323–385. Cambridge: Cambridge University Press. http://www.ipcc.ch/publications_and_data/ ar4/wg3/en/ch5.html.

Righelato, R. 2007. Biofuels or forests? *Scitizen*, August 23. http://scitizen.com/ authors/Renton-Righelato-a-809_s_5c85358b289831dc490e29b55191d47e. html.

Rodolfi, L., Zittelli, G. C., Bassi, N., Padovani, G., Biondi, N., Bonini, G., and Tredici, M. R. 2009. Microalgae for oil: Strain selection, induction of lipid synthesis and outdoor mass cultivation in a low-cost photobioreactor. *Biotechnology and Bioengineering* 102(1): 100–112. http://www.biomedexperts.com/Abstract .bme/18683258/Microalgae_for_oil_strain_selection_induction_of_lipid_ synthesis_and_outdoor_mass_cultivation_in_a_low-cost_photobiore.

Rosenburg, J. N., Oyler, G. A., Wilkinson, L., and Betenbaugh, M. J. 2008. A green light for engineered algae: Redirecting metabolism to fuel a biotechnology revolution. *Current Opinion in Biotechnology*, 19(5): 430–436.

Rosenthal, E. 2007. Food stocks dwindling worldwide, UN says; agriculture chief calls situation "unforeseen and unprecedented." *International Herald Tribune*, December 18. http://www.stwr.org/food-security-agriculture/world-food-stocks-dwindling-rapidly-un-warns.html.

Roth, A. M., Sample, D. W., Ribic, C. A., Paine, L., Undersander, D. J., and Bartelt, G. A. 2005. Grassland bird response to harvesting switchgrass as a biomass energy crop. *Biomass and Bioenergy*, 28(5): 490–498.

Samson, R. 1991. Switchgrass: A living solar battery for the prairies. Ste-Anne-de-Bellevue, QC: McGill University. http://eap.mcgill.ca/MagRack/SF/ Fall%2091%20L.htm.

Samson, R., and Bailey Stamler, S. 2009. Going green for less: Cost-effective alternative energy sources. Commentary No. 282. Toronto: C. D. Howe Institute. http://www.cdhowe.org/pdf/commentary_282.pdf.

Samson, R., Duxbury, P., Drisdale, M., and Lapointe, C. 2000. Assessment of pelletized biofuels. Ste. Anne de Bellevue, QC: Resource Efficient Agricultural Production-Canada. http://www.reap-canada.com/online_library/feedstock_biomass/15%20Assessment%20of.PDF.

Samson, R., Duxbury, P., and Mulkins, L. 2004. Research and development of fibre crops in cool season regions of Canada. Ste. Anne de Bellevue, QC: Resource Efficient Agricultural Production-Canada. http://www.reap-canada.com/online_library/agri_fibres_forestry/2%20Research%20and.pdf.

Sandia. 2009. Biofuels can provide viable, sustainable solution to reducing petroleum dependence, say Sandia researchers. Sandia National Laboratories press release, February 10. http://www.sandia.gov/news/resources/news_releases/biofuels-can-provide-viable-sustainable-solution-to-reducing-petroleum-dependence-say-sandia-researchers/.

Sass, M., Amendt, E., Glelim, R., McLenegan, T., and Whitacre, T. 2005. Biodiesel from microalgae [PowerPoint presentation]. http://www.slidefinder.net/b/biodiesel_from_microalgae_solution_sustainable/7161496.

Sassner, P., Galbe, M., and Zacchi, G. 2005. Steam pretreatment of Salix with and without SO^2 impregnation for production of bioethanol. *Applied Biochemisty and Biotechnology*, 124(1–3): 1101.

Schmitz, T. G., Seale, J., and Schmitz, A. 2009. The economics of alternative energy sources and globalization: The road ahead. Presentation at the CNAS Economics of Alternative Energy Sources and Globalization Conference, Orlando, FL, November 15–17. http://cnas.tamu.edu/ConfPresentations/SchmitzSealeHiddenSubsidiesInBrazilPres.ppt.

Spillius, A. 2008. IMF alert on starvation and civil unrest. *Sydney Morning Herald*, April 14.

Stanton, T. L., and LeValley, S. 1999. Feed composition for cattle and sheep. Colorado State University Extension, Report No. 1,615. http://www.ext.colostate.edu/pubs/livestk/01615.html.

Sun, Y., and Cheng, J. 2002. Hydrolysis of lignocellulosic materials for ethanol production: A review. *Bioresource Technology*, 83(1): 1–11.

Supalla, R. 2007. Biofuels: An emerging water resources hazard. Lincoln: Agricultural Economics Department, University of Nebraska. http://digitalcommons.unl.edu/cgi/viewcontent.cgi?article=1040&context=ageconworkpap.

Sylvester-Bradley, R. 2008. Critique of Searchinger (2008) and related papers assessing indirect effects of biofuels on land-use change. Version 3.2, June 12. In *Gallagher Biofuels Review for Renewable Fuels Agency*, Department of Transport. Boxworth, Cambs., England: ADAS UK. http://www.globalbioenergy.org/uploads/media/0806_ADAS_-_Seachinger_critique.pdf.

Tilman, D., Hill, J., and Lehman, C. 2006. Carbon negative biofuels from low-input, high-diversity grassland biomass. *Science*, 314:1598–1600.

Trainer, T. 2010. Can renewables etc. solve the greenhouse problem? The negative case. *Energy Policy*, 38:4107–4114.

Troeh, F. E., Hobbs, J. A., and Donahue, R. L. 2004. *Soil and water conservation for productivity and environmental protection*. Upper Saddle River, NJ: Prentice Hall.

Tsukahara, K., and Sawayama, S. 2005. Liquid fuel production using microalgae. *Journal of the Japan Petroleum Institute*, 48(5): 251–259.

UNICA. 2010. Brazilian sugarcane industry responds to introduction of Pomeroy-Shimkus legislation that taxes clean, renewable energy. March 25, 2010.

UNICA - Brazilian Sugarcane Industry Association. http://english.unica. com.br/noticias/show.asp?nwsCode={2E5A63C6-01AE-4D0B-80E1-D98A7A14B4D5}.

USCB (U.S. Census Bureau). 2004. *Statistical abstract of the United States, 2004–2005.* Washington, DC: USCB.

————. 2007. *Statistical abstract of the United States, 2008.* Washington, DC: USCB.

————. 2009. *Statistical abstract of the United States, 2010.* Washington, DC: USCB.

USDA (U.S. Department of Agriculture). 1975. Nutritive value of American foods: In common units. Agricultural Handbook 456. Washington, DC: USDA.

————. 1991. Corn-state: Costs of production. Stock #94018. Washington, DC: USDA, Economic Research Service, Economics and Statistics System.

————. 1997. Farm and ranch irrigation survey, 1998. In *1997 census of agriculture,* Vol. 3, *Special studies,* Part 1. Washington, DC: USDA, National Agricultural Statistics Service. http://www.agcensus.usda.gov/Publications/1997/Farm_ and_Ranch_Irrigation_Survey/index.asp.

————. 2005. National Agricultural Statistics Service. Accessed February 12, 2010, from: http://usda.mannlib.cornell.edu.

————. 2008. *Agricultural statistics, 2008.* Washington, DC: USDA.

Valle, S. 2007. Losing forests to fuel cars. *Washington Post,* July 31. http:// www.washingtonpost.com/wp-dyn/content/article/2007/07/30/ AR2007073001484.html?hpid=moreheadlines.

Venkatarama Reddy, B. V., and Jagadish, K. S. 2003. Embodied energy of common and alternative building materials and technologies. *Energy and Buildings,* 35(2): 129–137.

Ververis C., Georghiou, K., Christodoulakis, N., Santas, P., and Santas, R. 2004. Fiber dimensions, lignin and cellulose content of various plant materials and their suitability for paper production. *Industrial Crops and Products,* 19:245–254.

Volk, T. A., Abrahamson, L. P., Nowak, C. A., Smart, L. B., Tharakan, P. J., and White, E. H. 2006. The development of short-rotation willow in the northeastern United States for bioenergy and bioproducts, agroforestry and phytoremediation. *Biomass and Bioenergy,* 30(8–9): 715–727.

Walker, D. A. 2010. Biofuels—for better or worse? *Annals of Applied Biology,* 156:319–327.

Walter, A. 2009. Bioethanol development(s) in Brazil. In *Biofuels,* ed. Soetaert, W., and Vandamme, E. J., 55–75. Chippenham, England: John Wiley & Sons.

Walton, M. 2008. Algae: The ultimate in renewable energy. CNN, April 1. http:// www.cnn.com/2008/TECH/science/04/01/algae.oil/index.html.

Wereko-Brobby, C., and Hagen, E. B. 1996. *Biomass conversion and technology.* Chichester, England: John Wiley & Sons.

West, T., Dunphy-Guzman, K., Sun, A., Malczynski, L., Reichmuth, D., Larson, R., Ellison, J., et al. 2009. Feasibility, economics, and environmental impact of producing 90 billion gallons of ethanol per year by 2030. Preprint. http:// www.sandia.gov/news/publications/white-papers/90-Billion-Gallon-BiofuelSAND2009-3076J.pdf.

WHO (World Health Organization). 2000. Turning the tide of malnutrition: Responding to the challenge of the 21st century. Geneva: WHO. http:// www.who.int/mip2001/files/2232/NHDbrochure.pdf.

Williams, P. J. le B., and Laurens, L. M. L. 2010. Microalgae as biodiesel and biomass feedstocks: Review and analysis of the biochemistry, energetics and economics. *Energy and Environmental Science,* 3:554–590.

Wolcott, G. N. 1937. An animal census of two pastures and a meadow in northern New York. *Ecological Monographs*, 7(1): 1–90.

Wood Resources International. 2007. Wood chip prices. Global wood prices. Accessed July 10, 2011. http://www.wri-ltd.com/marketBriefs.cfm.

Xu, Y. J., and Wang, F. 2009. Role of Louisiana's forest ecosystems in carbon sequestration. Baton Rouge: Louisiana State University Ag Center Research and Extension. http://www.lsuagcenter.com/en/communications/publications/agmag/Archive/2006/Spring/The+Role+of+Louisianas+Forest+Ecosystems+in+Carbon+Sequestration.htm.

Yu, H., Guo, G., Zhang, X., Yan, K., and Xu, C. 2009. The effect of biological pretreatment with the selective white-rot fungus *Echinodontium taxodii* on enzymatic hydrolysis of softwoods and hardwoods. *Bioresource Technology*, 100(21): 5170–5175.

Zoellick, Robert. 2008. Interview by Steve Inskeep. *Morning Edition*, National Public Radio, April 11. http://www.npr.org/templates/story/story.php?storyId=89545855.

chapter eleven

Biofuels and world food and society issues

Philip McMichael

Contents

11.1   Introduction

The link between biofuels and food became quite clear during the recent food crisis. At the close of 2007, the *Economist*'s food-price index reached its highest point since originating in 1845, food prices had risen 75% since 2005, and world grain reserves were at their lowest (Holt-Giménez and Kenfield, 2008, p. 3). Estimates of the impact of industrial biofuels vary: the United Nations Food and Agriculture Organization (FAO) estimated that biofuels accounted for 10% of the food price rise, while the International Monetary Fund (IMF) and International Food Policy Research Institute (IFPRI) claimed 30% and the World Bank calculated a 65–75% contribution (Phillips, 2008; Chakraborrty, 2008).

Under present circumstances and projections, the biofuels project, as the purported solution to the energy and climate crises, is more likely to *deepen* these crises as well as the food crisis. Politicians with short-term horizons—driven by the exigencies of a system of states that need to renew their legitimacy by stabilizing currencies, consumer/energy prices, and employment—appeal to notions of "green capitalism" as they mandate emissions targets and subsidies that empower the biofuel industry. This nexus between government and the private sector is embedded in a fuel-food complex that underlies such agflation and only compounds these three crises. Thus, palm oil, "now used widely in food products ranging from instant noodles to biscuits and ice cream, has become so integrated into energy markets that its price moves in tandem with crude oil prices" (Greenfield, 2007, p. 4). In 2008, price inflation of such a basic commodity as cooking oil, essential to even those who grow their own food, disproportionately disadvantaged the poor.

This chapter argues that the growing synchrony of these markets (for oil and food in general) registers two fundamental crises—food and climate—associated with industrial agriculture.* The irony of the biofuels project is that it exacerbates the former crisis in seeking to address the latter. In addition, it inextricably links these crises through a process of deepening "biophysical override" (Weis, 2007), whereby the separation of agriculture from its biological base requires continual application of technological inputs to compensate for environmental degradation. One key input—chemical applications to the soil, pest, and weed populations— degrades soil and natural pest and weed control mechanisms through the intensification of fossil-fuel dependent monocultures.† Further, this interruption of natural cycles of regeneration of soil and water is accompanied, and accomplished, by the dispossession of small-farming cultures more likely to reproduce local ecologies in sustainable fashion.

11.2   The food/fuel crisis

Popular perceptions of the underlying cause of food price inflation target the biofuels revolution, with one author noting that the unsustainable agriculture and biofuels policies of the United States and the European Union have led to "huge food trade deficits of both countries" and are "at the heart of the current explosion of agricultural commodity prices" (Berthelot, 2008, p. 26). Thus, food stocks in the global North were run down by ballooning food trade deficits in addition to highly subsidized biofuel policies, especially for U.S. corn ethanol, identified by international institutions as the chief culprit in the explosion of world food prices:

> U.S. corn ethanol explains one third of the rise in the world corn price according to the FAO, and 70% according to the IMF. The World Bank estimates that the U.S. policy is responsible for 65% of the surge in agricultural prices, and for ... the former USDA [U.S. Department of Agriculture] Chief economist,

---

* The industrial food system contributes 22% of greenhouse gas emissions (McMichael et al., 2007), with an additional 8% of emissions from land use (Metz et al., 2007).
† Most industrial biofuel feedstocks require nitrogen fertilizers—and nitrous oxide emissions outstrip carbon dioxide emissions by 300%.

it explains 60% of the price rise. The World Bank
states that: "Prices for those crops used as bio-fuels
have risen more rapidly than other food prices in the
past two years, with grain prices going up by 144%,
oilseeds by 157% and other food prices only up by
11%." The U.S., as a result of its corn ethanol production, is clearly responsible for the explosion of world
agricultural prices. The second largest world corn
exporter, Brazil, produces ethanol from sugarcane
and hence has not influenced world market prices
for corn. In addition to the U.S. corn ethanol program, the U.S. biodiesel program [soybeans] also
contributes to soaring prices. (Berthelot, 2008, p. 27)

Demand from U.S. ethanol distilleries for corn doubled the increase
in global demand for corn from 2006 to 2008 (Holt-Giménez and Kenfield,
2008, p. 3). The Bush administration bonanza, the 2007 Renewable Fuels
Standards legislation, encouraged agroindustrial giants "ADM, Bunge
and Cargill to diversify their monopsonistic (df., situation where the
entire market demand for a product or service consists of only one buyer)
purchases to include corn for fuel as well as corn for food" (Holt-Giménez
and Kenfield, 2008, p. 2). With corn prices rising more than 50% in 2006,
grain traders profited from the captive market created by ethanol targets,
with corn inputs to U.S. ethanol distilleries tripling during 2001–2006
(Holt-Giménez and Kenfield, 2008). About 30% of U.S. corn production is
now converted to ethanol (ActionAid, 2010, p. 13).

The corn market expresses the commercial subordination of agriculture, as corporate strategies come to treat food and fuel as indistinguish-
able. Corn's impact on the food system is extensive, via its significance as
a feed crop for beef, poultry, eggs, dairy, and pork production and as a
component of sweeteners for candy, cereals, soft drinks, and other super-
market staples (Philpott, 2007). Pollan (2002) has observed:

A Chicken McNugget is corn upon corn upon
corn, beginning with corn-fed chicken all the
way through the obscure food additives and the
corn starch that holds it together. All the meat at
McDonald's is really corn. Chickens have become
machines for converting two pounds of corn into
one pound of chicken.

That is, to the extent that corn has become an indispensable and pervasive
ingredient of the industrial food diet, its conversion into fuel has extensive
impact on the price of processed food and meats, and other foods.

Globally, the diversion of U.S. corn to fuel feedstock affects grain markets. U.S. corn accounts for 40% of global production of corn so that:

> as demand for corn increases, more is planted, push-
> ing out other food grains such as wheat and soy-
> beans. With less land available for cultivation, the
> price of these products goes up. Because corn and
> soy are major inputs for processed food and live-
> stock feed, the increase in corn prices dramatically
> increases food prices worldwide. (Holt-Giménez
> and Kenfield, 2008, p. 3)

Working from a conservative impact of biofuels on food prices (30%), calculations show that biofuels are responsible for 30 million more hungry people, with another 260 million at risk of hunger. The FAO has suggested there are an additional 100 million hungry as a result of food price inflation (2008–2009). The Organization for Economic Cooperation and Development claims that industrial biofuels will be responsible for one-third of the rise of food prices over the next decade, and IFPRI estimates an additional 16 million hungry for each 1% food price rise (ActionAid, 2010, pp. 15–16).

Under these circumstances, it is no wonder that in 2007 the UN human rights rapporteur, Jean Ziegler, labeled grain ethanol a "crime against humanity" (Lederer, 2007). The displacement of food by fuel crops to feed the energy-intensive demands of the global consumer class is not simply impoverishing of low-income populations but also impoverishing of the future.

On current projections and alternative energy targets, the International Energy Agency calculates that, along with other renewables, industrial biofuels are expected to supply no more than 9% of the world's global energy consumption (GRAIN, 2007, p. 6). Meanwhile, the stress on land and water, not to mention the atmosphere, of an agroindustrial project to meet mandatory biofuels targets in the G8[*] and G5[†] countries will threaten planetary sustainability even more than it is currently threatened. For example, the *Gallagher Review of the Indirect Effects of Biofuels Production* (Renewable Fuels Agency, 2008) estimated, using a midrange scenario of land use, that by 2020 about 500 million more hectares of land, one-third more than currently under cultivation, would be required to meet global demand for biofuels. The additional land would include forests,

---

[*] The G8 comprises the United Kingdom, Russia, the United States, Italy, France, Germany, Japan, and Canada.
[†] The G5 comprises India, Brazil, China, Mexico, and South Africa.

woodlands, wetlands, and particularly grasslands—all of which currently serve as carbon sinks and centers of biodiversity.

In these ways, the industrial biofuels project becomes the problem rather than the solution to climate change and peak oil, by exacerbating land-use pressures for no actual benefit other than to the (heavily subsidized) investors in biofuels. According to a recent study reported in *Science*, the conversion of rain forests, peatlands, savannas, or grasslands to produce biofuels "creates a 'biofuel carbon debt' by releasing 17 to 420 times more CO_2 than the annual greenhouse gas (GHG) reductions these biofuels provide by displacing fossil fuels" (Fargione et al., 2008, p. 1235). Rainforest Action Network (2007) claims that a ton of palm oil produces 33 tons of CO_2 (10 times more CO_2 than a ton of petroleum produces), and Nobel laureate Paul Crutzen argues that biofuels raise rather than lower emissions (Corbyn, 2007).

11.3   The socioenvironmental impact

Following the green revolution's misleading attempt to "feed the world" via chemicals and biotechnologies, the biofuels project deepens the subordination of agriculture to an energy-industrial complex. The so-called biofuels transition has discarded feeding the world to deepen industrial agriculture's contribution to climate change and ecological degradation. Under these conditions, agriculture becomes fetishized as a profit source, rather than valued as a life source. The consequences are that rising hunger is coupled with rising environmental harm.

Industrial agriculture is a significant source of GHG emissions (ignored by Al Gore [2006] and Jeffrey Sachs [2008]), and with industrial biofuels, it now threatens to convert the global South into a "world energy farm." Some argue that global warming is evidence of an intensified "planetary metabolic rift" (Clark and York, 2005) via the interruption of natural processes of sequestration of carbon—thus, "when the impacts of forestation of land is compared to the impact of growing and using biofuels, the forested lands were capable of sequestering anywhere from two to nine times more carbon over a 30-year period" (Smolker et al., 2008, p. 12). Using biofuels to mitigate global warming has the opposite effect. Deforestation and draining peatlands for biofuels are widely acknowledged to increase emissions and/or remove land from agricultural production, raising food prices and converting pristine habitats into croplands (Monbiot, 2008).

In Brazil, sugarcane expansion for biodiesel intensifies deforestation. By "usurping agricultural lands previously used for other purposes, cane expansion has pushed those other uses, especially cattleraising, into forest frontier areas" (Smolker et al., 2008, p. 20). The new biodiesel—soy—depletes soils and nutrients, generating "climate-damaging emissions of nitrous oxide" (Smolker et al., 2008, p. 20), an outcome only intensified by the food-fuel nexus. During 2007, Amazonian forests were razed by ranchers and settlers to clear land in Brazil, Paraguay, and Bolivia. Thus, "rising prices for both cattle and soy for animal feed appear to be the major factor driving the demand for more land" (Smolker et al., 2008, p. 24), following the U.S. shift from soy into corn production for ethanol. The intensity of deforestation of the Amazon region is now overtaken by the colonization of the Brazilian *cerrado* (a biodiverse woodland/grassland covering 20% of Brazil): "More than half of this biome has already been turned over to cattle grazing and soy production, and it is now being considered as a promising area for sugar cane as well" (Smolker et al., 2008, p. 25).

Southeast Asia concentrates the biofuels rush on palm oil; the largest producers Malaysia and Indonesia supply roughly 85% of the world market (chiefly in Europe and China). Indonesian cultivation of oil palms has risen from 3.6 million hectares in 1961 to 8.1 million ha in 2009 (Rist et al., 2009). Indonesia is now the world's largest producer of palm oil, with 18 million ha of cleared forest for timber and future biofuel expansion (Colchester et al., 2006, p. 11-12).

Oil palm is a key to rural development strategy, exercised mainly through nucleus estate and smallholder schemes, in which farmers allocate a proportion of their land to an oil palm firm's estate plantation, with the remaining land retained by farmers but planted with oil palms by the firm (Rist et al., 2009). The Indonesian Department of Agriculture claims approximately 27 million additional hectares of "unproductive forestlands" (post logging and cultivation) are available for conversion into plantations, and Sawit Watch reports almost 20 million ha have been proposed for biofuel development by local governments (Colchester et al., 2006, p. 25). Palm oil plantations are encouraged by tax breaks, subsidies, and huge investments by the China National Offshore Oil Corporation and by oil and agribusiness firms like Shell, Neste Oil, Greenergy International, BioX, Cargill, and Archer Daniels Midland (ADM). Destruction of peatland forests, concentrated in Indonesia, contributes about 8% of annual GHG, exceeded only by the United States and China (Colchester et al., 2006, p. 9).

In Kalimantan, Indonesia, ethnographic research on oil palm plantations demonstrates the combined ecological and social effect of biofuel expansion, as well as growing food insecurity:

> Forest and land availability have been greatly reduced, making it more difficult for the local communities to obtain NTFP's [non-timber forest products] and leading to a lack of farming lands. As there are not enough farming lands, farming has become more intensive. The same lands are used continuously, so that the soil does not have enough

> time to regain fertility. As there is not enough arable
> land, many people have given up rice farming and
> a linear regression can be seen in the diversity of
> crops cultivated in relation to the proximity of the
> plantation.... Availability of, and access to foods
> such as meat, vegetables and fruits has declined, so
> that more food has to be bought, leading to higher
> food expenses. (Orth, 2007, p. 51)

While there are clearly some forest dwellers and smallholders choosing to participate in the commercial biofuel economy, there are obstacles to their long-term success. The Indonesian constitution recognizes "customary law communities"; however, legally the state has the right to control and allocate natural resources in the name of its citizenry. Rist et al. (2009), for example, note that oil palm development agreements with communities are often concluded without commitment by the companies, with recurring problems of land grabbing and a lack of transparency and appropriate consent procedures, exacerbated by the absence of clear land rights.

Southeast Asia is emerging as an offshore export platform of biofuels for China, which is "already importing feedstocks from other countries, including Nigeria, Malaysia, Indonesia, and the Philippines, and investing in refineries in Indonesia and Malaysia." Japan, an investor in agrofuel supplies in Brazil, "has plans for a jatropha biodiesel plant in South Africa, a coconut biodiesel plant in the Philippines and cassava ethanol plants in Indonesia, Thailand and Vietnam" (Smolker et al., 2008, p. 32).

Meanwhile, Africa and its extensive land reserves (the "Green OPEC") is a frontier for Brazil, Saudi Arabia, China, the World Bank, the U.S. Agency for International Development, the European Commission, and private corporations to develop biofuels primarily for export. Of most concern are plans to open up the Central African Republic and the Democratic Republic of Congo to biofuel development. The Congo Basin rain forest represents 18% of the world's rain forest, contains about 70% of Africa's vegetation with 25–30 billion tons of carbon, and regulates rainfall and weather patterns in West Africa and the world as a whole (Smolker et al., 2008, p. 35).

Overall, energy needs and emission reduction targets set in the North amount to a global biofuels project (McMichael, 2008). Responsible for about 18% of global GHG, the European Union has set emissions targets that depend on offshore supplies of biofuels. As Smolker and colleagues note:

> The EU is reducing its own emissions by raising
> emissions in developing countries that produce the

feedstock oils (through increased deforestation and
land-use change, for example) and [that] are not
bound by emissions reduction targets, especially
Indonesia and countries in Latin America. (Smolker
et al., 2008, p. 39)

This relationship—represented as a new vehicle of development in the
global South—in fact registers an energy crisis as well as ecological crisis
in the North, as "peak soil" in the United States and Europe undermines
productivity increases. By converting 70% of European farmland and the
entire U.S. corn and soy harvest to fuel crops, Northern fuel needs could be
met now (Holt-Giménez, 2007). But with access to Southern resources, this
solution is unlikely, particularly given the example of the Kyoto Protocol's
Clean Development Mechanism, which encourages North-to-South emission offsets to enable continued emissions in the North.

11.4   The biofuel complex

Access to cheaper resources offshore, to address peak oil and peak soil in
the North, capitalizes on Southern dependencies (notably debt reschedul-
ing) as a solution to Northern needs. For Southern governments, carbon-
offset and biofuel projects generate foreign-exchange for debt repayment.
Further, new oil/energy, auto, food, and biotech industrial alliances invest
in Southern land and biofuel infrastructures, including crop development. Biofuels are regarded as critical alternative energy supplies, with
a DuPont official claiming, "the demand for corn (for ethanol) could be
so dramatic that it could change farming practices"—substituting crop
rotation with monoculture, which depletes soil and produces insect and
disease buildup (Padilla, 2007, p. 7) and, of course, provides opportunity
for genetically modified (GM) crops. Silvia Ribeiro from the Action Group
on Erosion, Technology, and Concentration (ETC) reports:

> All the companies which produce transgenic
> crops—Syngenta, Monsanto, Dupont, Dow, Bayer,
> BASF—have investments in crops designed specifi-
> cally for the production of biofuels such as ethanol
> and biodiesel. They also have collaboration agree-
> ments in a similar vein with Cargill, Archer Daniel
> [*sic*] Midland, Bunge, transnational companies that
> dominate the global trade in grains.... All this is
> creating new alliances. For example, Monsanto and
> Dow have just signed an agreement to create GM
> seeds that will combine in the same plant both resis-
> tance to eight herbicides as well as making them

> insecticides. This in part reflects the recognition that
> GM seeds create resistance to herbicides and there-
> fore require more and more. And if the seeds are not
> for human use it will be possible to use more toxic
> herbicides in greater quantities. (Padilla, 2007, p. 6)

Following the agribusiness model, biofuel investors organize their own feedstock supply chains: "Most agrofuel factories are being built with simultaneous investments in crop production. The clear trend is towards the formation of fully integrated transnational agrofuel networks, bringing together everything from seeds to shipping" (GRAIN, 2007, p. 12). Profits from biofuels not absorbed by agribusinesses already producing potential feedstocks (soya, corn, palm oil, and sugar) are financing "a wave of new alliances and business groupings, bringing together financial companies, shippers, traders, and producers. In some cases major investment funds, such as the Carlyle Group, are even setting up their own fully integrated agribusiness/energy networks" (GRAIN, 2007, p. 13). Outside the agribusiness complex, "BP and ConocoPhillips have struck deals with major meat processors for the supply of animal fats to produce biodiesel. BP, along with several other companies, is also developing jatropha as a feedstock" (GRAIN, 2007, p. 13), while Chinese and South Korean firms are busy making deals for large-scale cassava production in Nigeria and Indonesia.

The biofuels complex involves new private–public partnerships, which provide a captive market for biofuel investors, sanctioned by the proliferating mandates by governments to increase biofuel content of transport fuels. Biofuel subsidies in the European Union and the United States amount to $16–18 billion a year, four times all their agricultural aid to the South, which focuses on agro-exports (Seager, 2008). For the palm oil complex, the Indonesian palm oil trade is managed by a combination of Cargill (the world's largest private company), an ADM-Kuck-Wilmar alliance (the world's largest biofuels manufacturer), and Synergy Drive, while the Malaysian government firm is "soon to become the world's biggest palm oil conglomerate" (Greenpeace, 2007).

Given the private–public partnering, it is not surprising that some cautionary responses have appeared, as recent commissions and reports express misgivings about the ecological and social consequences of biofuel outsourcing. On June 30, 2008, the French ecology minister stated that developing a biofuels target was "probably a mistake" and that environmental and social criteria should have been developed prior to setting a target (Phillips, 2008). Concern about market decisions driving politics emerged in the European Parliament a week later, when its members voted for significantly reduced targets, given the biofuel impacts on food prices, people, and biodiversity and their inability to significantly reduce

GHG emissions. At the same time in Britain, the *Gallagher Review* was released, the Executive Summary stating:

> Biofuels have been proposed as a solution to several pressing global concerns: energy security, climate change and rural development. This has led to generous subsidies in order to stimulate supply. In 2003 ... the European Union agreed to the Biofuels Directive....
>
> Five years later, there is growing concern about the role of biofuels in rising food prices, accelerating deforestation and doubts about the climate benefits. This has led to serious questions about their sustainability....
>
> We have concluded that there is a future for a sustainable biofuels industry but that feedstock production must avoid agricultural land that would otherwise be used for food production. This is because the displacement of existing agricultural production, due to biofuel demand, is accelerating land-use change, and, if left unchecked, will reduce biodiversity and may even cause greenhouse gas emissions rather than savings. The introduction of biofuels should be significantly slowed until adequate controls to address displacement effects are implemented and are demonstrated to be effective. (Renewable Fuels Agency, 2008, p. 1)

Misgivings aside, biofuels continue to showcase alternative energy policy, with the *Gallagher Review* arguing that "it should be possible to establish a genuinely sustainable industry provided that robust, comprehensive and mandatory sustainability standards are developed and implemented" (Renewable Fuels Agency, 2008, p. 9). Establishing standards is problematic. The problem is not simply that, other than perhaps sugarcane, biofuels release more GHG emissions than they reduce. Biofuelwatch has concluded that most biofuel industry responses

> reject any mandatory safeguards which would ensure that the biofuels sold in Europe will have lower greenhouse gas emissions than the petrol or diesel which they will replace.... Many responses suggest that not enough is known about life-cycle greenhouse gas emissions from biofuels, but nonetheless demand government support for rapid

> market expansion. (quoted in Gilbertson et al., 2007,
> pp. 15–16)

The scientific community has begun to acknowledge this problem, which has as much to do with the life-cycle effect as with the fiction of a standard metric of carbon in a biodiverse world populated by varying deposits and flows of fossil fuel and biotic carbon (Lohmann, 2006). Additional emissions produce side effects, thus "much of the 'evidence' presented for biofuels to reduce greenhouse gas emissions ignores the larger picture of 'land use change' (usually deforestation), soil erosion and nitrous oxide emissions" (Biofuelwatch et al., 2007, p. 10). Nobel laureate Paul Crutzen observes that biofuels raise rather than lower emissions, and from research with colleagues on nitrous oxide emissions from crop fertilizers, he concludes:

> The replacement of fossil fuels by biofuels may not bring the intended climate cooling due to the accompanying emissions of N_2O.... Depending on N content, the use of several agricultural crops for energy production can readily lead to N_2O emissions large enough to cause climate warming instead of cooling by "saved fossil CO_2." (Crutzen et al., 2007)

According to Gilbertson et al. (2007, p. 39):

> There are currently no peer reviewed life-cycle greenhouse gas studies for biodiesel from palm oil, jatropha or soya, and peer reviewed studies on sugarcane ethanol are limited to those looking at energy gains and fossil fuel displacement, rather than total greenhouse gas balances.

For GHG balances, "current margins of uncertainty, even at the micro-level, are currently too high for meaningful certification based on life-cycle emissions" (Biofuelwatch et al., 2007, p. 31). Storm (2009) claims that lack of verifiability leads to carbon imperialism, converting the global South into a "carbon dump" to sustain Northern lifestyles—a perspective echoed by Martinez-Alier (2009), who notes that Kyoto enabled the North to obtain property rights on carbon sinks in the South and the atmosphere in return for reduced emission targets.

Given varying environments across distinct spaces, calculating and/ or equating GHG emissions from land-use change, soil erosion, and nitrous oxide release is nearly impossible. Biofuelwatch reports that existing calculations of agrofuel emissions from land use, deforestation, and soil organic carbon loss adopt different methodologies with incommensurable results (Gilbertson et al., 2007, p. 30). Storm (2009) notes, for example, that "carbon savings" from offset projects are unmeasurable, because they are based on an unrealized counterfactual. For Lohmann, offsets are a fiction created by "deducting what you hope happens from what you guess would have happened" (Storm, 2009, p. 1020). Similarly, Berkeley scientists claimed that "including incommensurable quantities such as soil erosion and climate change into a single metric requires an arbitrary determination of their relative value" (quoted in Gilbertson et al., 2007). The Intergovernmental Panel on Climate Change (IPCC) considers "CO_2 equivalences" gross oversimplifications: "The effects and lifetimes of different greenhouse gases in different parts of the atmosphere are so complex and multiple that any straightforward equation is impossible" (Lohmann, 2008).

Despite attention to accurate calculation, the controversy over certification methodologies is, arguably, a distraction from the incommensurability of an economic and an ecological calculus. While the former, standardizing calculus depends on a virtual fractionation of carbon units, the latter recognizes the actual interactive complexity of carbon cycles, both natural and unnatural. (Lohmann, 2008).[*] Such interactive complexity informs climate scientists' concern with "positive feedback: the self-acceleration of climate change. The IPCC noted that 'emission reductions ... might be underestimated due to missing carbon cycle feedbacks'" (quoted in Monbiot, 2007).

Even so, the biofuels project continues to present itself as a solution to GHG emissions and the energy crisis. Carbon markets estimates suggest that "the output of bioethanol and biodiesel could rise up to 120 and 24 billion litres respectively in 2020 if instruments such as the CDM [Clean Development Mechanism] support the implementation of biofuel markets in developing countries" (Gilbertson et al., 2007). Anticipating incorporation of biofuels into the CDM, the World Bank brokers carbon markets, recognizing (in a 2005 leaked memo) that with a 5% or more commission (13% by 2009), it could earn $100 million in one year. Through its networks, the World Bank negotiates eligibility for local carbon-capture projects for carbon credit exchanges with Northern governments and corporations. In addition, establishing a Community Development Carbon Fund connects "small-scale projects seeking carbon finance with companies, governments, foundations, and NGOs [nongovernmental organizations] seeking to improve the livelihoods of local communities and obtain verified emission reductions" (Wysham, 2005). In this way, the development industry contributes to the consolidation of the biofuels complex.

---

[*] See Lohmann (2006) for an extended discussion of this.

11.5   Biofuels, development, and global ecology

Biofuels, the new export crop, are encouraged as a way to generate employment through rural diversification. Oxfam concludes in its "Bio-fuelling Poverty" report:

> Biofuels need not spell disaster for poor people
> in the South—they should instead offer new mar-
> ket and livelihood opportunities. But the agro-
> industrial model that is emerging to supply the EU
> target poses little in the way of opportunities and
> much in the way of threats. (Oxfam, 2007)

Accordingly, Oxfam proposes a set of social principles to govern the development of a biofuels industry, where "feedstock cultivation does not adversely impact on local communities or indigenous people," without which the European Union "must accept that the ten percent target will not be reached sustainably, and therefore should not be reached at all" (Oxfam, 2007). The *Gallagher Review* (Renewable Fuels Agency, 2008) complements Oxfam's social vision, cautioning against displacing food crops, but noting that alternative energy crops can simultaneously provide new employment and local development opportunities to rural communities.

While the criterion of sustainability is open to serious question, claims of renewable alternative energy work to legitimize the biofuels project. Nevertheless, there is an important ontological issue, namely, the global projection of a development model whose beneficiaries are a minority of the world's population, most of whom consume energy unsustainably, whether they like it or not. Biofueling poverty, in this sense, is a euphemism for commodifying resources on behalf of this minority, often at the expense of natural habitats and local food security. By converting Southern habitats into supply zones for an unsustainable industry in the name of sustainability, industrial biofuels deepen the project of "global ecology" whereby Southern resources sustain Northern lifestyles (Sachs, 1993).

Southern resources, as export commodities, can become a magnet for financial investments at the expense of developing local capacity. Thus the Plantation Office head in Sanggau District declared:

> We believe that the oil palm estate has a good mul-
> tiplier effect. The financial benefits from oil palm
> estates are by farmers on the estates, through wages
> for employment, as well as through the oppor-
> tunities for the community to conduct business
> around the estate. These can contribute significantly

> to the development of the area. We are aware that
> the development of oil palm plantations can also
> impose high social and financial costs. Nonetheless,
> we still feel we are more fortunate compared with
> other districts [without oil palm]. *Due to the lack of*
> *financial support for [alternative] agricultural activities,*
> *particularly from the commercial banks, it is really hard*
> *to develop the agriculture sector in Sanggau District.*
> Therefore, the most feasible activities that can
> be conducted in Sanggau District are plantation
> activities especially oil palm estates. (Quoted in
> Colchester et al., 2006; emphasis added)

Certainly the biofuel industry claim of extending social development has materialized for some already marginalized rural and forest-dwelling people, even if they have no choice in the matter.[*]

Land laws are crucial to these projections. Despite customary law traditions of communal landownership, the state can abrogate land rights in the name of development, and/or allow land-grabbing by companies. Thus, a recent UN report noted:

> Experience with existing and extensive oil palm
> plantations in other parts of Indonesia conclusively
> demonstrates that Indigenous peoples' property
> and other rights are disregarded, their right to
> consent is not respected, some are displaced, and
> they left with no alternative but to become *de facto*
> bonded laborers gathering oil palm fruit for the
> companies that manage the plantations. (Quoted in
> Smolker et al., 2008, p. 30)

As a general rule, *adat* communities diversify agricultural production with subsistence crops, rice, coffee, fruit trees, and damar trees, which yield a valuable resin. Indonesian villagers have reported to Sawit Watch that companies offering benefits to participate in plantations encourage (or compel) land transfers, undermining customary agricultures and degrading the environment as "changes in the vegetation cover have caused changes in species' distributions and have led to uncontrolled pests booming"—even though some smallholder oil palm producers improve their income and gain access to markets

---

[*] Rist et al. (2009) notes, for example, that oil palm production contributes more than 63% of smallholder household incomes in two locations in Sumatra and that there is evidence of oil palm alleviating poverty.

via new road systems (Smolker et al., 2008, p. 99). Nevertheless, according to Sawit Watch, social costs tend to include alcohol abuse and the breakdown of communal traditions via profit seeking, leading to "everything being measured only in economic terms" (Smolker et al., 2008, p. 100).

In 2007, Malaysia signed the UN Declaration on the Rights of Indigenous Peoples, which requires states to "consult and cooperate in good faith with the indigenous peoples concerned through their own representative institutions in order to obtain their free and informed consent prior to the approval of any project affecting their lands or territories." However, there is evidence of patterns of disregard for community decision making, the integrity of communal lands, and compensating loss or damage. Such patterns include surprise occupation of territories, ignoring the fact that, as one resident observes, "Our livelihood is greatly dependent on the resources in our surroundings" (Colchester et al., 2007, p. 47). As reported in *Land Is Life:*

> The people first found out about the oil palm scheme when workers started work on their lands, clearing lands which included rubber trees and fruit trees belonging to the indigenous communities. As the oil-palm land clearing work continued, the rivers that supplied water to the people and the fish stock were affected. In addition to the crops, and polluted rivers, the people's burial ground and farm lands were also destroyed. People were then unable to hunt for the game which is an important element in their diet. There was no more rattan to harvest either, the raw material for handicrafts which had provided extra cash income to the communities. Jungle food sources, like vegetables, were also destroyed. (Colchester et al., 2007, p. 54)

Malaysia has a plural legal regime, just as Indonesia does, including constitutional support for custom. The Sarawak government limits the exercise of native customary rights, however, refusing to reveal the location of lands actually subject to these rights and retaining the right to implement its policy of developing natural resources. By this policy, native landowners must surrender their lands to the state for 60 years for development as joint ventures with private companies, despite the absence of clear principles regarding compensation to native landowners and reclamation of their lands when the leases expire (Colchester et al., 2007, pp. 1–2). Colchester and his colleagues report that:

> As a direct result of its restricted interpretation of the extent of customary rights, companies are being given leases for oil palm development over supposedly "vacant" and "idle" State lands, which are, in fact, quite obviously inhabited, encumbered with customary rights and being actively used by local communities in their daily lives. The result is that most palm oil projects are contested by local communities. (Colchester et al., 2007, pp. 77–78)

Cultural systems of social reproduction not based on private property have historically been vulnerable to invisiblization. Continuing this trend of expropriation, states and corporations identify "idle" land for commercial biofuel expansion, but

> growing evidence raises doubts about the concept of *idle* land. In many cases, lands perceived to be *idle, under-utilised, marginal* or *abandoned* by government and large private operators provide a vital basis for the livelihoods of poorer and vulnerable groups, including through crop farming, herding and gathering of wild products. (Cotula et al., 2008, pp. 22–23; emphasis in original)

The FAO in 2008 underlined the marginalizing effect of biofuels on women in rural areas, noting that marginal lands provide important subsistence functions to rural peoples and are farmed by women who are denied access to property (Gaia Foundation et al., 2008). In India, jatropha plantations target "waste lands" that sustain millions of people as "commons" and pasturelands. In addition to pastoralists,

> refugees from development projects, displaced persons, jobless labourers and small farmers facing crop failure often rely on these lands as places where they can put their cattle during an emergency. If these lands are enclosed, the lifelines of many already disadvantaged people will be jeopardized. (GRAIN, 2008, p. 20)

Ironically, such displacement processes displace ways of life potentially vital to planetary sustainability.

11.6   Conclusions

The land grab under way across the global South encloses territory for both food and biofuel production via agroindustrial methods. This kind of path dependency, in the name of food and energy security and renewable energy, deepens the threat to planetary sustainability. Not only does agriculture (and associated deforestation) account for almost a third of GHG, but agroindustrialization also proceeds at the expense of peasant agriculture, at a time when industrialization has been decoupled from urbanization, generating a planet of slums and global circuits of migrant labor (Davis, 2006). Human redundancy combines with the elimination of biodiversity and vital food provisioning cultures, upon which the future of humanity depends.

As the food sovereignty movement rhetoric goes, small farming diversity feeds the world and cools the planet, significantly. Pretty et al. (2006) have estimated that industrial agriculture uses 6–10 times more energy than agroecological methods (Goldsmith/ISIS, 2004), while agroecology restores soils, reducing emissions up to 15% and restoring grasslands, and wetlands reduce emissions up to another 20% (Apfelbaum and Kimble, 2007). Altieri and Nicholls (2008, p. 474) claim that small farms "cool the climate," using organic fertilizer that absorbs and sequesters carbon more effectively than industrial agriculture. They summarize the advantage of small diversified farming thus:

> In polycultures developed by smallholders, productivity, in terms of harvestable products, per unit area is higher than under sole cropping with the same level of management. Yield advantages range from 20 to 60%, because polycultures reduce losses due to weeds, insects and diseases, and make more efficient use of the available resources of water, light and nutrients. In overall output, the diversified farm produces much more food. (Altieri and Nicholls, 2008, p. 474)

Associated with the smallholder vision (for roughly two-fifths of the world's inhabitants) is not only food sovereignty to grow food for local and regional populations while stewarding the land but also the possibility of energy sovereignty. Brazil has shown us one such possibility (over and above the energy self-reliance built into stable biodiverse practices of recycling waste). Brazil's biofuel project subdivides into two sectors: The first is a relatively unregulated agroindustrial ethanol program, centered in the São Paulo region, which has seen sugarcane expansion at the expense of dairy farming, orange groves, and other staple crops

(Wilkinson and Herrera, 2008, p. 24). Second, there is the government-regulated biodiesel program, decentralized and designed to promote regional development and social inclusion. The National Petroleum Agency organizes auctions through which firms, on acquiring a Social Seal provided by the Ministry for Agrarian Development, gain access to the biodiesel market. These firms must "demonstrate that a given percentage of their raw material or crude oil has been contracted with family farms in agreement with the rural trade unions" (Wilkinson and Herrera, 2008, p. 8).

In a detailed report examining the operation of this biodiesel program by region, Wilkinson and Herrera (2008) observe that its economic viability is threatened by problems of land access and competition from agroindustry, especially in the north and northeast. The south, however, with its cooperative traditions, saw a recent surge of locally focused cooperatives and associations experimenting with ethanol from sugarcane (and manioc and sorghum) and biodiesel from jatropha and tung (tree crops) via a variety of intercropping systems, both with tree crops and short-cycle food crops, and joint processing and farming activities. Each producer is allowed to plant only two hectares of biofuel, in order to ensure adequate food supply. Wilkinson and Herrera (2008) view this program as a "radical challenge to the dominant agribusiness model," noting that together systems of food and energy production

> are seen as strategies for increasing the autonomy of the less favoured family farm sector, an important feature of which includes the production of ethanol for local consumption. These projects are still at an early stage of development and so definitive conclusions cannot be drawn as to their feasibility. (p. 57)

The serious conjuncture of food, energy, and climate crises demands a new direction in how to harness agriculture to the joint tasks of energy saving, emissions reduction, and food security. There is mounting evidence of the adaptability of smallholder communities to ecological changes under conditions of global warming, provided they are sufficiently valued, adequately heard, and generously supported. Unfortunately, the biofuels project is premised on the enclosure of smallholder lands and addressing the needs of the global consumer-class minority for fuel and all-season foodstuffs from around the world.

Food security through the global market has been an illusion, underlined during the food crisis of 2007–2008. Nevertheless, partly in consequence, corporate alliances are forming that profit from flexible portfolios in food and fuel crops. Under these conditions, it is unsurprising to find declarations from Indonesian smallholders like, "It's as if we were ghosts on our own land" (Forest Peoples Programme and Sawit Watch, 2006), and Amazonian indigenous peoples referring to biofuel plantations as the "devil's orchards." While they speak for themselves, they could well speak for humanity.

References

ActionAid. 2010. Meals per gallon: The impact of industrial biofuels on people and global hunger. London: ActionAid. http://www.actionaid.org.uk/doc_lib/meals_per_gallon_final.pdf.

Altieri, M., and Nicholls, C. 2008. Scaling up agroecological approaches for food sovereignty in Latin America. *Development*, 51(4): 472–480.

Apfelbaum, S. I., and Kimble, J. 2007. A dirty, more natural way to fight climate change. *Ithaca Journal*, December 6.

Berthelot, J. 2008. The food crisis explosion: Root causes and how to regulate them. *Kurswechsel*, 3:23–31.

Biofuelwatch et al. 2007. Biofuels: Towards a reality check in nine key areas. June. http://archive.corporateeurope.org/docs/AgrofuelsRealityCheck.pdf.

Chakraborrty, A. 2008. Exclusive: We publish the biofuels report they didn't want you to read. *Guardian*, July 10. http://www.guardian.co.uk/environment/blog/2008/jul/10/exclusivethebiofuelsreport.

Clark, B., and York, R. 2005. Carbon metabolism: Global capitalism, climate change, and the biospheric rift. *Theory & Society*, 34:391–428.

Colchester, M., Jiwan N., Andiko, M., Sirait, M., Firdaus, A.Y., Surambo, A., and Pane, H. 2006. *Promised land: Palm oil and land acquistion in Indonesia; implications for local communities and indigenous peoples*. Moreton-in-Marsh, England: Forest Peoples Programme; Bogor, Indonesia: Perkumpulan Sawit Watch. http://www.forestpeoples.org/documents/prv_sector/oil_palm/promised_land_eng.pdf

Colchester, M., Pang, W. A., Chuo, W. M., and Jalong, T. 2007. *Land is life: Land rights and oil palm development in Sarawak*. Bogor, Indonesia: Perkumpulan Sawit Watch; Moreton-in-Marsh, England: Forest Peoples Programme. http://www.forestpeoples.org/documents/asia_pacific/sarawak_land_is_life_nov07_eng.pdf.

Corbyn, Z. 2007. Biofuels could boost global warming, finds study. *Chemistry World*, September 21. http://www.rsc.org/chemistryworld/News/2007/September/21090701.asp.

Cotula, L., Dyer N., and Vermeulen, S. 2008. *Fuelling exclusion? The biofuels boom and poor people's access to land*. London: International Institute for Environment and Development (IIED); Rome: FAO. http://www.iied.org/pubs/pdfs/12551IIED.pdf.

Crutzen, P. J., Mosier, A. R., Smith, K. A., and Winiwarter, W. 2007. N_2O release from agro-biofuel production negates global warming reduction by replacing fossil fuels. *Atmospheric Chemical Physics Discussion*, 7:11,191–11,205.

Davis, M. 2006. *Planet of slums*. London: Verso.

Fargione, J., Hill, J., Tilman, D., Polasky, S., and Hawthorne, P. 2008. Land clearing and the biofuel carbon debt. *Science*, 319:1235–1238.

Forest Peoples Programme and Sawit Watch. 2006. *Ghosts on our own land: Indonesian oil palm smallholders and the Roundtable on Sustainable Palm Oil.* Moreton-in-Marsh, England: Forest Peoples Programme; Bogor, Indonesia: Perkumpulan Sawit Watch. http://www.forestpeoples.org/documents/ prv_sector/oil_palm/ghosts_on_our_own_land_txt_06_eng.pdf.

Gaia Foundation, Biofuelwatch, the African Biodiversity Network, Salva La Selva, Watch Indonesia, and EcoNexus. 2008. Agrofuels and the myth of the marginal land: A briefing. http://www.africanbiodiversity.org/ media/1221812708.pdf.

Gilbertson, T., Holland, N., Semino, S. and Smith, K. 2007. *Paving the way for biofuels: EU policy, sustainability criteria and climate calculations.* Amsterdam: Transnational Institute. http://archive.corporateeurope.org/docs/agrofuelpush.pdf.

Goldsmith, E. 2007. Feeding the world under climate change. Institute of Science in Society, June. www.i-sis.org.uk/FTWUCC.php?printing=yes.

Gore, A. 2006. *An inconvenient truth.* New York: Rodale.

GRAIN. 2007. Agrofuels special issue. *Seedling,* July. http://www.grain.org/seedling_ files/seed-07-07-en.pdf.

———. 2008. Agrofuels in India, private unlimited. *Seedling,* April. http://www .grain.article/categories/22-seedling-april-2008.

Greenfield, H. 2007. Rising commodity prices and food production: The impact on food & beverage workers. International Union of Food, Agricultural, Hotel, Restaurant, Catering, Tobacco and Allied Workers' Associations (IUF), December.

Greenpeace. 2007. How the palm oil industry is cooking the climate. Amsterdam: Greenpeace International. http://www.greenpeace.org/international/ Global/international/planet-2/report/2007/11/cooking-the-climate-full.pdf.

Holt-Giménez, E. 2007. Biofuels: Myths of the agro-fuels transition. [Video.] *Food First Backgrounder,* 13(2). http://www.foodfirst.org/en/node/2638.

Holt-Giménez, E., and Kenfield, I. 2008. When "renewable isn't sustainable": Biofuels and the inconvenient truths behind the 2007 U.S. Energy Independence and Security Act. Policy Brief No. 13. Oakland: Institute for Food and Development Policy. http://www.foodfirst.org/en/node/2063.

Lederer, E. M. 2007. UN expert calls biofuels "crime against humanity." *Live Science,* October 27. http://www.livescience.com/4692-expert-calls-biofuel-crime-humanity.html.

Lohmann, L. 2006. *Carbon trading: A critical conversation on climate change, privatisation and power.* Development Dialogue No. 48. Uppsala, Sweden: Dag Hammarskjold Centre. http://www.scribd.com/doc/8172014/Carbon-Trading-A-Critical-Conversation-on-Climate-Change-Privatization-and-Power-.

———. 2008. Carbon trading, climate justice and the production of ignorance: Ten examples. *Development,* 51:359–365.

Martinez-Alier, J. 2009. Socially sustainable economic de-growth. *Development and Change,* 40(6): 1099–1119.

McMichael, A. J., Powles, J. W., Butler, C. D., and Uauy, R. 2007. Food, livestock production, energy, climate change, and health. *Lancet,* 370(9594): 1253–1263.

McMichael, P. 2008. Agro-fuels, food security, and the metabolic rift. *Kurswechsel,* 3:14–22.

Metz, B., Davidson, O. R., Bosch, P. R., Dave, R., and Meyer, L. A., eds. 2007. *Climate change, 2007: Mitigation of climate change: Contribution of Working*

Group III to the fourth assessment report of the Intergovernmental Panel on Climate Change*. Geneva: Intergovernmental Panel on Climate Change; Cambridge: Cambridge University Press. http://www.ipcc.ch/ipccreports/ar4-wg3.htm.

Monbiot, G. 2007. This crisis demands a reappraisal of who we are and what progress means. *Guardian*, December 4. http://www.guardian.co.uk/commentisfree/2007/dec/04/comment.politics.

———. 2008. These objects of contempt are now our best chances of feeding the world. *Guardian*, June 10. http://www.guardian.co.uk/commentisfree/2008/jun/10/food.globaleconomy.

Orth, M. 2007. *Subsistence foods to export goods: The impact of an oil palm plantation on local food sovereignty, North Barito, Central Kalimantan, Indonesia*. Wageningen, The Netherlands: Van Hall Larenstein. http://www.biofuelwatch.org.uk/docs/foodsovereigntyindonesia.pdf.

Oxfam. 2007. Bio-fuelling poverty: Why the ET renewable-fuel target may be disastrous for poor people. *Oxfam Briefing Note*, October 29. http://www.oxfam.org/en/node/217.

Padilla, A. 2007. *Biofuels: A new wave of imperialist plunder of Third World resources*. People's Coalition on Food Sovereignty. http://www.foodsov.org/resources/resources_000006.pdf.

Phillips, L. 2008. EU biofuels target "probably a mistake," France says. *EU Observer*, June 30. http://euobserver.com/9/26419.

Philpott, T. 2007. Bad wrap: How Archer Daniels Midland cashes in on Mexico's tortilla woes. *Grist*, February 22. http://www.grist.org/article/tortillas/.

Pollan, M. 2002. The life of a steer. *The New York Times*, March 31. http://www.nehbc.org/pollan1.html.

Pretty, J., Noble, A. D., Bossio, D. Dixon, J., Hine, R. E., Penning de Vries, F. W. T., and Morison, J. I. 2006. Resource conserving agriculture increases yields in developing countries. *Environmental Science & Technology*, 40(4): 1114–1119.

Rainforest Action Network. 2007. Getting real about biofuels. http://ran.org/content/fact-sheet-getting-real-about-biofuels.

Renewable Fuels Agency. 2008. *The Gallagher review of the indirect effects of biofuels production*. St Leonards-on-Sea, East Sussex, England: Renewable Fuels Agency. http://www.globalbioenergy.org/uploads/media/0807_RFA_-_Report_of_the_Gallagher_review.pdf.

Rist, L., Feintrenie, L., and Levang, P. 2009. The livelihood impacts of oil palm: Smallholders in Indonesia. *Biodiversity and Conservation*, 19(4): 1009–1024.

Sachs, J. 2008. *Commonwealth*. New York: Penguin.

Sachs, W. 1993. *Global ecology*. London: Zed.

Seager, A. 2008. Poverty: 260m driven into hunger by push for biofuel—ActionAid. *Guardian*, July 3. http://www.guardian.co.uk/environment/2008/jul/03/biofuels.food.

Smolker, R., Tokar, B., Petermann, A. and Hernandez, E. 2008. *The real cost of agrofuels: Impacts on food, forests, peoples and the climate*. Global Forest Coalition and Global Justice Ecology Project.

Storm, S. 2009. Capitalism and climate change: Can the invisible hand adjust the natural thermostat? *Development and Change*, 40(6): 1011–1038.

Weis, T. 2007. *The global food economy: The battle for the future of farming*. London: Zed.

Wilkinson, J., and Herrera, S. 2008. *Agrofuels in Brazil: What is the outlook for its farming sector?* Rio de Janeiro: Oxfam International. http://www.globalbio energy.org/uploads/media/0811_WilkinsonHerrera_-_Agrofuels_in_ Brazil.pdf.

Wysham, D. 2005. Carbon rush at the World Bank. *Durbin Group for Climate Justice,* February 28. http://www.durbanclimatejustice.org/articles/a-carbon-rush-at-the-world-bank.html.

chapter twelve

The potential of algae and jatropha as biofuel sources

Robert Rapier

Contents

12.1 Introduction

The world presently consumes approximately 4.25 billion metric tons of petroleum-derived products per year (EIA, 2010, Table 3D).* Distillates, the class of fuels that includes diesel, jet fuel, and fuel oil, account for almost 30% of worldwide petroleum consumption (EIA, 2004). As global petroleum supplies deplete, economical and sustainable distillate substitutes are needed.

Fuels derived from the glycerides contained in vegetable oils or animal fats are capable of substituting for petroleum-derived distillates. Renewable distillate fuels have historically been produced with edible vegetable oils such as soybean oil, cottonseed oil, or rapeseed (canola) oil. However, the potential for growing jatropha and algae on marginal land has led to increasing interest in using their respective plant oils as biofuel feedstock.

12.2 Sustainability

As unsustainable sources of energy like fossil fuels are depleted, the world must move to sustainable sources. Sustainable energy sources should meet certain criteria. The first is that the energy yielded from the process must exceed the energy inputs required to grow, harvest, and process the biomass into usable energy. If that criterion is met, then the excess energy yielded could be classified as renewable, in most cases based on the captured solar energy. For example, if the production of 10 units of renewable energy requires 8 units of fossil energy for its production, then the source is only 20% renewable on the basis of the energy balance.

However, certain other criteria must be met as well. Even if a system efficiently captures solar energy and converts it into usable energy with few external energy inputs, the system must not deplete other resources in an unsustainable fashion. For example, a system that yields 2 units of energy for every 1 unit of external input cannot be classified as sustainable if soil degradation is taking place as a result of loss of certain nutrients. It may be that the system can *become* sustainable, however, by modifying the process to address any unsustainable aspects (e.g., if soil depletion is the issue in question, leaving more biomass in the field when harvesting may be necessary).

Some have expanded the concept of sustainability far beyond simply the environmental implications. In 2006, the Copernicus Institute in the Netherlands published an assessment of sugarcane ethanol in Brazil (Smeets et al., 2006). In their study, the authors not only covered sustainability basics such as waste management and soil and groundwater

---

* This is equivalent to 85 million barrels of petroleum-derived products per day.

preservation but also included the welfare of the labor force and the local population, a more comprehensive approach to quantifying sustainability. Fuels that are considered sustainable replacements for petroleum should at a minimum be evaluated for the basic environmental implications and ideally would meet the criteria set forth in the Dutch assessment of sugarcane ethanol.

12.3   Renewable diesel

Renewable distillates may be produced from vegetable oils via two primary routes.[*] In the first, the oils are converted to biodiesel. Biodiesel is a distillate substitute, but distinct from a hydrocarbon distillate. The second route for renewable distillate production is by hydrotreating glyceride feedstocks. The hydrotreating process removes the oxygen from the glyceride molecules. The products of this reaction when glycerides are the feed source are a hydrocarbon diesel and a propane by-product (Hodge, 2006). Synthetic hydrocarbon diesel produced from biomass in this way is often referred to as "green diesel."

12.3.1   Biodiesel

Biodiesel is defined as the mono-alkyl ester product derived from the glycerides (long-chain fatty acids contained in lipids),[†] primarily triglycerides, in vegetable oils or animal fats (Knothe 2001).[‡] The chemical structure of biodiesel is distinctly different from petroleum diesel, and it has different physical and chemical properties.

Biodiesel is normally produced by reacting glycerides with an alcohol in a base-catalyzed reaction (Sheehan et al., 1998). Methanol[§] is usually the preferred alcohol, due to its lower cost and faster reaction relative to ethanol and higher alcohols. The primary reaction products are the alkyl ester (e.g., methyl ester if methanol is used) and glycerol.[¶] A key advantage of processing straight vegetable oil into biodiesel is that the viscosity is

---

[*] Renewable distillates could also be produced via gasification and the Fischer-Tropsch reaction, but this would be less efficient usage of triglycerides than the two routes discussed.

[†] Lipids are oils obtained from recently living biomass. Examples are soybean oil, jatropha oil, algal oil, and animal fats.

[‡] The technical definition for biodiesel under the standard ASTM D 6751 is a fuel comprised of mono-alkyl esters of long-chain fatty acids derived from vegetable oils or animal fats, designated B100, and meeting the requirements of ASTM D 6751.

[§] Methanol is usually produced from natural gas, although some is commercially produced from light petroleum products or from coal. Methanol therefore represents a frequently overlooked fossil fuel input to the biodiesel process.

[¶] Glycerol is also commonly referred to as *glycerin* or *glycerine*.

greatly reduced relative to unprocessed vegetable oil, albeit at the cost of additional processing and a lower-value glycerol by-product.

Biodiesel is reported to be nontoxic and biodegradable (Sheehan et al., 1998). An Environmental Protection Agency (EPA) study published in 2002 showed a mostly favorable impact of biodiesel on exhaust emissions (EPA, 2002). Relative to petroleum diesel, pure biodiesel was estimated to increase the emission of nitrogen oxides (NO_x) by 10%, but to reduce hydrocarbon emissions by almost 70%.

12.3.2   Green diesel

Hydrotreating is the process of reacting a feedstock with hydrogen under elevated temperature and pressure in order to remove oxygen from the molecules. The technology has long been used in the petroleum industry to fracture very large organic molecules into smaller organic molecules, ranging from those suitable for liquid petroleum gas (LPG) applications through those suitable for use as distillate fuels.

Hydrotreating technology has recently been used to convert lipid feedstocks into distillate fuels that are sometimes referred to as "green diesel" or "green jet fuel." The resulting products are propane and a hydrocarbon distillate fuel with properties very similar to petroleum diesel (Hodge, 2006). Renewable distillates produced in this way contain no aromatics, oxygen, or sulfur. Like biodiesel production, which normally utilizes fossil-fuel-derived methanol, hydrotreating generally relies on fossil fuel inputs due to the hydrogen required for the process.[*]

There are five primary advantages of hydrotreating relative to first-generation biodiesel technology:

1. The cold weather properties are superior.
2. The propane by-product is preferable over glycerol by-product.
3. The energy density is greater.
4. The cetane number is higher.
5. Capital costs and operating costs are lower (Arena et al., 2006).

A number of companies are developing renewable distillate projects based on hydrotreating technology. Neste Oil in Finland began developing its NExBTL technology in 2002, and in May 2007 it inaugurated a plant with a capacity of 170,000 metric tons per year of renewable diesel fuel from a feedstock of vegetable oil and animal fat (Neste Oil, 2007). In 2009, Neste inaugurated a second plant, and it has two more under construction (Thalén, 2009). Italy's Eni has announced plans for a facility in Livorno, Italy, that will hydrotreat vegetable oil for supplying European markets.

[*] Hydrogen is commercially produced almost exclusively from natural gas.

Brazil's Petrobras also produces renewable diesel, using its own patented hydrotreating technology (NREL, 2006).

12.4 Jatropha curcas *as a source of biofuel*

12.4.1 *Jatropha description*

Production of biodiesel or green diesel requires a glyceride feedstock. *Jatropha curcas*, also known as the physic nut, produces glycerides that can be converted into fuel. Jatropha is a hardy deciduous shrub indigenous to Central America. Portuguese explorers in the 15th and 16th centuries helped spread jatropha throughout the tropical and subtropical regions of Africa and Asia (Augustus et al., 2002). Jatropha is tolerant of arid conditions (Sirisomboon et al., 2007) and may be used to reclaim marginal, desert, or degraded land (Wood, 2005).

Jatropha has several apparent advantages as a renewable diesel feedstock. Because it is nonedible and can be grown on marginal lands, it is potentially a sustainable biofuel that will not compete with food crops. This is not the case with biofuels derived from soybeans, rapeseed, or palm.

Jatropha has historically been grown for two purposes: as a hedge and for its oil content. It contains toxic compounds such as lectins, phorbol esters, and curcin. As a result, animals do not browse on jatropha, and the thick foliage and drought tolerance have made it an ideal "living fence" for protection of crops and livestock. On the other hand, the toxicity issue led the Australian government to ban the plant in 2006. It was declared an invasive species and too risky for Western Australian agriculture and the environment (*Biofuel Review*, 2006).

Commercially, jatropha has long been grown for its oil. Cape Verde in Africa exported jatropha nuts for production of lamp oil and soaps beginning in the mid-1800s for the European market. Commercial production declined with the arrival of cheap fossil oil, but renewed interest in jatropha as an alternative to petroleum has resulted in a number of new commercial ventures around the world.

12.4.2 *Cultivation*

Jatropha may be established by direct seeding or via direct planting of cuttings (Henning, 2000; Openshaw, 2000). Cuttings are the preferred method if erosion control is the primary objective, as the roots develop primarily near the surface (Heller, 1996). With direct seeding, a taproot develops that helps to anchor the plant and reach nutrients in the subsoil. Thus, for commercial cultivation as a source of biofuel, establishing plants from seeds would be preferred.

Jatropha can grow in a wide variety of soils, but well-drained soils are preferred for better root development (Heller, 1996). Jatropha is drought tolerant, but during extended drought conditions, it responds by shedding its vegetation. Although it can grow in marginal soils, jatropha has a high demand for nitrogen and phosphorus to thrive and accumulate significant biomass. On the other hand, if the moisture and nutrient levels are *too* high, foliage is accumulated at the expense of fruits (Foidl et al., 1996). Optimal levels of water and nutrients for high oil yields have yet to be determined.

Once planted outdoors, survival rates in tropical climates have been reported to be high (Zahawi, 2005). Jatropha has been grown at altitudes from sea level to 1,800 meters (Foidl et al., 1996). However, jatropha is very sensitive to frost, which causes immediate damage to the plants (Gour, 2006).

Jatropha can yield fruit in the first year (Lutz, 1992). Seed yields have been reported over a very large range—from 0.5 tons per hectare under arid conditions to 12 tons per hectare under optimum conditions (Francis et al., 2005). Likewise, the oil concentration in the seeds has been reported to range from 30% to 50% (Pramanik, 2003). The upper end of both seed and oil yields appears to be the basis of extrapolations of yields from single trees to hectare yields per year (Achten et al., 2008).

Achten et al. (2008) surveyed the yields from jatropha crops in Nicaragua and Paraguay and found positive correlations to the amount of rainfall the plants received, as well as the age of the plant. For five-year-old plants receiving 1,200 mm of rainfall per year, they reported a dry seed yield of 5 t/ha. However, most of the reported yields were in the 1–3 t/ha range. While jatropha is drought tolerant, in order to thrive it must receive adequate rainfall. Contrary to its popular image of a plant that can produce fuel with minimal water inputs, one study estimated that it requires 20,000 liters of water on average to produce one liter of biodiesel from jatropha (University of Twente, 2009).

12.4.3   Harvesting and processing

Like coffee beans, the fruits of jatropha ripen at different times. This has made mechanization of jatropha harvesting impossible to date, so harvesting takes place primarily by handpicking the fruits from the tree or from the ground. A worker harvesting in this manner can harvest approximately 5 kg of fruit per hour (Riedacker and Roy, 1998).

There are two major by-products from jatropha oil production. The husk comprises approximately 40% of the weight of the fruit. After processing to remove the kernels, the husk remains. The husks have an energy content comparable to wood and could be used for combustion or gasification (Achten et al., 2008).

After removal of the husk, jatropha oil can be extracted from the kernels via mechanical or chemical means. The products of the extraction process are the jatropha oil and a seed cake—the second major by-product. The seed cake contains the toxic compounds of the jatropha fruits and is therefore inedible. However, it also contains many of the nutrients and could be recycled back to the soil as an organic fertilizer (Francis et al., 2005).

12.4.4 Jatropha oil

The primary glycerides contained in jatropha oil are oleic acid (36–64%), linoleic acid (18–45%), palmitic acid (10–17%), stearic acid (5–10%), and myristic acid (1.4%) (Münch and Kiefer, 1989). These glycerides contain carbon chains of 15–18 carbon atoms. Lighter distillates like kerosene are composed of hydrocarbons containing 10–18 carbon atoms, while heavier distillates like diesel and gas oil contain 12–20 carbon atoms. Thus, processing jatropha's glycerides into biofuel results in carbon chain lengths that are suitable for kerosene and diesel replacements.

The average oil content of jatropha seeds has been reported to be 34.4%, and even the unmodified oil has been shown to perform adequately as a 50/50 blend with petroleum diesel (Achten et al., 2008; Pramanik, 2003). If biodiesel is produced from the glycerides, then the product will be the corresponding fatty acid (e.g., methyl ester if methanol is used as the alcohol). In the case of palmitic acid triglycerides, the resulting biodiesel would be palmitic acid methyl ester and glycerol. If a hydrotreating process is used, then the primary product would be paraffinic[*] hydrocarbons in the C_{16}–C_{17} range and a propane by-product (Hodge, 2006).

12.4.5 Jatropha biofuel feasibility calculations

12.4.5.1 Economics of harvesting jatropha

Consider a jatropha plantation with a typical annual yield of 2 dry tons of seeds per hectare. The average oil content of dry seeds, as determined by Achten et al. (2008) is 34.4%. The highest obtainable oil extraction efficiency with an engine-driven screw press is approximately 80% (Tewari, 2007), and therefore the maximum amount of oil extracted per hectare in this example would be 544 kg. This is the energy equivalent of 476 kg of crude oil, or 3.3 barrels of crude oil equivalent per hectare.

A worker harvesting jatropha by hand can harvest approximately 5 kg of fruit per hour (Riedacker and Roy, 1998), or harvest and shell up

---

[*] Paraffinic hydrocarbons are defined by the general formula C_nH_{2n+2}.

to 15 kg of fruit per day (EcoWorld, 2008)*. With an oil content of 34.4% and an extraction efficiency of 80%, a worker can harvest and shell fruits containing approximately 5.2 kg of jatropha oil per day. At an approximate weight of 146 kg of jatropha oil per barrel, it would take a single worker 28 days to harvest and shell enough fruits by hand to equal one barrel of oil.

If jatropha oil is valued at roughly the present price of petroleum— say, $75 per barrel—then a worker harvesting 5.2 kg of oil per day will have collected only $2.68 of jatropha oil. Additional costs will be incurred in the growing phase, the oil extraction phase, and the conversion of the oil into a distillate substitute. Therefore, jatropha harvesting must be mechanized, or labor costs will have to be much lower than $2.68 per day for hand harvesting to be commercially viable.

12.4.5.2   Land availability

Land availability will also be a factor if jatropha is to displace a mean- ingful amount of global petroleum consumption. The world presently consumes approximately 4.25 billion metric tons of petroleum-derived products per year. To replace 10% of that demand with jatropha oil, a land area of around 781 million hectares would be required, given a dry seed yield of 2 tons per hectare.

The total amount of arable land in the world is estimated to be 1.5 billion hectares (CIA, 2010). Of that, permanent crops presently occupy 155 million hectares. Thus, replacing 10% of the world's oil consumption with jatropha oil would require a land area approximately five times greater than that presently occupied by permanent crops—a land area equal to half of all the arable land in the world.

An estimated 2 billion acres of land is considered to be degraded and perhaps suitable for jatropha cultivation (Oldeman et al., 1991). However, on degraded land the yield of dry seeds would be expected to be signifi- cantly less than the 2 t/ha cited above.

12.4.6   Conclusions on the near-term viability of jatropha-based biofuel

A notable commercial venture in the Western world was announced in June 2007 between BP and D1 Oils to develop jatropha biodiesel (BP, 2007). The companies announced that they would invest $160 million with the stated intent of becoming the largest jatropha biodiesel producer in the world. The target was production volumes of up to 2 million tons of biodie- sel per year. However, BP abandoned the venture when it determined that

---

* The commercial grower in Madagascar who reported the harvesting rate concluded, "We initially approached the production of *Jatropha curcas* with great enthusiasm but now have grave doubts as to the commercial viability of the exercise."

a commercial jatropha operation required much more water and land than expected (Fahey, 2010).

India has already invested billions of dollars into establishing a jatropha industry. However, recent reports indicate that commercial efforts there have stalled, placing "nationwide investments in jatropha of more than $5 billion" at risk (UPI, 2010). India intended to grow jatropha on dry wasteland, but one of the cited reasons for the difficulties was the discovery that jatropha needs moderate amounts of water in order to thrive.

Given present jatropha yields, as well as the fact that harvesting has not been mechanized, commercial production of jatropha does not appear to be economically viable at this time. With the relatively slow speed of handpicking the fruits, mechanization will be required to produce scalable quantities of fuel from jatropha. Jatropha does have the potential to supply small quantities of fuel for local consumption, though. If farmers use jatropha as fencing, then the plant oil could be a by-product that could be used for local energy needs.

While jatropha has intriguing potential, it must be properly domesticated before it can be a scalable contributor to the global energy supply. It is still essentially a wild plant that has not undergone major domestication efforts. One area of opportunity of improving the traits of jatropha lies in hybridization. *Jatropha curcas* may be crossed with most of the 160 species of the genus *Jatropha* (Münch and Kiefer, 1989), which means there is potentially fertile ground for researchers to explore in improving traits through a dedicated breeding program.

The potential for detoxification should be studied (Heller, 1996), because an edible seedcake could be a valuable by-product. Developing varieties more amenable to mechanical harvest is critical for large scale jatropha ventures. Furthermore, a systematic study of the factors influencing oil yields should be undertaken, because higher yields are probably needed before jatropha can contribute significantly to world energy supplies. Finally, studying the potential for jatropha varieties that thrive in more temperate climates may be worthwhile, as jatropha is presently limited to tropical climates.

12.5   Algae as a source of biofuel

12.5.1   Algae description

The production of biofuels from land-based biomass will ultimately run into limitations regarding the amount of land available for growing biofuel crops. Further, there will continue to be concerns over the competition for land to grow food versus bioenergy crops. Algae-based biofuels are an attractive prospect because of the potential to grow biomass in areas wholly unsuitable for crop production.

Algae* are defined as nonvascular plants; they lack roots and obtain their nutrition directly through their tissues. They range from the microscopic (microalgae) to giant kelp more than 100 feet in length. Algae are found in all geographic regions from the Arctic to deserts, but more commonly live in aqueous environments throughout the world. Most algae produce their food via photosynthesis (autotrophic algae), although some strains are capable of growing in the dark and using sugars as a food source (heterotrophic algae). Both autotrophic and heterotrophic algae are being researched for the possibility of producing biofuels.

From 1978 to 1996, the U.S. Department of Energy funded a study by the National Renewable Energy Laboratory (NREL) on the feasibility of producing renewable fuels from algae (Sheehan et al., 1998). After initially investigating closed photobioreactors, the bulk of this program focused on the production of algae in outdoor ponds. Many different strains of algae were studied, and the major variables affecting the production of biomass and lipids were identified. Important metabolic pathways in the production of lipids from algae were also identified, and the research laid the foundation for present commercialization efforts.

12.5.2 Algal oil

Certain strains of algae are capable of producing lipids that can be extracted and converted to renewable fuel. Many different species of algae are capable of producing lipids; examples include *Chlorella vulgaris, Botryococcus braunii, Scenedesmus dimorphus,* and *Haematococcus pluvialis.* Under conditions of stress, certain species are capable of producing 40–50% of their dry weight in lipids (Sheehan et al., 1998). The fatty acids produced are similar to those found in other oil crops, including palmitic, stearic, oleic, linoleic, and linolenic acids (Damiani et al., 2010).

Estimates of the oil production potential from algae vary wildly because there are no commercial algal fuel operations and estimates have been extrapolated from lab data. Theoretical estimates based on the photosynthetic efficiency of algae have been as high as 30,000–50,000 L/ha—more than 100 times the oil yield of soybeans (Hu et al., 2008). However, to date, reported yields have fallen far short of the theoretical estimates.

12.5.3 Algae production systems

Three primary systems for producing triglycerides or hydrocarbons from algae have been utilized. The first is an open pond system, typically in the form of a raceway system in which water is constantly circulated by means

---

* The word *algae* encompasses a wide variety of organisms covering many taxonomic divisions.

of large paddlewheels or other devices. The primary advantage of open pond systems is that they are relatively simple, so capital and operating costs are lower than for other algae production systems. Disadvantages to the open pond approach include:

- Variations in the weather can stunt algal growth.
- Wild strains of algae can invade and overtake strains that have been developed for high oil yields.
- Evaporation in open ponds may necessitate the addition of freshwater to maintain the required water volume.
- Light penetration into the ponds is minimal, limiting the photosynthetic potential to only the algae that have been circulated to the surface.

The second type of system is a closed system, commonly known as a *photobioreactor* (PBR). A PBR is a device that contains the algae, water, and nutrients and is transparent to allow light to reach the algae. PBRs can take the form of cylindrical tubes, flat plates, or bags of various types. One primary advantage of the PBR is that invasive strains are easier to control, so a high-yielding oil strain of algae could be grown without the likelihood that it will be outcompeted by wild strains. The risk of contamination for PBRs is also lower than for open pond systems. Conservative cost estimates of PBRs are $150 per square meter of surface area, approximately tenfold higher than the open pond systems (Alabi et al., 2009). Some systems are more complex, using some type of light guide, optical fibers, or even artificial light to introduce more light into the cultures.

Both open pond systems and PBRs are ultimately limited by the amount of light that can be utilized by the algae. The third type of algal production system utilizes heterotrophic algae. Heterotrophic algae production systems usually take the form of industrial fermenters. The algae must receive nutrients and a supply of food and are not directly dependent upon solar energy like the autotrophic algae are. Advantages of using fermenters include:

- The ability of the algae to grow in the dark, which means productivity can be greater than in photosynthetic-based algal systems
- Process conditions that can be more carefully controlled, leading to a more reproducible product

The primary disadvantage is that a source of food in the form of sugars must be added. The agricultural area required for producing the sugars would be manyfold higher than the land required to produce algal oil from ponds or PBRs.

To support their growth, autotrophic algae require carbon dioxide; water; nutrients such as nitrogen, phosphates, and sulfates; and sunlight. During the growth phase, algae accumulate fatty acids primarily for esterification into glycerol-based membrane lipids (Hu et al., 2008). Under conditions optimal for production of biomass, lipids comprise generally less than 20% of the algal cells. However, when stressed, the algae switch from a growth mode into a lipid production mode in which the lipid concentration in the cells may reach 50% of the dry algal mass.

Nitrogen is one of the most crucial elements for algal growth, and nitrogen deficiency has the strongest influence on stressing algae into a lipid production mode (Hu et al., 2008). Other stressors include deficiencies of phosphate and sulfate, temperature changes, and low light conditions. The primary disadvantage of these stressors is that while the oil content in the biomass increases, algal growth, and hence biomass yield, is stunted. Maximizing oil content while maximizing productivity has not yet been achieved.

12.5.4   Algae life cycle assessment

In 2010, researchers at the University of Virginia published a life cycle assessment (LCA) on algae cultivation systems (Clarens et al., 2010).* While algal systems were found to utilize land more efficiently than corn, canola, or switchgrass, the overall greenhouse gas emissions for algae were determined to be higher than for the land-based crops. In fact, the researchers argued that algae cultivation emits more greenhouse gases than it sequesters, which would imply that algae cultivation acts as an energy sink as opposed to an energy producer.

There are multiple reasons that this could be the case. Algae production requires the algae to be constantly circulated via pumps or paddlewheels, and algae must be separated from water when it is cultivated. Algae have high nutritional demands as well, and if this takes the form of fertilizer additions, that can comprise a signification energy input. In open ponds, evaporation can be a problem that necessitates makeup water. Further, even though the University of Virginia LCA examined only the cultivation of algae, extraction of oil and subsequent conversion into fuel are two more energy-intensive steps that must be considered for algae-to-fuel schemes.

Still, none of these factors is either alone or in combination decisive. There are techniques for reducing the energy requirements. Pond mixing (at 20–25 cm/s) can be achieved with less than 2 kW/ha (Benemann, pers. comm., 2010), algae harvesting can be accomplished by spontaneous

---

* The assessment was "cradle to gate" and did not consider the subsequent extraction of oil and conversion into fuel.

settling of the algae once removed from the ponds, and fertilizers could be recycled after oil extraction. Makeup water for blow-down and evaporation losses could be supplied by water sources not otherwise usable in agriculture (seawater, brackish waters, various wastewaters, etc.).

Furthermore, the University of Virginia LCA made many assumptions and calculations that have been called into question. Others who have examined the entire process, including the extraction of oil and subsequent conversion into fuel, came to different conclusions, finding that algae oil production can generate lower greenhouse gas emissions than land-based crops. Those more optimistic projections were based in part on the idea that after extraction of the oil, the residual algal biomass could be digested to produce enough methane gas to generate energy that can not only power the entire process but also export surplus energy. However, this analysis was based on process assumptions that need to be verified in an actual operating process. Until then, this must be considered purely theoretical.

12.5.5   Algal biofuel feasibility calculations

Algae have been cultivated for about 40 years, mostly for food supplements (nutraceuticals). *Spirulina* and *Chlorella* are the two major algae genera mass-produced today in the United States, China, India, and Southeast Asia as a food supplement. The lowest cost to produce *Spirulina* in open pond systems in the United States for nutraceuticals is currently estimated to be approximately $5,000 per dry ton of biomass (Benemann, 2009).

Under the optimistic assumption that algae can be mass-cultivated with an average oil content of 40%, then $5,000 of production costs could produce 400 kg of oil. The embedded production cost of the oil, if the residual biomass has little value, is more than $12.50 per kilogram of contained algal oil. This translates into a cost of more than $40 per gallon—more than an order of magnitude above current petroleum diesel prices—*before* considerations for extracting the oil and converting it into biofuel.

Other cost estimates have also placed algal fuel at least an order of magnitude above present petroleum prices. One study was commissioned by the British Columbia Innovation Council et al. to examine the viability of an algal biodiesel industry in British Columbia (Alabi et al., 2009). The study looked at photobioreactors, open raceways, and fermenters. The estimated net cost of production per liter for PBRs was $24.60 (more than US$90/gallon); for open raceways, it was $14.44 per liter, and for fermenters, $2.58. Another estimate was provided in 2009 by the algal biofuel company Solix, which indicated that its production costs were approximately $33 per gallon (Kanellos, 2009). High energy costs were cited as a factor in the high cost of production, which is another indicator that the energy balance for conventional algal fuel production may be unfavorable.

However, it should not be presumed that algae would be grown for fuel with the same methods used commercially today for food-grade algae. When cultivating algae for biofuels, the scale would be much larger, and higher-yielding strains might be used (*Spirulina* has very low productivity, no more than 30 mt/ha/yr dry weight). Further, wastewater— unsuitable for nutraceutical algae—could be used when producing algae for fuel. Some studies for open pond systems (Benemann and Oswald, 1996) have estimated very low costs of algae oil production, based on favorable assumptions on future achievable production rates (e.g., over 100 mt/ha/yr) and oil content (e.g., over 40%), as well as low-cost harvesting and so forth.

12.5.6   Conclusions on the near-term viability of algal biofuel

While the production of biofuels from a raw material like algae has obvious appeal, many cost estimates are presently more than an order of magnitude too high to be economically viable. The NREL closeout report (2006) concluded that there are many technical challenges to be overcome. A major challenge is to increase oil yields without stressing the algae and causing them to shift out of a growth mode; this shift ultimately results in lower overall oil yields because of the slower algal growth. Furthermore, because of lack of data on continuous lipid production from algae, the energy return on the process is unknown.

The NREL report suggested that algae could potentially supply the equivalent of a large fraction of U.S. demand, but costs must come down and technical challenges must be solved. On the subject of costs, the report noted, "Even with aggressive assumptions about biological productivity, we project costs for biodiesel which are two times higher than current petroleum diesel fuel costs" (NREL, 2006, p. 4).*

The report concluded that even with aggressive assumptions about technical breakthroughs, the ultimate cost of algal fuels would be several times higher than that of petroleum-based fuel. Recent studies on the commercial viability of algal fuels indicate that this is still the case, although a number of studies have identified potential areas for substantial cost reduction.

12.6   Conclusions

While it is technically viable to convert lipids extracted from jatropha and algae into either biodiesel or green diesel, there are multiple economic hurdles. Problems with commercialization of jatropha to date have been due to the lack of mechanization for harvesting the fruits, the variable

---

* Albeit at a time when oil prices were a fraction of today's prices.

yields of the fruits, and the low oil yield relative to the cost of harvesting and extracting the oil. Further, the water requirements for jatropha to thrive appear to be much higher than initially believed.

The primary problem with fuel from algae is that it is currently expensive to grow, harvest, and process algae. There are a number of commercial algae operations around the world today, primarily for the purpose of supplying the small food supplement market (the total world market is less than 10,000 tons). Thus, production costs are based on very small systems, the largest of which are not much over 20 hectares. They are consequently limited by small scale, high costs of inputs (e.g., merchant CO_2 costs of more than $1,000/ton of algae), low productivity (due to the strains cultivated), and high processing costs (e.g., spray-drying the biomass). Extrapolating from current commercial production to future biofuels production may therefore not be appropriate.

Both jatropha and algae have the potential to supply energy, either through future technical breakthroughs or via integrated systems. Jatropha serves a purpose as a living fence in many tropical countries, so any fuel extracted from the fruits can be a secondary benefit. Likewise, algae used in wastewater treatment may overcome the financial challenges if the oil is a by-product of these systems and could be a relatively near-term application.

12.7   Conversion factors

While mostly SI units are used in this chapter, imperial/English units are commonly used in the United Kingdom and the United States. Therefore, a number of common conversion factors are listed here, which should enable to reader to convert between SI and imperial units. A number of measures in the text have been converted from imperial units, but the conversion factors listed should enable the reader to reproduce all figures.

   1 barrel of oil = 42 gallons = 158.984 liters = 0.140 metric tons
   1 barrel of oil = 5.8 million BTUs of energy = 6.1 gigajoules (GJ)
   1.0 hectare = 10,000 m² = 2.47 acres

   The specific gravity of crude oil is 0.88.
   The specific gravity of diesel oils is 0.84.
   The specific gravity of biodiesel is 0.88.
   The specific gravity of jatropha oil is 0.92 (Münch and Kiefer 1989).

The lower heating value (LHV) is the heat released by combusting a substance without recovering the heat lost from vaporized water. The LHV is a more accurate representation of actual heat utilized during combustion, as vaporized water is rarely recovered.

The LHV for crude oil is 138,100 BTU/gallon = 38.5 MJ/liter = 45.3 GJ/t
The LHV for distillates is 130,500 BTU/gallon = 36.4 MJ/liter = 42.8 GJ/t
The LHV for biodiesel is 117,000 BTU/gallon = 32.6 MJ/liter = 37.8 GJ/t
The LHV for jatropha oil is 120,700 BTU/gallon = 33.6 MJ/liter = 39.6 GJ/t

References

Achten, W. M. J., Verchot, L., Franken, Y. J., Mathijs, E., Singh, V. P., Aerts, R., and Muys, B. 2008. Jatropha bio-diesel production and use. *Biomass and Bioenergy*, 32(12): 1063–1084.

Alabi, O. A., Tampier, M., and Bibeau, E. (2009). Microalgae technologies and processes for biofuels/bioenergy production in British Columbia. N.p.: Seed Science. http://www.bcic.ca/images/stories/publications/lifesciences/microalgae_report.pdf.

Arena, B., Holmgren, J., Marinangeli, R., Marker, T., McCall, M., Petri, J., Czernik, S., Elliot, D., and Shonnard, D. 2006. *Opportunities for biorenewables in petroleum refineries*. Paper presented at the Rio Oil & Gas Expo and Conference, Instituto Braserileiro de Petroleo e Gas, Rio de Janeiro.

Augustus, G. S., Jayabalan, M., and Seiler, G. J. 2002. Evaluation and bioinduction of energy components of *Jatropha curcas*. *Biomass and Bioenergy*, 23:161–164.

Benemann, J. 2009. Microalgae biofuels: A brief introduction. Walnut Creek, CA: Benemann Associates and MicroBio Engineering. http://www.adelaide.edu.au/biogas/renewable/biofuels_introduction.pdf.

Benemann, J., and Oswald, W. 1996. Systems and economic analysis of microalgae ponds for conversion of CO_2 to biomass: Final report. Washington, DC: USDOE-NETL.

Biofuel Review. 2006. Western Australia bans *Jatropha curcas*. Retrieved April 27, 2010, from http://www.biofuelreview.com/content/view/28/2/.

BP. 2007. BP and D1 Oils form joint venture to develop jatropha biodiesel feedstock. Press release, June 29. http://www.bp.com/genericarticle.do?categoryId=2012968&contentId=7034453.

CIA (Central Intelligence Agency). 2010. *The world factbook*. [Online database accessed April 26, 2010.] https://www.cia.gov/library/publications/the-world-factbook/geos/xx.html.

Clarens, A., Resurreccion, E., White, M., and Colosi, L. 2010. Environmental life cycle comparison of algae to other bioenergy feedstocks, *Environmental Science & Technology*, 44(5): 1813–1819.

Damiani, C., Popovich, C., Constenla, D., and Leonardi, P. 2010. Lipid analysis in *Haematococcus pluvialis* to assess its potential use as a biodiesel feedstock. *Bioresource Technology*, 101(11): 3801–3807.

EcoWorld. 2008. Jatropha reality check. http://www.ecoworld.com/energy-fuels/jatropha-reality-check.html.

EIA (U.S. Energy Information Administration). 2004. World output of refined petroleum products, 2003. Table 3.2 of *International Energy Annual, 2004*. Washington, DC: EIA. http://www.eia.doe.gov/pub/international/iea2004/table32.xls.

———. 2010. *Short term energy outlook*. Washington, DC: EIA. April 7. http://www.eia.doe.gov/emeu/steo/pub/contents.html.

EPA (U.S. Environmental Protection Agency). 2002. Modeling and inventories. Accessed September 23, 2010. http://epa.gov/oms/models.htm.

Fahey, J. 2010. A new life for jatropha. *Forbes*, February 24. http://www.forbes.com/2010/02/23/sg-biofuels-technology-ecotech-jatropha.html.

Foidl, N., Foidl, G., Sanchez, M., Mittelbach, M., and Hackel, S. 1996. *Jatropha curcas L.* as a source for the production of biofuel in Nicaragua. *Bioresource Technology*, 58:77–82.

Francis, G., Edinger, R., and Becker, K. 2005. A concept for simultaneous wasteland reclamation, fuel production, and socio-economic development in degraded areas in India: Need, potential and perspectives of jatropha plantations. *Natural Resources Forum*, 29(1): 12–24.

Gour, V. K. 2006. Production practices including post-harvest management of *Jatropha curcas*. In *Proceedings of the biodiesel conference toward energy independence—focus of Jatropha, Hyderabad, India, June 9–10*, ed. Singh, B., Swaminathan, R., and Ponraj, V., 223–251. New Delhi: Rashtrapati Bhawan.

Heller, J. 1996. *Physic nut* Jatropha Curcas L.: *Promoting the conservation and use of underutilized and neglected crops*. Vol. 1. Gatersleben, Germany: Institute of Plant Genetics and Crop Plant Research; Rome: International Plant Genetic Resources Institute.

Henning, R. K. 2000. The Jatropha booklet: A guide to the jatropha system and its dissemination in Zambia. Weissensberg, Germany: GbR.

Hodge, C. 2006. *Chemistry and emissions of NExBTL*. Presentation at the University of California, Davis. http://bioenergy.ucdavis.edu/downloads/Neste_NExBTL_Enviro_Benefits_of_paraffins.pdf.

Hu, Q., Sommerfeld, M., Jarvis, E., Ghirardi, M., Posewitz, M., Seibert, M., and Darzins, A. 2008. Microalgal triacylglycerols as feedstocks for biofuel production: Perspectives and advances. *Plant Journal*, 54:621–639.

Kanellos, M. 2009. Algae biodiesel: It's $33 a gallon. GreentechMedia, February 3. http://www.greentechmedia.com/articles/read/algae-biodieselits-33-a-gallon-5652/.

Knothe, G. 2001. Historical perspectives on vegetable oil-based diesel fuels. *Inform*, 12(11): 1103–1107.

Lutz, A. 1992. Vegetable oil as fuel. *GTZ Gate*, 1992(4): 38–46.

Münch, E., and Kiefer, J. 1989. Purging nut *(Jatropha curcas L.)*: Multi-use plant as a source of fuel in the future. *Shriftenreihe der GTZ*, No. 209: pp. 1–32.

Neste Oil. 2007. *Neste Oil inaugurates new diesel line and biodiesel plant at Porvoo*. Press release, May 31. http://www.nesteoil.com/default.asp?path=1,41,540,1259,1260,7439,8400.

NREL (National Renewable Energy Laboratory). 2006. *Biodiesel and other renewable diesel fuels*. NREL/FS-510-40419. Golden, CO: NREL.

Oldeman, L. R., Hakkeling R. T. A., and Sombroek, W. G. 1991. *World map of the status of human-induced soil degradation: An explanatory note*. Wageningen, The Netherlands: International Soil Reference and Information Centre; Nairobi: United Nations Environment Program.

Openshaw, K. 2000. A review of *Jatropha curcas*: An oil plant of unfulfilled promise. *Biomass and Bioenergy*, 19:1–15.

Pramanik, K. 2003. Properties and use of *Jatropha curcas* oil and diesel fuel blends in compression ignition engine. *Renewable Energy Journal*, 28(2): 239–248.

Riedacker, A., and Roy, S. 1998. Jatropha (physic nut). In *Energy plant species: Their use and impact on environment and development*, ed. Bassam, N. E., 162–166. London: James & James.

Sheehan, J., Dunahay, T., Benemann, J., and Roessler, P. 1998. A look back at the U.S. Department of Energy's Aquatic Species Program: Biodiesel from algae. NREL/TP-580-24190. Golden, CO: NREL. http://www.nrel.gov/docs/legosti/fy98/24190.pdf.

Sirisomboon, P., Kitchaiya, P., Pholpho, T., and Mahuttanyavanitch, W. 2007. Physical and mechanical properties of *Jatropha curcas L.* fruits, nuts and kernels. *Biosystems Engineering*, 97(2): 201–207.

Smeets, E., Junginger, M., Faaij, A., Walter, A., and Dolzan, P. 2006. *Sustainability of Brazilian bio-ethanol*. Utrecht, Netherlands: Copernicus Institute, University of Utrecht.

Tewari, D. N. 2007. *Jatropha and biodiesel*. New Delhi: Ocean Books.

Thalén, S. 2009. NExBTL diesel outdoes its rivals. Helsinki: Tekes. http://www.tekes.fi/en/community/Success%20stories/416/Success%20story/667?name=NExBTL+diesel+outdoes+its+rivals.

University of Twente. 2009. Bioenergy makes heavy demands on scarce water supplies. *ScienceDaily*, 4 June. http://www.sciencedaily.com/releases/2009/06/090603091737.htm.

UPI (United Press International). 2010. Indian biofuel efforts falter. April 27. http://www.upi.com/Science_News/Resource-Wars/2010/04/27/Indian-biofuel-efforts-falter/UPI-49151272395583/.

Wood, P. 2005. Out of Africa: Could jatropha vegetable oil be Europe's biodiesel feedstock? *Refocus*, 6(4): 40–44.

Zahawi, R. A. 2005. Establishment and growth of living fence species: An overlooked tool for the restoration of degraded areas in the tropics. *Restoration Ecology*, 13:92–102.

chapter thirteen

Crop residues for biofuel and increased soil erosion hazards

Rattan Lal

Contents

13.1 Introduction

It is now widely accepted that global warming beyond 2°C is not acceptable (Group of 8, 2009; Meinshausen et al., 2009). Therefore, global emissions must peak and the declining trend must begin in the near future (Meinshausen et al., 2009). With roughly 8.7 Gt of CO_2-C emissions from fossil fuel combustion and another 1.6 Gt from deforestation (Global Carbon Project, 2010), anthropogenic emissions are likely to be exacerbated by rising energy demand from the increasing world population and the rapidly industrializing economies such as the BRIC nations (Brazil, Russia, India, China; Hoogwijk et al., 2005). The drastic changes in atmospheric constituents and in radiative forcing during the 20th century (Foster and Ramaswamy, 2007) indicate that the time to take a strong action on climate change is now (Bowman et al., 2010), through stabilizing the atmospheric concentration of CO_2 and other trace gases (Koonin, 2008). In this regard, sustainably produced biofuels can play an important role. For example, ethanol production from sugarcane in Brazil is a success story (Lynd and Brito Cruz, 2010). The question is: Can the Brazilian experience be replicated elsewhere?

The emphasis on sustainably produced biofuel implies minimal or no adverse impacts on global food production, because diversion of corn (*Zea mays*) and soybeans (*Glycine max*) to biofuels has increased food prices (Piesse and Thirtle, 2009) and increased the food-insecure population to more than 1 billion (FAO, 2009). Thus, biofuel must only be produced without jeopardizing global food security. Removal of crop residues can also adversely impact soil quality, exacerbate soil degradation, and jeopardize food security (Blanco-Canqui et al., 2007; Lal, 2007b, 2009).

The objective of this chapter is to discuss the positive feedbacks of harvesting crop residues for biofuels on emissions of greenhouse gases (GHGs) through a negative soil carbon budget and an increase in CO_2 emissions from erosion-induced decomposition of soil organic matter (SOM). The specific focus of this chapter is on the effectiveness of using crop residues as mulch to conserve soil and water, improve soil quality, and enhance the favorable rhizospheric processes.

13.2   Global food security

The world population of 0.2 billion in A.D. 1 increased to 7 billion by 2011 and is projected to be 9 billion by 2045 (see Table 13.1). Thus, enhancing and sustaining agricultural production are essential to meeting the demands of the growing world population for food, feed, fiber, and, increasingly, modern (liquid) biofuels (bioethanol, biodiesel). Between 1960 and 2007, the global yield of corn (*Zea mays*) and wheat (*Triticum aestivum*) increased by a factor of 2.6, of rice (*Oryza sativa*) by 2.2, and of soybean (*Glycine max*) by 2.0 (Alston et al., 2009). This surge in agronomic production was caused

Table 13.1 Increase in World Population

Year	Population (billion)
A.D. 1	0.2
1800	1
1930	2
1960	3
1974	4
1987	5
1999	6
2011	7
2024	8
2045	9

Source: Kunzig, R., Population 7 billion, *National Geographic*, 219, 32–69, 2010.

by growing input-responsive varieties on irrigated soils with liberal use of chemical fertilizers and pesticides. While the focus was on crop yields, soil quality management was not an integral component of the agronomic package of the Green Revolution technology used in South Asia and elsewhere.

With a minimal need of about 330 kilograms of grains per person per year, the world's total grain requirement will be about 4.6 Gt by 2050 (70% more than that in 2011) for the projected population of 9.2 billion. The challenge is how food security can be advanced without degrading natural resources (Godfray et al., 2010; Lal, 2010b). To meet the future demand, global average cereal yields are estimated to be required to increase from 2.64 Mg/ha in 2000 to 3.60 Mg/ha (+36%) by 2025 and 4.30 Mg/ha (+67%) by 2050, assuming no major dietary change. With a projected increase in the animal-based diet of the populations in emerging economies, the average cereal yield would have to be 4.40 Mg/ha (+67% over 2000) by 2025 and 6.0 Mg/ha (+127%) by 2050 (Wild, 2003). Indeed, the increasing world consumption of beef is a major driver of regional and global climate change (McAlpine et al., 2009). While recognizing the strong link between livestock and land (Naylor et al., 2005), judicious management of crop residues (controlled grazing) and using dung as manure are important strategies to minimize risks of soil degradation.

The challenge of enhancing cereal yields will be exacerbated by the adverse impacts of projected climate change (Cline, 2007; Battisti and Naylor, 2008) and the attendant risks of soil degradation (Sugden et al., 2004). In contrast to the demand, the global yield and agricultural productivity growth rates are declining (Alston et al., 2009). The average global increase in crop yield annually between 1961 and 1990 was 2.9% for wheat, 2.25% for corn, 2.2% for rice, and 1.8% for soybeans. By comparison, the rate of growth between 1990 and 2007 was 0.5% for wheat, 1.75% for corn, 1.1% for soybeans, and 1.0% for rice (Alston et al., 2009). This slowing in the rate of increase in crop yield since the 1990s is to a large extent due to degradation of soil quality (Lal, 2009, 2010a). While easier access to an improved seed bank is important (Finkel, 2009), the yield potential of even the elite varieties cannot be realized unless grown under optimal soil and agronomic conditions (Lal, 2007a, 2010a), which requires retention of crop residues as mulch, as well as land application of other biosolids (i.e., compost, manure).

13.3 Agriculture and the global carbon cycle

Agriculture has influenced and will continue to influence the global carbon cycle, even when a switch to noncarbon fuel sources occurs. Agricultural activities have been source of GHGs ever since the dawn of settled agriculture (Ruddiman, 2003). As much as 480 Gt of carbon may have been lost from the biosphere (Ruddiman, 2003), including 78 Gt from soils (Lal, 1999). In comparison, emissions from fossil fuel combustion are estimated at about 300 Gt (Holdren, 2008).

Of the 11.5 Gt of carbon emissions worldwide in 2007 (Koonin, 2008), 18% (2 Gt of carbon equivalent or Ce) is attributed to land use conversion, 14% (1.6 Gt Ce) to agriculture, and 3% (0.34 Gt Ce) to waste and wastewater management (Holdren, 2008). Global energy demand and fossil fuel consumption are increasing. The per capita energy use in China is only 30% of that of the United States and 50% that of Japan. Per capita energy use in India is even smaller, only 7% of the United States and 13% of the Japanese amounts. There is a strong energy–poverty–climate nexus (Casillas and Kammen, 2010). With an increase in energy use in emerging economies to alleviate poverty, there will have to be a larger reliance on biofuels (Maoris, 2006; Regalbuto, 2009). However, whether the production of biofuels will reduce net emissions of GHGs remains to be seen.

13.4   The biofuel conundrum

Humans have used biofuels ever since the discovery of fire. Traditional biofuels used over the millennia include firewood, crop residues, animal dung, and so forth. These biofuels are still an important energy source, especially in the developing countries of Asia and Africa. Yet, modern biofuels (bioenergy) pose a dilemma because of their competition with food availability and demands for water (Gerbens-Leenes et al., 2009; Bennett, 2008; Wolf et al., 2003). Nonetheless, there is a wide range of modern biofuels being used in developed countries (see Figure 13.1). These include the so-called first generation, such as the production of bioethanol from starch (corn) and sugar (sugarcane) and of biodiesel from plant-derived oils (e.g., soybean, oil palm, jatropha). Ethanol from sugarcane in Brazil is a success story among the first generation of biofuels (Goldemberg and Guardabassi, 2009). However, first-generation biofuels compete for food availability and access with poor people (Piesse and Thirtle, 2009).

Regardless of the food demand, ethanol production from corn grain is increasing in the United States. It was 10.5 billion gallons in 2009 and is projected to be 15 billion gallons in 2015. The increase in biofuel production globally will heighten the competition for cropland (Popp, 2009; Searchinger et al., 2008), and an increase in demands for food and biofuel will accentuate emissions from agricultural ecosystems and land use conversion. It is in this context that set-asides are considered better options than corn-based ethanol (Piñeiro et al., 2009). Even the biorefinery feedstock production on Conservation Reserve Program (CRP) land (Mapemba et al., 2007) and harvesting biomass from CRP land (Tilman et al., 2006) are considered viable options.

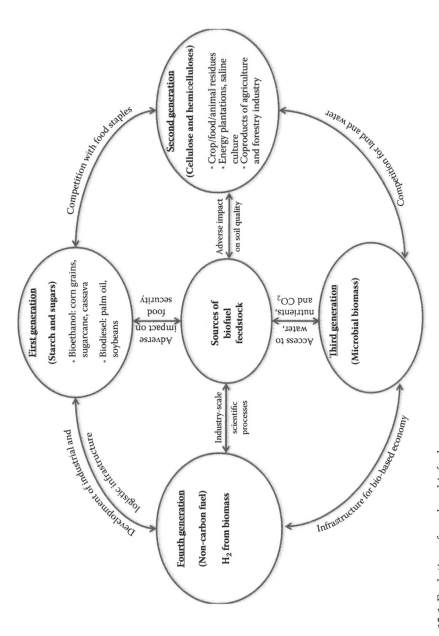

Figure 13.1 Evolution of modern biofuels.

There is no spare cropland to be converted to the production of bio-fuel feedstock (Young, 1999), especially considering the sensitivity to climate change (Ramankutty et al., 2002; Foley et al., 2005). Therefore, sustainable agriculture (Pretty, 2008; Pretty et al., 2010) must be integral to any solution addressing climate and other environmental issues and global economic development (Bierbaum, 2010). There is an urgent need for radically rethinking agriculture for the 21st century (Fedoroff et al., 2010). Maintaining soil quality in agroecosystems is crucial to sustainability (Kibblewhite et al., 2008).

The second-generation biofuels involve those in which bioethanol and sugars are produced from celluloses and hemicelluloses. Between 2010 and 2030, there will be a large-scale demand for the production of second-generation biofuels. However, the production of biomass feed-stock can compete with cropland, and there is an additional need for water and nutrients (Searchinger et al., 2008). Harvesting biomass from prairies, the low-density and high-diversity biomass, has also been pro-posed (Tilman et al., 2006). Regardless of the political emphasis, the life cycle analysis of such systems remains to be done to assess the carbon, nutrient, and water footprints and ecological impacts of bioenergy pro-duction (Firbank, 2008).

Use of cropland for biofuels can increase emissions of GHGs from land use change (Searchinger et al., 2008). Furthermore, harvesting of crop residues to produce cellulosic ethanol can also adversely impact soil qual-ity (Blanco-Canqui et al., 2007). Thus, future research must be focused on the third-generation biofuels produced from microalgae (Sayre, 2010; Chisti, 2008; Hielman et al., 2010) and the fourth-generation biofuels based on production of hydrogen (H_2) from biomass rather than from fossil fuel (see Figure 13.1).

Strategies to mitigate climate change by adopting recommended management practices, outlined by Smith and Martino (2007) and Watson et al. (2000), are based on the goals to:

1. Produce more food grains and biomass from water, energy, and nutrients by enhancing the use-efficiency of inputs and increas-ing agronomic yields and net primary productivity from existing croplands
2. Enhance environmental sustainability in agriculture by increasing soil/ecosystem and social resilience through conservation, seques-tration, substitution, and recycling of the ecosystem carbon, water, and nutrient pools
3. Improve adaptation to climate change through identification and conversion to ecologically compatible agroecosystems that can buf-fer against extreme events and the attendant changes in growing-season characteristics

4. Take advantage of new opportunities and niches created by climate change for adoption of new crops and cropping systems and other options through shifts in the biome
5. Adopt innovative agricultural practices with the goal of exploiting the benefits of biodiversity at the genetic, variety, species, and ecosystem levels at every point of agronomic production through improvement in soil quality

These five goals can be realized through retention of crop residues as mulch and the application of other biosolids to croplands to create a positive ecosystem carbon budget. These goals are realized through improvements in rhizospheric processes (see Figure 13.2). Principal rhizospheric

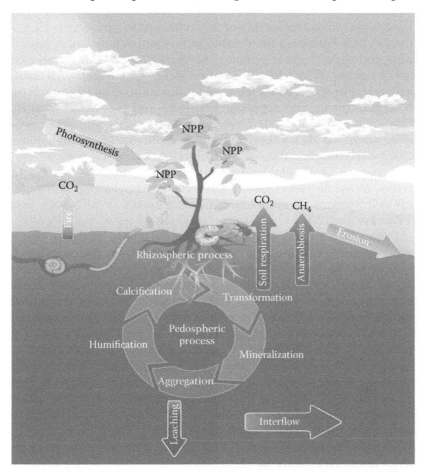

Figure 13.2 Rhizospheric processes (aggregation, humification, mineralization, transformations, and calcification) are impacted by soil surface characteristics such as the residue mulch and cover crops.

processes accentuated by mulch farming are soil aggregation, biomass humification, recycling of nutrients, and restoration of soil quality and its long-term productivity.

13.5 Agriculture and emission of greenhouse gases

Agricultural activities are the primary source of three gases: carbon dioxide (CO_2), methane (CH_4), and nitrous oxide (N_2O) (EPA, 2006). Further, agriculture is a principal contributor of CH_4 with its global warming potential (GWP) of 21 and of N_2O, with a GWP of 310.

Deforestation, biomass burning, land use conversion, drainage of wetlands (peat soils), and soil tillage all contribute to emissions of CO_2. All sources of CO_2 (including fossil fuel combustion, deforestation, land use change, and agricultural activities) contribute 63.5% to the overall global radiative forcing (WMO, 2008; Foster and Ramaswamy, 2007). All energy-based agricultural operations (plowing, cultivation, spraying, harvesting, and manufacturing and applying fertilizers and pesticides) contribute to CO_2 emissions (Lal, 2004). Drainage of peat lands and other poorly drained soils, and their subsequent cultivation, exacerbate the emission of CO_2 by accentuating the rate of mineralization of soil organic matter due to changes in soil temperature and moisture regimes. Agricultural activities, especially deforestation and land use conversion, contribute as much as 2.5 tonnes of carbon equivalents annually (WMO, 2008; Foster and Ramaswamy, 2007).

Methane contributes 18.2% to the overall global radiative forcing. The atmospheric concentration of CH_4 increased from 700 parts per billion (ppb) in the preindustrial era to 1,797 ppb in 2008 (WMO, 2008), rising by 13 ppb/yr during the late 1980s and by 7 ppb in 2008. Among principal agricultural sources of CH_4 are raising livestock (ruminants), including the production and management of manure; rice paddy cultivation; and biomass burning. Reduction of enteric CH_4 emissions from ruminants involves diets based on grains rather than straw and the use of flaxseed and other oils (Veerasamy et al., 2011). Technological options for decreasing CH_4 emissions from livestock are described in detail by Smith and Martino (2007) and may in some cases reduce animal productivity. Conversion of rice paddies to aerobic rice is important to reducing CH_4 emissions (Bouman et al., 2006, 2007). Capacity of soils to oxidize CH_4 is accentuated by improvement of soil quality, for which the use of crop residues as mulch is essential.

Nitrous oxide contributes 6.2% to the overall global radiative forcing (WMO, 2008; Foster and Ramaswamy, 2007). N_2O concentrations in the atmosphere increased from 270 ppb in the preindustrial era to 322 ppb in 2008 and continue to increase at the rate of 0.8–0.9 ppb/yr (WMO, 2008). Principal agricultural sources of N_2O are fertilizer use, biomass burning, and land use conversion. Enhancing fertilizer use efficiency is an important strategy of reducing N_2O emissions. Conversion of plow-tilled to no-till farming may in some cases accentuate N_2O emissions (Lal, 2004). N_2O released from agro-biofuel production negates the global warming reduction from lower CO_2 emissions (Crutzen et al., 2007). Moderation of nitrification and denitrification processes by improving soil and water conservation through use of mulch farming is the basic strategy of decreasing N_2O emissions from agroecosystems (Smith and Martino 2007).

13.6   Mulch farming and soil conservation

Retention of crop residue mulch at the soil–air interface improves soil quality and enhances pedospheric processes in the rhizosphere. Favorable impacts of mulching on soil quality are attributed to the following:

- Moderation of soil temperature regime
- Increase in water infiltration rate
- Decrease in soil evaporation
- Conservation of water in the root zone
- Decrease in soil erosion
- Reduction in nonpoint-source pollution
- Decrease in sedimentation
- Reduction in risks of flooding and siltation of reservoirs and waterways
- Decline in risks of anoxia of coastal ecosystems
- Increase in soil carbon sequestration and offsetting anthropogenic emissions
- Improvement in soil structure, water stable aggregation, and aggregate stability
- Increase in use-efficiency of inputs (e.g., fertilizer)
- Adaptation to extreme events (e.g., drought) resulting from climate change
- Sustainable and stable productivity
- High economic profitability

Long-term and indiscriminate removal of crop residues for biofuel and other purposes would jeopardize these benefits and marginalize other ecosystem services.

Conservation effectiveness of crop residue retention as mulch is attributed to soil protection against raindrop detachment and

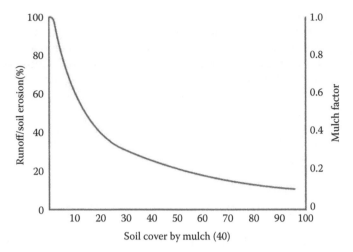

Figure 13.3 Soil cover by mulch and the exponential decline of runoff, and mulch factor. The mulch factor (M) is the ratio of runoff and soil erosion from a mulched soil to that of a bare or plowed soil.

erosivity (attributed to kinetic energy) of surface runoff and blowing wind. The erosivity of wind-driven rain—higher than that of rain falling in no-wind or low-wind conditions—is buffered by the mulch layer. Consequently, soil erosion decreases exponentially with increase in the mulch cover (see Figure 13.3). In general, the magnitude of decrease with increase in mulch cover is more for erosion than surface runoff. The effectiveness of mulch for erosion control is attributed to:

- Protection against direct raindrop impact, which causes aggregate disruption and particle detachment
- Reduction in the velocity of runoff due to increase in resistance
- Decline in sediment-carrying capacity of overland flow because of the decrease in its velocity

Soil erosion is a four-stage process:

1. Detachment
2. Transport
3. Redistribution
4. Deposition

The soil organic carbon (SOC) pool is preferentially removed by water (and wind), and the enrichment ratio (relative amount of SOC in sediment compared with that in the surface layers of the eroding soil) can be 1.5:1

to 50:1. The disruption of aggregates and breakdown of particles exposes the SOC hitherto encapsulated and protected against microbial processes. Therefore, erosion-induced transport of SOC increases the rate of mineralization and emission of CO_2 and other GHGs. The emission is also exacerbated by the erosion-induced alterations in soil temperature and moisture regimes.

In addition to CO_2 (mineralization under aerobic conditions), CH_4 is emitted because of methanogenesis under anaerobic environments, and N_2O because of acceleration in either nitrification or denitrification process. While the carbon buried in depressional sites and that in the aquatic ecosystems may be protected against mineralization (Van Oost et al., 2007), that being transported and redistributed over the landscape is more prone to mineralize (Lal, 2003). Erosion-induced emissions are estimated at about 1.1 Gt of carbon per year on a global scale (Lal, 2003). A continuous use of crop residue mulch, applied in conjunction with no-till farming and cover cropping, can minimize erosion-induced emissions of GHGs. Some mechanisms of reducing gaseous emission by mulching are outlined in Figure 13.4.

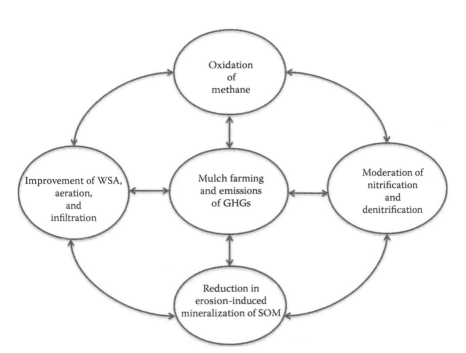

Figure 13.4 Effects of mulch farming on emission of greenhouse gases from soil into the atmosphere.

13.7   On-site and off-site benefits of crop residue retention as surface mulch

The on-site benefits of residue retention are attributed to improvements in pedological, edaphological, biological, and ecological processes (see Figure 13.5). Some pedological processes influenced by residue mulch are:

- Favorable changes in physical, chemical, and biological soil quality
- An increase in the rate of soil formation
- Strengthening of elemental/nutrient cycling
- Improvement in interactive soil, hydrological, atmospheric, and vegetational processes

Edaphological processes influenced by mulch farming that are of significance to plant growth comprise:

- Favorable conditions of soil temperature and moisture regimes
- Favorable rhizospheric processes
- Creation of disease-suppressive soils
- Conditions conducive to soil–plant interactions

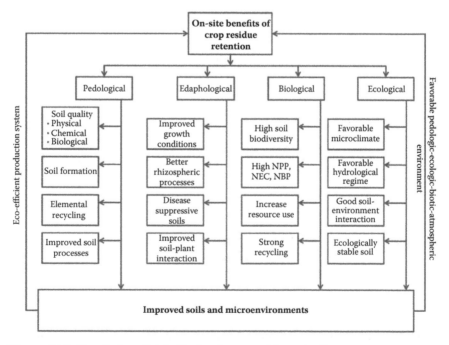

Figure 13.5 On-site beneficial effects of crop residue retention

Mulching strongly influences soil biological processes such as activity and biodiversity of soil fauna, high ecosystem and biome productivity, and increased resource use. Some ecological processes influenced by mulching include favorable micro- and mesoclimate and creation of an ecologically stable soil in dynamic equilibrium with its environment. In essence, mulching creates eco-efficient production systems under a favorable pedologic, ecologic, hydrologic, and atmospheric environment (Figure 13.5).

Mulching also creates numerous off-site benefits (see Figure 13.6). Because of the reduction in runoff and erosion, there is a favorable effect on the quantity and quality of water resources, including seasonality of stream flow and quality of aquatic/coastal ecosystems (anoxia). Mulching also affects the nature (type) and magnitude of the emission of GHGs (see the previous section). The runoff and its dissolved and suspended loads also impact the biomass production in depressional sites, thereby impacting its biodiversity and productivity. Mulch farming influences landscape and watershed processes and increases ecosystem resilience (Figure 13.6).

13.8   Biofuels from crop residues

There is a strong energy–poverty–climate nexus, and thus mitigating climate change, increasing energy access, and alleviating rural poverty can

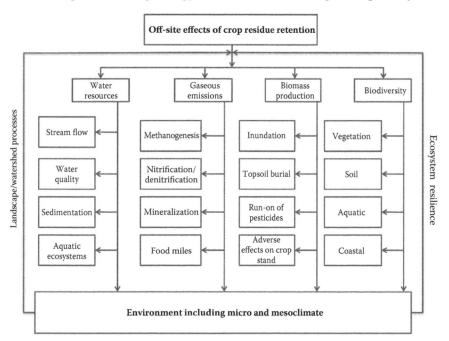

Figure 13.6 Off-site effects of the retention of crop residues on cropland soils.

all be complementary (Casillas and Kammen, 2010). An estimated two-thirds of the world's poor live in rural areas (Bierbaum, 2010). Mulch farming has a direct impact on rural poverty through its effects on agronomic production, profitability, and sustainability. Agriculture being an engine of economic development, mulching can reduce or eliminate poverty by increasing crop yields and sustaining production.

Removal of crop residue mulch can set in motion a downward spiral exacerbating the problem of soil degradation and strengthening the poverty trap, as is already the case of resource-poor farmers in South Asia and Sub-Saharan Africa. Indeed, there is no such thing as a free biofuel from crop residues (Lal, 2007b), because of the long-term adverse impacts on soil quality and productivity.

Using biochar produced from coproducts of agri-silvo-pastoral industry (rice husk, coconut shells, food packing) may be a viable option if it produces energy through pyrolysis and also generates biochar that can be used as a soil amendment (Gaunt and Lehmann, 2008; Chan et al., 2007, 2008). Black (biochar) is considered by some to be the "new green" (Harvey, 2009).

Conversion of biomass into biochar, with production of energy through a low-temperature pyrolysis, and its application to cropland soils is being widely discussed for environmental management (Lehmann and Joseph, 2009; Bruges, 2009). The answer lies in conducting a life cycle assessment with long-term field plot studies under a range of practices including the use of biochar at feasible rates (4–6 Mg/ha). Rather than subsidies, important strategies can include payments to farmers for ecosystem services (Gibbs et al., 2008) and development of a certification system for sustainable bioenergy trade (Lewandowski and Faaij, 2006).

13.9   Conclusions

Biofuels can be an important strategy to meet the growing energy demands of the world population. However, production of biofuel feedstocks must neither compete with food (corn, grains, soybeans) nor with land area, water, and nutrients required for food production. Removal of crop residues for co-combustion (with coal) and for production of cellulosic ethanol can adversely impact carbon sequestration in terrestrial ecosystems. It can also degrade soil quality by exacerbating soil erosion hazards, and water quality by increasing risks of nonpoint-source pollution. Converting crop residues into biochar can also adversely impact soil quality. Thus, the decision for production of biofuel feedstocks must be based on life cycle analyses and assessment of the net energy gains. Production of third-generation biofuels (from algae) and fourth-generation biofuels (H_2 from algae and other biomass) are options that must be explored.

References

Alston, J. M., Beddow, J. M., and Pardey, P. G. 2009. Agricultural research, productivity and food prices in the long run. *Science*, 325:1209–1210.

Battisti, D. S., and Naylor, R. L. 2008. Historical warnings of future food insecurity with unprecedented seasonal heat. *Science*, 323:240–244.

Bennett, K. 2008. Food or fuel? The bioenergy dilemma. Washington, DC: World Resources Institute. http://www.wri.org/stories/2008/08/food-or-fuel-the-bioenergy-dilemma.

Bierbaum, M. F. 2010. *World development report, 2010: Development and climate change*. Washington, DC: World Bank.

Blanco-Canqui, H., Lal, R., Owens, L. B., Post, W. M., and Shipitalo, M. J. 2007. Soil hydraulic properties influenced by stover removal from no-till corn in Ohio. *Soil & Tillage Research*, 92:144–154.

Bouman, B. A. M., Humphreys, E., Tuong, T. P., and Baker, R. 2007. Rice and water. *Advances in Agronomy*, 92:187–237.

Bouman, B. A. M., Yang, X., Wang, H., Zhao, J., and Chen, B. 2006. Performance of aerobic rice varieties under irrigated conditions in north China. *Field Crops Research*, 97:53–55.

Bowman, T. E., Maibach, E., Mann, M. E., Somerville, R. C. J., Seltser, B. J., Fischhoff, B., Gardiner, S. M., Gould, R. J., Leiserowitz, A., and Yohe, G. 2010. Time to take action on climate communication. *Science*, 330:1044.

Bruges, J. 2009. *The biochar debate: Charcoal's potential to reverse climate change and build soil fertility*. White River Junction, VT: Chelsea Green.

Casillas, C., and Kammen, D. M. 2010. The energy–poverty–climate nexus. *Science*, 330:1181.

Chan, K. Y., Van Zwieten, L., Meszarus, I., Downie, A., and Joseph, S. 2007. Agronomic value of greenwaste biochar as soil amendment. *Australian Journal of Soil Research*, 45:629–634.

———. 2008. Using poultry litter biochars as soil amendments. *Australian Journal of Soil Research*, 46:427–444.

Chisti, Y. 2008. Biodiesel from microalgae beats bioethanol. *Trends in Biotech*, 26:126–131.

Cline, W. R. 2007. *Global warming and agriculture*. Washington, DC: Center for Global Development, Peterson Institute for International Economics.

Crutzen, P. J., Mosier, A. R., Smith, K. A., and Winiwarter, W. 2007. N_2O release from agro-biofuel production negates global warming reduction by replacing fossil fuels. *Atmospheric Chemistry and Physics*, 7:11,191–11,205.

EPA (Environmental Protection Agency). 2006. *Global mitigation of non-CO_2 greenhouse gases*. EPA-430-R-06-005. Washington DC: EPA. http://www.epa.gov/climatechange/economics/downloads/GlobalMitigationFullReport.pdf.

FAO (Food and Agriculture Organization). 2009. *Food security and agricultural mitigation in developing countries: Options for capturing synergies*. Rome: FAO. http://www.fao.org/docrep/012/i1318e/i1318e00.pdf.

Fedoroff, N. V., Battisti, D. S., Beachy, R. N., Cooper, P. J. M., Fischhoff, D. A., Hodges, C. N., Knauf, V. C., et al. 2010. Radically rethinking agriculture for the 21st century. *Science*, 327:833–834.

Finkel, E. 2009. Scientists seek easier access to seed banks. *Science*, 324:1376.

Firbank, L. G. 2008. Assessing the ecological impacts of bioenergy projects. *BioEnergy Research*, 1:12–19.

Foley, J. A., DeFries, R., Asner, G. P., Barford, C., Bonan, G., Carpenter, S. R., Chapin, F. S., et al. 2005. Global consequences of land use. *Science*, 309:570–574.

Foster, P., and Ramaswamy, V. 2007. Changes in atmospheric constituents and in radiative forcing. In *Climate change, 2007: The Physical Science Basis: Contribution of Working Group I to the fourth assessment report of the Intergovernmental Panel on Climate Change*, ed. Solomon, S., Qin, D., Manning, M., Chen, Z., Marquis, M., Averyt, K. B., Tignor, M., and Miller, H. L. 129–234. Cambridge: Cambridge University Press.

Gaunt, J. L., and Lehmann, J. 2008. Energy balance and emissions associated with biochar sequestration and pyrolysis bioenergy production. *Environmental Science and Technology*, 42:4152–4158.

Gerbens-Leenes, W., Hoekstra, A. Y., and van der Meer, T. H. 2009. The water footprint of bioenergy. *Proceedings of the National Academy of Sciences*, 106:10,219–10,223.

Gibbs, H. K., Johnston, M., Foley, J. A., Holloway, T., Monfreda, C., Ramankutty, N., and Zaks, D. 2008. Carbon payback times for crop-based biofuel expansion in the tropics: The effects of changing yield and technology. *Environmental Research Letters*, 3:034001.

Global Carbon Project. 2010. *Carbon budget 2009: An annual update of the global carbon budget and trends*. http://www.globalcarbonproject.org/carbonbudget/.

Godfray, H. C. J., Beddingon, J. R., Crute, I. R., Haddad, L., Lawrence, D., Muir, J. F., Pretty, J., Robinson, S., Thomas, S. M., and Toulmin, C. 2010. Food security: The challenge of feeding 9 billion people. *Science*, 327:812–818.

Goldemberg, J., and Guardabassi, P. 2009. Are biofuels a feasible option? *Energy Policy*, 37:10–14.

Group of 8 (G8). 2009. *Responsible leadership for a sustainable future*. L'Aquila, Italy: G8. http://www.g8italia2009.it/static/G8_Allegato/G8_Declaration_08_07_09_final,0.pdf.

Harvey, F. 2009. Black is the new green. *Financial Times*, February 27. http://www.ft.com/cms/s/2/67843ec0-020b-11de-8199-000077b07658.html#axzz19Q6Ckl7L.

Hielman, S. M., Davis, H. T., Jader, L. R., Lefebvre, P. A., Sadawski, M. J., Schendel, F. J., Kietz, M. G. von, and Valentas, K. J. 2010. Hydrothermal carbonization of microalgae. *Biomass and Bioenergy*, 34:875–882.

Holdren, J. P. 2008. *Meeting the climate-change challenge*. Eighth Annual John H. Chafee Memorial Lecture on Science and Environment, National Council for Science and the Environment, Washington, DC, January 17. http://ncseonline.org/conference/Chafee08final.pdf.

Hoogwijk, M., Faaij, A., Eickhout, B., de Vries, B., and Turkenburg, W. 2005. Potential of biomass energy out to 2100 for four IPCC SRES land-use scenarios. *Biomass Bioenergy*, 29:225–257.

Kibblewhite, M. G., Ritz, K., and Swift, M. J. 2008. Soil health in agricultural systems. *Philosophical Transactions of the Royal Society B*, 363:685–701.

Koonin, S. E. 2008. The challenge of CO_2 stabilization. *Elements*, 4:294–295.

Kunzig, R. 2010. Population 7 billion. *National Geographic*, 219:32–69.

Lal, R. 1999. Soil management and restoration for carbon sequestration to mitigate the greenhouse effect. *Programs in Environmental Science*, 1:307–326.

———. 2003. Soil erosion and the global C budget. *Env. Intl.* 29: 437–450.

———. 2004. Carbon emission from farm operations. *Environment International*, 30:981–990.

————. 2007a. Technology without wisdom. *CSA News*, 52(8): 12–13.

————. 2007b. There is no such thing as a free biofuel from crop residues. *CSA News*, 52(5): 12–13.

————. 2009. Soil degradation as a reason for inadequate human nutrition. *Food Security*, 1:45–57.

————. 2010a. Enhancing eco-efficiency in agroecosystems through soil carbon sequestration. *Crop Science*, 1:30–40.

————. 2010b. Managing soils and ecosystems for mitigating anthropogenic carbon emissions and advancing global food security. *Bioscience*, 60:708–721.

Lehmann, J., and Joseph, S., eds. 2009. *Biochar for environmental management: Science and technology.* London: Earthscan.

Lewandowski, I., and Faaij, A. 2006. Steps towards the development of a certification system for sustainable bio-energy trade. *Biomass Bioenergy*, 30:83–104.

Lynd, L. R., and Brito Cruz, C. H. de. 2010. Make way for ethanol. *Science*, 330:1176.

Maoris, E. 2006. Drink the best and drive the rest. *Nature*, 444:670–672.

Mapemba, L. D., Epplin, F. M., Taliaferro, C. M., and Huhnke, R. L. 2007. Biorefinery feedstock production on Conservation Reserve Program Land. *Review of Agricultural Economics*, 29:227–246.

McAlpine, C. A., Etter, A., Feearnside, P. M., Seabrook, L., and Laurance, W. F. 2009. Increasing world consumption of beef as a driver of regional and global change: A call for policy action based on evidence from Queensland (Australia), Colombia and Brazil. *Global Environmental Change*, 19:21–33.

Meinshausen, M., Meinshausen, N., Hare, W., Raper, S. C. B., Frieler, K., Knutti, R., Frame, J., and Allen, M. R. 2009. Greenhouse-gas emission targets for limiting global warming to 2°C. *Nature*, 458:1158–1162.

Naylor, R., Steinfeld, H., Falcon, W., Galloway, J., Smil, V., Bradford, E., Alder, J., and Mooney, H. 2005. Losing the links between livestock and land. *Science*, 310:1621–1622.

Piesse, J., and Thirtle, C. 2009. Three bubbles and a panic: An explanatory review of recent food commodity price events. *Food Policy*, 34:119–129.

Piñeiro, G., Jobbágy, E. G., Esteban, G., Baker, J., Murray, B. C., and Jackson, R. B. 2009. Set-asides can be better climate investment than corn ethanol. *Ecological Applications*, 19:277–282.

Popp, J. 2009. Economic balance on competition for arable land between food and biofuel. In *Biofuel and soil quality*, ed. Lal, R., and Stewart, B. A., 151–182. New York: Taylor & Francis.

Pretty, J. 2008. Agricultural sustainability: Concepts, principles and evidence. *Philosophical Transactions of the Royal Society B*, 363:447–465.

Pretty, J., Sutherland, W. J., Ashby, J., Auburn, J., Baulcombe, D., Bell, M., Bentley, J., et al. 2010. The top 100 questions of importance to the future of global agriculture. *International Journal of Agricultural Sustainability*, 8:219–236.

Ramankutty, N., Foley, J. A., Norman, J., and McSweeney, K. 2002. The global distribution of cultivable lands: Current patterns and sensitivity to possible climate change. *Global Ecology and Biogeography*, 11:377–392.

Regalbuto, J. R. 2009. Cellulosic biofuels: Got gasoline? *Science*, 325:822–824.

Ruddiman, W. F. 2003. The anthropogenic greenhouse era began thousands of years ago. *Climatic Change*, 61:261–293.

Sayre, R. 2010. Microalgae: The potential for carbon capture. *BioScience*, 60:722–727.

Searchinger, T., Heimlich, R., Houghton, R. A., Dong, F., Elobeid, A., Fabiosa, J., Tokgoz, S., Hayes, D., and Yu, T. H. 2008. Use of U.S. croplands for biofuels increases greenhouse gases through emissions from land use change. *Science*, 319:1238–1240.

Smith, P., and Martino, D. 2007. Agriculture. In *Climate change, 2007: Mitigation of climate change: Contribution of Working Group III to the fourth assessment report of the Intergovernmental Panel on Climate Change*, ed. Metz, B., Davidson, O. R., Bosch, P. R., Dave, R., and Meyer, L. A., 497–541. Cambridge: Cambridge University Press.

Sugden, A., Stone, R., and Ash, C. 2004. Ecology in the underworld—introduction. *Science*, 304:1613.

Tilman, D., Hill, J., and Lehman, C. 2006. Carbon-negative biofuels from low-input high diversity grassland biomass. *Science*, 314:1598–1600.

Van Oost, K., Quine, T. A., Govers, G. et al., 2007. The impact of agriculture soil erosion on the global carbon cycle. *Science*, 318: 626–629.

Veerasamy, S., Lal, R., Lakritz, J., and Ezeji, T. 2011. Measurement and prediction of enteric methane emission. *International Journal of Biometeorology*, 55(1): 1–16.

Watson, R. T., Noble, I. R., Bolin, B., Ravindranath, N. H., Verardo, D. J., and Dokken, D. J., eds. 2000. *Land use, land-use change and forestry: A special report of the International Panel on Climate Change*. Cambridge: Cambridge University Press.

Wild, A. 2003. *Soils, land and food: Managing the land during the 21st century*. Cambridge: Cambridge University Press.

WMO (World Meteorological Organization). 2008. *Greenhouse gas bulletin: The state of greenhouse gases in the atmosphere using global observations through 2007*. Geneva: WMO.

———. 2009. *Greenhouse gas bulletin*. Geneva: WMO.

Wolf, A. T., Yoffe, S. B., and Giordano, M. 2003. International waters: Identifying basins at risk. *Energy Policy*, 5:29–60.

Young, A. 1999. Is there really spare land? A critique of estimates of available cultivable land in developing countries. *Environment, Development, and Sustainability*, 1:3–18.

chapter fourteen

Biofuels, foods, livestock, and the environment

David Pimentel

Contents

14.1   Introduction

Shortages of fossil energy are encouraging the heavy use of biomass for biofuel production (Santa Barbara, 2007). Emphasis on biofuels has developed globally to include the use of crops such as corn, sugarcane, and soybeans as renewable energy sources. Wood and crop residues also are being used as fuel by some (Pimentel, 2008a). Though it may seem beneficial to use plant materials for biofuels, the use of crop residues and other biomass for biofuels raises many environmental and ethical concerns (Pimentel, 2006).

Diverse conflicts exist in the use of land, water, energy, and other environmental resources for food, livestock products, and biofuel production. Foods, livestock, and biofuels are dependent on the same resources for production: land, water, and energy. In the United States, about 19% of all fossil energy is utilized in the food system: about 7% for agricultural production, 7% for processing and packaging foods, and 5% for distribution and preparation of food (Pimentel, Marklein et al., 2009). In developing countries, about 50% of wood energy is used for cooking (Nonhebel, 2005).

The objectives of this chapter are to analyze, first, the use of and inter-dependencies among land, water, and fossil energy resources in food versus biofuel and livestock production, and second, the environmental impacts and ethical issues related to livestock and biofuel production.

14.2   Food and malnourishment

The Food and Agricultural Organization (FAO) of the United Nations confirms that worldwide the amount of food available per capita has been declining, based on availability of cereal grains from 1983 to 2006 (FAO, 2011a). Cereal grains make up 80% of the world's food supply (Pimentel and Pimentel, 2008). Although grain yields per hectare in both developed and developing countries are still gradually increasing, the rate of increase is slowing, while the world population and its food needs are rising (FAO, 2011a; PRB, 2011). For example, from 1950 to 1980, U.S. grain yields increased at about 3% per year. Since 1980, the annual rate of increase for corn and other grains has been only approximately 1% (USDA, 1980, 2007). Worldwide, the rate of increase in grain production is not keeping up with the 1.2% rate of world population growth (PRB, 2011).

The result of decreasing food supply is widespread malnutrition. There are more deaths from malnutrition than any other cause of death in the world today (Pimentel et al., 2007). The World Health Organization (WHO, 2000) reports that more than 3.7 billion people (56% of the global population) are currently malnourished, with 2 billion cases related to iron deficiencies. In addition, the FAO (2009) reports an additional 700 million malnourished due to protein/calorie deficiencies. Thus the total malnourished population is 4.4 billion, or 66% of the total world population. Although much of the cropland worldwide is occupied by grains and other crops, malnutrition is still globally prevalent and increasing (WHO, 2000; FAO, 2011b).

14.3   World cropland and water resources

More than 99.7% of human food comes from the terrestrial environment, while less than 0.3% comes from the oceans and other aquatic ecosystems (FAO, 2011b). Worldwide, of the total of 13 billion hectares of land area on Earth, cropland accounts for 11%, pastureland 27%, forests 32%, and urban areas 9%. Most of the remaining land area (21%) is unsuitable for crops, pasture, or forests because the soil is too infertile or shallow to support plant growth or the climate or topography is too harsh, cold, dry, steep, stony, or wet (FAO, 2011). Most of the suitable cropland is already in use.

As the human population continues to increase rapidly, diverse human activities expand, dramatically reducing available cropland and pastureland. Much vital cropland and pastureland has been covered by transportation systems and urbanization. In the United States, about 0.4 ha (1 acre) of land per person is covered with urban areas and roads (USCB, 2008). In 1960, when the world population numbered only 3 billion, approximately 0.5 ha was available per person for the production of a diverse, nutritious diet of plant and animal products (Giampietro and Pimentel, 1994). It is widely agreed that 0.5 ha is essential for a healthy, diverse diet.

China's recent explosion in development provides an example of rapid declines in the availability of per capita cropland (Pimentel and Wen, 2004). The current available cropland in China is only 0.08 ha per capita. This relatively small amount of cropland provides the people of China with a predominantly vegetarian diet, which requires less energy, land, water, and biomass than a typical American diet.

In addition to land, water is a vital controlling factor in crop production (Gleick, 1996). The production of 9 tons per hectare of corn requires about 7 million liters of water (about 700,000 gallons of water per acre; Pimentel et al., 2004). Other crops also require large amounts of water. Irrigation provides much of the water for world food production. For example, 17% of the crops that are irrigated provide 40% of the world food supply (Postel, 1999, 2000). A major concern is that world irrigation declined about 10% from 1987 to 1997 (Postel, 1997).

14.4   Energy resources and use

Since the industrial revolution of the 1850s, the rate of energy use from all sources has been growing much faster than world population growth. For example, from 1970 to 1995, energy use increased at a rate of 2.5% per year (doubling every 30 years) compared with the worldwide population growth rate of 1.7% per year (doubling every 40–60 years; Pimentel and Pimentel, 2008). Developed nations annually consume about 70% of the fossil energy worldwide, while the developing nations, which have around 75% of the world population, use only 30% (DOE, 2006).

Although roughly half of all the solar energy captured by worldwide photosynthesis is used by humans for food, fiber, forest products, and other materials, it is still inadequate to meet all of human needs— namely, food production needs (Pimentel, 2001). To make up for this shortfall, about 463 quads (1 quad = 1×10^{15} BTU) of fossil energy—mainly oil, gas, and coal—and a small amount of nuclear energy are utilized worldwide each year (USCB, 2008). Of this 463 quads, about 100 quads of the world's total energy is utilized just in the United States (USCB, 2008). In other words, the United States, with only 4.5% of the world's population, consumes about 22% of the world's fossil energy output (USCB, 2008).

Each year, the U.S. population uses three times more fossil energy than the total solar energy captured by all harvested U.S. crops, all forests,

Table 14.1 Total Amount of Biomass and Solar Energy Captured Each Year in the United States

	Million ha	Tons/ha	$\times 10^6$ tons	Total energy collected $\times 10^{15}$ BTU
Crops	160	5.5	901	3,600
Pasture	300	1.1	333	1,332
Forests	264	2.0	527	2,108
Total	*724*		*1,758*	*7,040*

An estimated $7,040 \times 10^{15}$ BTU of sunlight reaching the United States per year suggests that the green plants (crops, grasses, and forests) in the United States are collecting 0.1% of the solar energy reaching these plants (Pimentel, Marklein et al., 2009).

and all grasslands (see Table 14.1). Industry, transportation, home heating and cooling, and food production account for most of the fossil energy consumed in the United States (USCB, 2008). Per capita use of fossil energy in the United States amounts to about 9,500 liters of oil equivalents per year—more than seven times the per capita use in China (Pimentel and Pimentel, 2008). In China, most fossil energy is used by industry, although approximately 25% is now used for agriculture and in the food production system (Pimentel and Wen, 2004).

The world supply of oil is projected to last approximately 40 years, if use continues at current production rates (EIA, 2009). Worldwide, the Earth's natural gas and coal supplies are considered adequate for about 40 and 100 years, respectively (BP, 2005; Youngquist, 1997). In the United States, natural gas supplies are already in short supply; projections are that the United States will deplete its natural gas resources in about 20 years (depletion might take place at a later date because of natural gas extract via fracking) (Youngquist and Duncan, 2003). Many agree that the world reached peak oil reserves in 2007; after this, oil resources will decline slowly and continuously until there is little or no oil left (Youngquist and Duncan, 2003; Campbell, 2006; Ruppert, 2009).

Youngquist (1997) reports that earlier excessively optimistic estimates of the oil and gas resources in the United States were based on exploratory drilling. Both the U.S. oil production rate and existing reserves are continuing to decline. Domestic oil and natural gas production have been decreasing for more than 30 years (USCB, 2008). Approximately 90% of U.S. oil resources have already been exploited (W. Youngquist, pers. comm., 2002). At present, the United States is importing about 70% of its oil (USCB, 2008).

14.5 Biomass resources

The total sustainable biomass energy potential in the United States has been estimated to be about 7,040 quads (1 quad = 10^{15} BTU) per year (see

Table 14.2 Total Amount of Biomass and Solar Energy Captured Each Year in the World

	Billion ha	Tons/ha	× 10⁹ tons	Total energy collected × 10¹⁸ BTU
Crops	1.5	5.0	750	3,000
Grasses	3.5	1.1	385	1,540
Forests	3.0	1.6	480	1,920
Total	*8.0*		*1,615*	*6,460*

An estimated $6,460 \times 10^{18}$ BTU of sunlight reaching the Earth is captured as biomass by green plants (crops, grasses, and forests) each year in the world, suggesting that green plants are collecting about 0.1% of the solar energy reaching these plants (Pimentel, Marklein et al., 2009).

Table 14.1), which represents 7% of total global energy use. The total forest biomass produced in the United States is 2.1 quads per year, which represents 2% of total energy use. In the United States, only 1–2% of home heating is achieved with wood (USCB, 2008).

The total sustainable world biomass energy potential in the world has been estimated to be about $6,460 \times 10^{18}$ BTU per year (see Table 14.2), which represents nearly 100% of total global energy produced. The total forest biomass produced worldwide is 480 billion tons per year, which represents about 99% of total energy produced globally.

The global forest area removed each year totals 15 million ha (Sundquist, 2007). About 90% of fuelwood is utilized in developing countries (Parikka, 2004). A significant portion (26%) of all forest wood is converted into charcoal (Arnold and Jongma, 2007), but the production of charcoal causes between 30% and 50% of the wood energy to be lost (Demirbas, 2001) and produces large quantities of smoke. Charcoal is cleaner burning, however, and thus produces less smoke during cooking than burning wood fuel directly (Arnold and Jongma, 2007); charcoal is dirty to handle but lightweight.

In developing countries, about 2 kcal of wood is utilized in cooking 1 kcal of food (Fujino et al., 1999). Thus, more biomass, land, and water are needed to produce the biofuel for cooking than to produce the food.

Worldwide, most biomass is burned for cooking and heating; however, it can also be converted into electricity. Assuming an optimal yield globally of 3 dry metric tons per hectare per year of woody biomass can be harvested sustainably (Ferguson, 2001, 2003), this would provide a gross energy yield of 13.5 million kcal/ha. Harvesting this wood biomass requires an energy expenditure of approximately 30 liters of diesel fuel per ha, plus the embodied energy for cutting and collecting wood for transport to an electric power plant. Thus, the energy input per output ratio for such a system is calculated to be 1:25 (Hendrickson and Gulland, 1993).

Woody biomass has the capacity to supply the United States with about 5 quads (1.5 × 10^12 kWh thermal) of its total gross energy supply by 2050, provided the amount of forest area stays constant (Pimentel, 2008b). A city of 100,000 people using the biomass from a sustainable forest (3 t/ha per year) for electricity requires approximately 200,000 ha of forest area, based on an average electrical demand of slightly more than 1 billion kWh (860 kcal = 1 kWh; Pimentel, Marklein et al., 2009).

Air quality impacts from burning biomass are less harmful than those associated with coal but more harmful than those associated with natural gas (Pimentel, 2001). Biomass combustion releases more than 200 different chemical pollutants, including 14 carcinogens and 4 co-carcinogens, into the atmosphere (Burning Issues, 2006). As a result of this, approximately 4 billion people worldwide suffer from continuous exposure to smoke (Smith, 2006). In the United States, wood smoke kills 30,000 people each year (Burning Issues, 2003), although many of the pollutants from electric plants that use wood and other biomass can be mitigated using the same scrubbers that are frequently installed on coal-fired plants.

14.6 Energy inputs in livestock product production

Livestock production is a major consumer of water, because grains and forage biomass consumed by livestock require significant amounts of water for biomass growth and production. To produce 1 kg of grain requires about 1,000 liters of water. Based on grain and forage consumption, about 43,000 liters of water are required to produce 1 kg of beef. Each year, an estimated 50 million tons of plant protein is fed to U.S. livestock to produce approximately 8 million tons of animal protein for human consumption (USDA, 2007). In addition, about 28 million tons of plant protein from grain and 17 million tons of plant protein from forage are fed to the animals. Thus, for every kilogram of high-quality animal protein, livestock are fed nearly 6 kg of plant protein. In the conversion of plant protein into animal protein, there are two principal costs:

1. The direct costs of production of the harvested animal, including the grain and forage
2. The indirect costs for maintaining the breeding animals (mother and father)

The major fossil energy inputs for grain and forage production include fertilizers, farm machinery, fuel, irrigation, and pesticides (Pimentel et al., 2008). The energy inputs vary according to the particular crop and forage being grown. When these inputs are balanced against their energy and protein content, grains and some legumes like soybeans are produced more efficiently in terms of energy inputs than fruits, vegetables, and animal products (Pimentel and Pimentel, 2008). In the United States, the average protein yield of the five major grains (plus soybeans) fed to livestock is about 700 kg/ha. To produce a kilogram of plant protein requires approximately 10 kcal of fossil energy (Pimentel and Pimentel, 2008).

Forage can be fed to ruminant animals, like cattle and sheep, because they can convert forage cellulose into usable nutrients through microbial fermentation. The total plant protein produced on good U.S. pasture and fed to ruminants is 60% of the amount produced by grains. The current yield of beef protein from productive pastures is about 66 kg/ha, while the energy input per kilogram of animal protein produced is 3,500 kcal (Pimentel and Pimentel, 2008). Therefore, animal protein production on good pastures is less expensive in terms of fossil energy inputs than grain protein production.

Of the livestock systems evaluated in this investigation, chicken-broiler production is the most efficient, with an input of 4 kcal of fossil energy per 1 kcal of broiler protein produced (see Table 14.3). Broilers are a grain-only system. Turkey production, also a grain-only system, is next in efficiency with a ratio of 10:1. Milk production based on a mixture of grain and forage also is relatively efficient, with a ratio of 14:1. For milk production, about two-thirds of the feed protein consumed by the cattle is grain (see Table 14.3); of course, 100% of milk production could be produced on forage. Both pork and egg production also depend upon grain. Pork has a 14:1 ratio, whereas egg production is relatively more costly in terms of feed energy, requiring a 39:1 ratio.

Table 14.3 Grain and Forage Inputs per Kilogram of Animal Product Produced, and Fossil Energy Inputs (kcal) Required to Produce 1 kcal of Animal Protein

Livestock	Grain (kg)[a]	Forage (kg)[b]	kcal input/kcal protein
Lamb	21	30	57:1
Beef cattle	13	30	40:1
Eggs	11	—	39:1
Beef cattle	—	200	20:1
Swine	5.9	—	14:1
Dairy (milk)	0.7	1	14:1
Turkeys	3.8	—	10:1
Broilers	2.3	—	4:1

[a] USDA, 2001
[b] Heischmidt, R. K., Short, R. E., and Grings, E. E., 1996, Ecosystems, sustainability and animal agriculture, *Journal of Animal Science*, 74(6): 1395–1405; Morrison, F. B., *Feeds and feeding* (Ithaca, NY: Morrison, 1956)

The two livestock systems that depend most heavily on forage, but still use significant amounts of grain, are the beef and lamb production systems (see Table 14.3). The lamb system with a ratio of 57:1 and the beef system with a ratio of 40:1 are the two highest. If these animals were fed only on good-quality forage, the energy inputs could be reduced by about half, depending on the conditions of the pasture-forage and the management practices. Beef cattle fed 200 kg of forage and no grain had an energy input per kcal protein (beef) output ratio of 20:1 (see Table 14.3). Rainfall is critical for all productive pasture systems.

Per kilogram of animal product foods, broiler chicken flesh has the largest percentage of protein and milk the lowest (Pimentel and Pimentel, 2008). Beef has the highest calorie content because of its high fat content and relatively low water content. Of all the animal products, milk has the highest water content with 87%.

The average fossil energy input for all animal protein production systems studied is about 25 kcal of fossil energy input per kcal of animal protein produced (see Table 14.3). This energy input is more than 10 times greater than the average input/output ratio for grain protein production, which was about 2.5 kcal per kcal of protein produced. As a food for humans, however, animal protein has about 1.4 times the biological value as a food compared with grain protein.

14.7   Biofuels, environment, and biomass

Biofuel production is increasing starvation in the world because of the increasing use of grains and other food crops to produce bioethanol and biodiesel (Ford Runge and Senauer, 2007; Monbiot, 2009). Most of the biofuels are produced using corn, soybeans, and sugarcane. All are grown on the best agricultural cropland. For example, in the United States, 34% of all corn goes into ethanol production. The use of corn has resulted in price increases for milk, meat, and eggs (Pimentel, 2009).

The Congressional Budget Office (2009) reports that corn ethanol accounted for 10–15% of the food price increases between April 2007 and April 2008, costing families between $6 billion and $9 billion per year in higher grocery bills. An additional $12 billion is invested by the U.S. government each year to subsidize corn ethanol, along with an added $6 billion to subsidize biodiesel (Koplow and Steenblik, 2008).

Using food crops such as corn and soybeans to produce ethanol or biodiesel raises major nutritional and ethical concerns. As mentioned, about 66% of humans in the world now are currently malnourished, so the need for grains and other basic foods is critical. Growing crops for fuel squanders land, water, and energy resources vital for the production of food for people (Pimentel et al., 2009). The president of the World Bank has reported that biofuels have increased world food prices 75%

(Chakrabortty, 2008). Jacques Diouf, director general of the FAO, agrees that biofuels increase world starvation (Diouf, 2008). The UN secretary general, Ban Ki-moon, has also raised serious doubt about switching crops to biofuels because it would increase world starvation (Vidal, 2008). The Western countries' appetite for biofuels is increasing world malnutrition (Monbiot, 2007).

Corn production causes more soil erosion than any other crop grown in the United States (NAS, 2002). In Brazil, sugarcane production also causes more soil erosion than any other crop grown in Brazil (Pimentel and Patzek, 2008). Soil erosion is reported as the major threat to food production in the United States and, indeed, in the world (Pimentel, 2006).

Water is a critical resource for agricultural production, and biofuel production utilizes enormous amounts of it. For example, the production of 1 liter of ethanol requires 1,700 liters of freshwater for the corn grain production and processing of the grain into ethanol (Pimentel and Patzek, 2008a). Most of the water is utilized in the production of the corn. A hectare of corn during the three-month growing season requires about 7 million liters of water, or nearly 10 million liters if the corn is irrigated (Pimentel and Pimentel, 2008).

Corn production utilizes more insecticides and more herbicides than any crop grown in the United States (Pimentel and Patzek, 2008a). These pesticides reduce the biodiversity on the cornland, and pesticides that drift or are washed from the treated land area have negative consequences for biological species that come in contact with them (Pimentel, 2005). Pesticides are also a major threat to public health. For example, in the United States, the General Accounting Office (GAO, 1992) reported there were more than 300,000 nonfatal pesticide poisonings annually. Worldwide, the situation is far more serious, with an estimated 26 million human pesticide poisonings and some 220,000 deaths (Richter, 2002).

Furthermore, biofuel production significantly contributes to the global warming problem by adding enormous amounts of CO_2 to the atmosphere (Pimentel and Patzek, 2008). From 50% to 170% more fossil energy is required to produce 1 liter of ethanol by various means than the energy in the 1 liter of ethanol, and this significantly adds to the atmospheric CO_2. In addition, because corn and sugarcane production cause enormous amounts of soil erosion, soil organic matter is lost from the eroded soil as CO_2 to the atmosphere (Pimentel, 2006; Lal, 2004).

Several groups in the U.S. Congress and private groups in the United States oppose congressional efforts to prohibit or delay the inclusion of carbon emission from indirect land use change in the life cycle greenhouse gas emissions from biofuels in the Renewable Fuel Standard (Coalition Advocates for the Scientific Integrity in Federal Ethanol Policy, 2009).

14.8   Conclusions

The rapidly increasing world population and growing consumption of fossil fuels are increasing demand for food, livestock products, and biofuels. This higher demand will exaggerate food and fuel shortages. Producing food, livestock products, and biofuels all require huge amounts of fossil energy resources. Clearly, this intensifies the conflicts among these products for resources.

Utilizing various grain and other food crops to produce ethanol and biodiesel raises major nutritional and ethical concerns. Nearly 60% of the world population is currently malnourished (WHO, 2000), so the need for grains and other basic foods is critical. Growing grains for fuel and livestock takes away cropland, water, and fossil energy resources vital for human society.

Meat, milk, and egg production in the United States relies on significant quantities of fossil energy, cropland, pastureland, and water resources. Grain-fed livestock systems use large quantities of fossil energy because of the grain crops cultivated, although cattle grazed on pastures use considerably less energy than grain-fed cattle. An average of 25 kcal of fossil energy is required to produce 1 kcal of animal protein and represents about 10 times the energy expended to produce 1 kcal of plant protein.

Nearly one-third of the U.S. land area is devoted to livestock production. Of this, around 10% is devoted to grain production for livestock and the remainder is used for forage and rangeland production. The pastureland and rangeland are marginal in terms of productivity, because there is too little rainfall and nutrients for major biomass production.

Acknowledgments

This research was supported in part by the Podell Emeriti Award at Cornell University.

References

Arnold, J. E. M., and Jongma, J. 2007. Fuelwood and charcoal in developing countries: An economic survey. Rome: Forestry Department, FAO. http://www.fao.org/docrep/l2015e/l2015e01.htm.

BP. 2005. *BP statistical review of world energy.* London: BP. http://www.bp.com/liveassets/bp_internet/globalbp/globalbp_uk_english/publications/energy_reviews_2005/STAGING/local_assets/downloads/pdf/statistical_review_of_world_energy_full_report_2005.pdf.

Burning Issues. 2003. References for EPA wood smoke brochure [from 2003]. http://burningissues.org/car-www/science/wsbrochure-ref-3-03.htm.

Campbell, C. 2006. Regular conventional oil production to 2100 and resource based production forecast [Excel spreadsheets]. *The coming global oil crisis.* http://www.oilcrisis.com/campbell/.

Chakrabortty, A. 2008. Secret report: Biofuel caused food crisis. Thursday 3 July 2008. *The Guardian* http://www.guardian.co.uk/environment/2008/jul/03/biofuels.renewableenergy.

Coalition Advocates for the Scientific Integrity in Federal Ethanol Policy. 2009. Letter to the Senate Committee on Appropriations. September 22. http://www.usclimatenetwork.org/resource-database/letter-to-chairman-inouye-and-vice-chairman-cochran-urging-opposition-to-legislative-prohibition-or-delay-of-the-inclusion-of-carbon-emission-from-indirect-land-use-in-the-lifecycle-of-ghg-emisiions-from-biofuels.

Congressional Budget Office. 2009. The impact of ethanol use on food prices and greenhouse gas emissions. Washington, DC: U.S. Congress, Congressional Budget Office. http://www.cbo.gov/ftpdocs/100xx/doc10057/04-08-Ethanol.pdf.

Demirbas, A. 2001. Biomass resource facilities and biomass conversion processing for fuels and chemicals. *Energy Conversion & Management*, 42(11): 1357–1378.

Diouf, J. 2008. Statement by the director-general on the occasion of the launch of SOFA 2008. October 7. Food and Agriculture Organization of the United Nations, Rome. http://www.fao.org/newsroom/common/ecg/1000928/en/sofalaunch.pdf.

DOE (U.S. Department of Energy). 2006. *International energy annual.* Washington, DC: DOE.

EIA (U.S. Energy Information Administration). 2009. World proved reserves of oil and natural gas: Most recent estimates. Washington, DC: DOE. http://www.eia.gov/international/reserves.html.

FAO. 2009. *World hunger facts, 2009.* Rome: FAO.

———. 2011a. FAOSTAT. [Online database accessed June 2, 2011.] http://faostat.fao.org/site/339/default.aspx.

———. 2011b. Food balance sheets. [Online database accessed June 23, 2011.] http://faostat.fao.org/site/368/default.aspx.

Ferguson, A. R. B. 2001. Biomass and energy. *Optimum Population Trust Journal*, 4(1): 14–18.

———. 2003. Implications of the USDA 2002 update on ethanol from corn. *Optimum Population Trust Journal*, 3(1): 11–15. http://populationmatters.org/journal/j31.pdf.

Ford Runge, C., and Senauer, B. 2007. How biofuels could starve the poor. *New York Times*, May 7. http://www.nytimes.com/cfr/world/20070501faessay_v86n3_runge_senauer.html.

Fujino, J., Yamaji, K., and Yamamoto, H. 1999. Biomass-balance table for evaluating bioenergy resources. *Applied Energy*, 63(2): 75–89.

GAO (U.S. General Accounting Office). 1992. Hired farmworkers: Health and well-being at risk. GAO/HRD-92-46. Washington, DC: GAO. http://archive.gao.gov/t2pbat7/145941.pdf.

Giampietro, M., and Pimentel, D. 1994. Energy utilization. In *Encyclopedia of agricultural science*, Vol. 3, ed. Arntzen, C. J., and Ritter, E. M., 63–76. San Diego, CA: Academic Press.

Gleick, P. H. 1996. Basic water requirements for human activities: Meeting basic needs. *Water International*, 21:83–92.

Heischmidt, R. K., Short, R. E., and Grings, E. E. 1996. Ecosystems, sustainability and animal agriculture. *Journal of Animal Science*, 74(6): 1395–1405.

Hendrickson, O. Q., and Gulland, J. 1993. Residential wood heating: Forest, the atmosphere and public consciousness. Presentation at the Air and Waste Management Association Conference, New Orleans. http://woodheat.org/public-conciousness.html.

Koplow, D., and Steenblik, R. 2008. Subsidies to ethanol in the United States. In *Biofuels, solar and wind as renewable energy systems: Benefits and risks*, ed. Pimentel, D., 79–108. Boca Raton, FL: CRC Press.

Lal, R. 2004. Soil carbon sequestration impacts on global climate change and food security. *Science*, 34:1623–1627.

Monbiot, G. 2007. The Western appetite for biofuels is causing starvation in the poor world; Developing nations are being pushed to grow crops for ethanol, rather than food—all thanks to political expediency. *Guardian*, November 6. http://www.guardian.co.uk/commentisfree/2007/nov/06/comment.biofuels.

———. 2009. U.S. car manufacturers plough a lonely furrow on biofuels [Blog posting]. *Guardian.co.uk*, July 22. http://www.guardian.co.uk/environment/georgemonbiot/2009/jul/22/biofuels.

Morrison, F. B. 1956. *Feeds and feeding*. Ithaca, NY: Morrison.

NAS (National Academy of Sciences). 2002. *Frontiers in agricultural research: Food, health, environment, and communities*. Washington, DC: National Academies Press.

Nonhebel, S. 2005. Renewable energy and food supply: Will there be enough land? *Renewable and Sustainable Energy Reviews*, 9(2): 191–201.

Parikka, M. 2004. Global biomass fuel resources. *Biomass Bioenergy*, 27:613–620.

Pimentel, D. 2001. Biomass utilization, limits of. In *Encyclopedia of physical science and technology*, 3rd ed., ed. Meyers, R. A., 2:159–171. San Diego, CA: Academic Press.

———. 2005. Environmental and economic costs of the application of pesticides primarily in the United States. *Environment, Development and Sustainability*, 7:229–252.

———. 2006. Soil erosion: A food and environmental threat. *Environment, Development and Sustainability*, 8:119–137.

———. 2008a. The ecological and energy integrity of corn ethanol production. In *Reconciling human existence with ecological integrity: Science, ethics, economics and law*, ed. Westra, L., Bosselmann, K., and Westra, R., 245–256. London: Earthscan.

———. 2008b. Renewable and solar energy technologies: Energy and environmental issues. In *Biofuels, solar and wind as renewable energy systems: Benefits and risks*, ed. Pimentel, D., 1–17. Dordrecht, The Netherlands: Springer.

———. 2009. Corn ethanol as energy: The case against US production subsidies. *Harvard International Review*, 31(2): 50–52.

Pimentel, D., Berger, B., Filiberto, D., Newton, M., Wolfe, B., Karabinakis, E., Clark, S., Poon, E., Abbett, E., and Nandagopal, S. 2004. Water resources: Agricultural and environmental issues. *Bioscience*, 54(10): 909–918.

Pimentel, D., Cooperstein, S., Randell, H., Filiberto, D., Sorrentino, S., Kaye, B., Yagi, C. J., et al. 2007. Ecology of increasing diseases: Population growth and environmental degradation. *Human Ecology*, 35:653–668.

Pimentel, D., Marklein, A., Toth, M. A., Karpoff, M., Paul, G. S., McCormack, R., Kyriazis, J., and Krueger, T. 2009. Food versus biofuels: Environmental and economic costs. *Human Ecology*, 37:1–12.

Pimentel, D., and Patzek, T. 2008. Ethanol production: Energy and economic issues related to U.S. and Brazilian sugarcane. In *Biofuels, solar and wind as renewable energy systems: Benefits and risks*, ed. Pimentel, D., 357–371. Dordrecht, The Netherlands: Springer.

Pimentel, D., and Patzek, T. 2008a. Ethanol production using corn, switchgrass and wood; Biodiesel production using soybean. In *Biofuels, solar and wind as renewable energy systems: Benefits and risks*, ed. Pimentel, D., 373–394. Dordrecht, The Netherlands: Springer.

Pimentel, D., and Pimentel, M. 2008. *Food, energy and society*. 3rd ed. Boca Raton, FL: CRC Press.

Pimentel, D., and Wen, D. 2004. China and the world: Population, food and resource scarcity. In *Dare to dream: Vision of 2050 agriculture in China*, ed. Tso, T. C., and He Kang, 103–116. Beijing: China Agricultural University Press.

Pimentel, D., Williamson, S., Alexander, C. E., Gonzalez-Pagan, O., Kontak, C., and Mulkey, S. E. 2008. Reducing energy inputs in the U.S. food system. *Human Ecology*, 36(4): 459–471.

Postel, S. 1997. *Last oasis: Facing water scarcity*. New York: Norton.

———. 1999. *Pillar of sand: Can the irrigation miracle last?* New York: Norton.

———. 2000. Redesigning irrigated agriculture. In *Proceedings of the 4th Decennial National Irrigation Symposium*, ed. Evans, R. G., Benham, B. L., and Trooien, T. P., 1–12. St. Joseph, MI: American Society of Agricultural Engineers.

PRB (Population Reference Bureau). 2011. 2010 population data sheet. Washington, DC: PRB. http://www.prb.org/Publications/Datasheets/2010/2010wpds.aspx.

Richter, E. D. 2002. Acute human pesticide poisonings. In *Encyclopedia of pest management*, ed. Pimentel, D., 1:3–6. New York: Marcel Dekker.

Ruppert, M. C. 2009. *Confronting collapse: The crisis of energy and money in a post peak oil world: A 25-point program for action*. White River Junction, VT: Chelsea Green.

Santa Barbara, J. 2007. The false promise of biofuels. San Francisco: International Forum on Globalization; Washington, DC: Institute for Policy Studies. http://www.ifg.org/pdf/biofuels.pdf.

Smith, K. R. 2006. Health impacts of household fuelwood use in developing countries. *Unasylva* 224(57): 41–44. http://ehs.sph.berkeley.edu/krsmith/?p=113.

Sundquist, B. 2007. Degradation data. In *Forest lands degradation: A global perspective*. http://home.windstream.net/bsundquist1/df4.html.

USCB (U.S. Census Bureau). 2008. *Statistical abstract of the United States, 2009*. Washington, DC: USCB.

USDA (U.S. Department of Agriculture). 1980. *Agricultural statistics, 1980*. Washington, DC: USDA.

———. 2001. *Agricultural statistics, 2001*. Washington, DC: USDA.

———. 2007. *Agricultural statistics, 2007*. Washington, DC: USDA.

Vidal, J. 2008. Crop switch worsens food price crisis. *Guardian*, April 5. http://www.guardian.co.uk/environment/2008/apr/05/food.biofuels.

WHO (World Health Organization). 2000. *Nutrition for health and development: A global agenda for combating malnutrition.* Department of Nutrition for Health and Development. Geneva: WHO. http://whqlibdoc.who.int/hq/2000/WHO_NHD_00.6.pdf.

Youngquist, W. 1997. *GeoDestinies: The inevitable control of earth resources over nations and individuals.* Portland, OR: National Book.

Youngquist, W., and Duncan, R. C. 2003. North American gas: Data show supply problems. *Natural Resources Research,* (12): 229–240.

Index

The letter *t* following a page number denotes a table; the letter *f* following a page number denotes a figure.

Milton Keynes UK
Ingram Content Group UK Ltd.
UKHW031138141024
449569UK00024B/1230